GENERAL CLIMATOLOGY

Dr. Jaya Arora

Toronto Academic Press

GENERAL CLIMATOLOGY

Dr. Jaya Arora

Toronto Academic Press
4164 Lakeshore Road
Burlington ON L7L 1A4
Canada
www.tap-books.com

Email: orders@arclereducation.com

ISBN: 978-1-77956-309-5

ABOUT THE AUTHOR

Dr. Jaya Arora holds a doctorate in Environmental Studies and specialises in abiotic disturbances such as landslides. She has a varied background in ecology, climate change, biodiversity, soil microbiology, cytogenetics, and eco-genomics. She currently serves as an Assistant Professor at O.P Jindal Global University in Sonipat, Haryana, India, where she presents her teaching and content delivery skills in the field of Environmental Biology. Appointed as an Assistant Professor in the Delhi School of Journalism, Dr. Jaya is specialised in guiding federal environmental laws, including the National Environmental Policy Act (NEPA), Endangered Species Act (ESA), Clean Water Act (CWA), and Clean Air Act (CAA) with over two years of professional experience in teaching Environmental Biology, Policy and Law with ability to collect, analyze, and interpret field data and biological information.

She had genuine interest in field work in the broad area of ecology and the environment. She chose to work on a highly challenging and a relatively new field of research namely, 'Ecogenomics' that integrates ecology and molecular biology. Dr. Jaya worked on two contrasting research areas, viz., field research which focuses on basic ecological investigations of the disturbed Himalayan ecosystems and molecular biology of the plant species involved in the colonisation of these habitats. She has performed an in-depth analysis of plant phytosociological properties, soil physicochemical parameters, genomic investigations of plants by flow cytometry and plant rhizobiome/metagenomics analysis using high throughput next-generation sequencing techniques. Considering the challenges of working in a relatively pristine area – both in terms of the study site as well as the subject area – Dr. Arora conducted vital experiments in order to conclude her research work and bring it to a logical conclusion. Professionally, she has over two years of expertise in ecological restoration, land management, and environmental consulting. As a restoration ecologist, she is dedicated to protecting and recovering damaged ecosystems.

Jaya earned a Ph.D. from the University of Delhi "Department of Environmental Studies" Delhi, India. Committed to constant study, Jaya documented ethnobotanical notes on a list of therapeutic plants in Chaubatia (Ranikhet), in the Indian Himalayan highlands. Jaya, who is interested about working in plant sectors, collaborated with WIMCO Seedlings Limited (ITC) on a project to analyse the variety and distribution of Poplar trees and their genotypes in the Udham Singh Nagar district of Uttarakhand, India.

Table of Contents

List of Figures

List of Tables

PREFACE

Climatology and meteorology are closely intertwined yet operate on different time scales. Weather pertains to immediate atmospheric conditions encountered day-to-day, whereas climate refers to long-term patterns and expectations. Climate can be conceptualized as a probability distribution dictating the range of weather variations over time. It is governed by Earth's system properties, resembling a boundary value problem, contrasting with weather's moment-to-moment evolution akin to an initial value problem in physics, particularly in hydraulics.

Weather forecasting remains reliable for only a few days to a week due to the inherent complexity and dynamical nature of atmospheric processes. In contrast, climate forecasting is challenged by modeling limitations and insufficient observational data, reflecting our ongoing efforts to fully grasp the intricacies of Earth's climate system. This complexity is compounded by gaps in understanding the hydrologic cycle, encompassing phenomena like cloud formation, ocean dynamics, and other small-scale processes, alongside the inherent unpredictability within climate dynamics.

The Earth's climate is a complex system interwoven with multiple feedback mechanisms, exemplified by the cryosphere—the icy realms including polar ice sheets, glaciers, and sea ice. These components play pivotal roles in regulating global climate through mechanisms such as solar radiation reflection (albedo), heat exchange with oceans and the atmosphere, and influencing sea levels and greenhouse gas dynamics.

For students specializing in fields like water resources, hydrology, or environmental engineering, climatology provides a foundational understanding of long-term weather patterns. It distinguishes itself from meteorology by focusing on historical climate trends and their impacts on natural systems and human activities. By studying climatology, engineers can better predict hydrological scenarios, understand solar radiation impacts, and anticipate climatic changes' effects on water resource management.

This book aims to introduce engineers to the fundamentals of climate science, equipping them to comprehend how climatic factors influence water resource systems from a beginner's perspective. It emphasizes the importance of climatological knowledge in addressing contemporary environmental challenges and engineering solutions.

Content/Structure of the Book

This textbook organized into ten chapters, each focusing on a critical aspect of climatology to provide a comprehensive understanding of the subject.

Chapter 1 serves as an introduction to the field of climatology, offering a foundational understanding of the science. It covers the definition, scope, and significance of climatology, providing historical context and an overview of key concepts and principles.

In Chapter 2, the book examines into the composition and structure of the Earth's atmosphere. It explores the various layers of the atmosphere, the gases present, and their respective roles in atmospheric processes. This chapter also discusses the vertical and horizontal variations in atmospheric composition.

Chapter 3 focuses on the sun as the primary source of energy for the Earth's climate system. It examines the principles of solar radiation, the Earth's energy budget, and the processes of energy absorption, reflection, and transmission. It also discusses the significance of the greenhouse effect and its impact on global temperatures.

In Chapter 4, the textbook explores the factors influencing temperature distribution across the globe. It discusses the concepts of heat transfer, temperature gradients, and the effects of latitude, altitude, and ocean currents on temperature variations. The chapter also covers diurnal and seasonal temperature changes.

Chapter 5 addresses the role of moisture in the atmosphere, including the processes of evaporation, condensation, and precipitation. It examines the hydrological cycle, types of precipitation, and the factors affecting precipitation patterns. The chapter also discusses humidity, dew point, and the formation of clouds.

In chapter 6, the textbook then focuses on atmospheric pressure and the dynamics of wind systems. It explains the causes and effects of pressure variations, the formation of high and low-pressure systems, and the Coriolis effect. The chapter also covers global wind patterns, including trade winds, westerlies, and polar easterlies.

Chapter 7 explores the characteristics and movements of air masses, as well as the formation and behavior of weather fronts. It discusses the classification of air masses, the processes leading to their development, and the impact of fronts on weather patterns. The chapter also examines cyclones and anticyclones.

Chapter 8 presents various methods of climate classification and the identification of climatic zones. It covers the Köppen climate classification system, as well as other classification schemes. The chapter provides detailed descriptions of different climatic zones, including tropical, temperate, and polar climates.

Chapter 9 addresses the critical issues of climate change and variability. It discusses the evidence for historical and contemporary climate change, the factors driving these changes, and their potential impacts on natural and human systems. The chapter also explores the concepts of climate variability and the role of human activities in altering the climate.

Chapter 10 focuses on the practical applications of climatology and future climate projections. It examines the use of climatic data in various fields such as agriculture, urban planning, and disaster management. The chapter also discusses climate modeling, scenarios of future climate change, and the implications for policy and adaptation strategies.

This organized approach provides readers with a deep and comprehensive understanding of climatology, from basic concepts to advanced applications and future challenges.

Audience

This book spans a diverse range of readers interested in gaining a comprehensive understanding of climate science and its broader implications. Primarily, the book targets students and researchers in environmental science, geography, atmospheric science, and related disciplines who seek a solid foundation in climatological principles and methodologies. It serves as a fundamental resource for academic courses exploring climate dynamics, climate change, and environmental sustainability.

Additionally, professionals and practitioners in fields such as agriculture, urban planning, and natural resource management will find the book invaluable for enhancing their knowledge of climate patterns and their practical applications. Dealing with topics like atmospheric composition, solar radiation, temperature variations, precipitation patterns, and climate classification, the book equips readers with essential tools to analyze and interpret climatic data in various contexts.

Policy makers and environmental consultants will also benefit from the book's insights into climate variability, climate modeling, and the potential impacts of climate change on ecosystems and human societies. Its accessible yet detailed approach to complex climatological concepts makes it suitable for both introductory readers and those seeking deeper insights into contemporary issues in climate science and its interdisciplinary connections. Overall, this book aims to empower its audience with the knowledge needed to navigate and address the challenges posed by a changing climate in the modern world.

Learning Features

- Start-of-chapter learning objectives
- Key terms in each chapter
- Summary Points at the end of each chapter
- Review Questions at the end of each chapter
- Case Studies and Activities in each chapter for solving real problems
- References at the end of each chapter

CHAPTER

1 Introduction to Climatology

LEARNING OBJECTIVES

After studying this chapter, you will be able to:

- Understand the meaning and scope of climatology;
- Explain the history of climatology;
- Define meteorology and climatology;
- Explain scales in climatology;
- Describe sub-fields of climatology.

KEY TERMS FROM THIS CHAPTER

Agricultural climatology	Atmosphere	Bioclimatology
Climate	Cryosphere	Dynamic climatology
Evaporation	Frequencies	Global hydrologic cycle
Human bioclimatology	Lithosphere	Local scale
Mesoscale	Meteorology	Paleoclimatology
Physical climatology	Proxy evidence	Regional climatology

INTRODUCTION

An important area of atmospheric science called climatology examines the long-term trends and variations in local weather. It includes examining historical documents, computer models, atmospheric data, and a range of other methods to comprehend climate dynamics and how they relate to the environment. Planning for sustainable development, forecasting weather patterns, and creating strategies and policies to lessen the effects of climate change all depend heavily on climatology. The definition, nature, and application of climatology will be covered in this article, with an emphasis on the discipline's multidisciplinary nature and importance in tackling the problems brought on by climate change.

1.1. MEANING AND SCOPE OF CLIMATOLOGY

The scientific study of climate and its long-term trends and variations is known as climatology. To comprehend how the climate has changed over time and to forecast future trends, it entails the analysis of data on atmospheric conditions, such as temperature, precipitation, humidity, wind, and atmospheric pressure. The multidisciplinary field of climatology incorporates concepts and methods from some academic fields, including geography, biology, physics, chemistry, geology, and mathematics. Understanding the causes and effects of climate change as well as creating mitigation plans for its effects on human societies and the environment require a thorough understanding of climatology.

1.1.1. Nature and Scope of Climatology

The goal of the large, multidisciplinary field of climatology is to comprehend the trends, origins, and effects of climate variability and change. Climate science includes a wide range of disciplines, such as meteorology, oceanography, ecology, and mathematics, in addition to overlapping fields like geology, physics, chemistry, biology, and mathematics.

Understanding the intricate relationships between atmospheric variables—such as temperature, precipitation, wind, and humidity—and how they affect the environment and human societies is one of climatology's main objectives. This covers both naturally occurring and man-made variables, such as emissions of greenhouse gases, modifications to land use, and volcanic activity that can affect climate.

In addition, the study of climatology attempts to create models and instruments for forecasting future climates and evaluating the possible effects of climate change. These models project future scenarios and assist decision-makers in making plans by utilizing information from historical and current weather conditions as well as highly developed computer simulations.

Climatology has practical uses in numerous domains, in addition to its scientific and technical uses. For instance, knowing how the climate will change over time can help water managers prepare for future floods and droughts, or farmers choose which crops to plant and when. Climatic studies can be used by urban planners to create cities that are more resistant to extreme weather, and by energy companies to forecast supply and demand. Last but not least, climatology has important social and ethical implications since social justice and sustainable development are severely hampered by climate change. To address these issues, innovative solutions that take into consideration the requirements and values of various stakeholders must be found, along with interdisciplinary and international cooperation.

1.1.2. Importance of Climatology

Here are some key points highlighting its importance:

1. Understanding Climate Change: Climatology helps us understand the causes and impacts of climate change, including global warming, sea level rise, extreme weather events, and shifts in precipitation patterns. This understanding is vital for developing strategies to mitigate and adapt to climate change.
2. Weather Prediction: Climatology provides the foundation for weather prediction models by analyzing long-term atmospheric patterns and trends. By understanding how different climate variables interact, meteorologists can make more accurate short-term weather forecasts, which are crucial for disaster preparedness, agriculture, transportation, and many other sectors.
3. Impacts on Ecosystems: Climate influences the distribution and behavior of plant and animal species, as well as ecosystems as a whole. Climatological research helps us understand how changes in temperature, precipitation, and other climate variables affect biodiversity, ecosystem health, and ecosystem services like pollination, water purification, and carbon sequestration.
4. Agriculture and Food Security: Climatology plays a vital role in agriculture and food security by informing decisions related to crop selection, planting schedules, irrigation practices, and pest management. Farmers rely on climate data and forecasts to maximize yields, minimize losses due to extreme weather events, and adapt to changing climatic conditions.
5. Water Resources Management: Climate influences the availability, distribution, and quality of water resources, including rivers, lakes, groundwater, and glaciers. Climatological studies help policymakers and water managers anticipate changes in water supply and demand, develop sustainable water management strategies, and mitigate the impacts of droughts, floods, and water scarcity.
6. Energy Planning: Climatology informs energy planning and policy by assessing the potential of renewable energy sources like solar, wind, hydroelectric, and biomass. Understanding regional climate patterns and variability helps optimize the design and operation of renewable energy systems, as well as assess the risks associated with climate-related hazards to energy infrastructure.
7. Public Health: Climate influences the spread of vector-borne diseases, air quality, heat-related illnesses, and other public health issues. Climatological research helps public health agencies anticipate and respond to climate-related health risks, develop early warning systems for heat waves and epidemics, and design interventions to protect vulnerable populations.
8. Policy and Decision Making: Climatology provides the scientific basis for climate policy and decision making at local, national, and international levels. Policymakers rely on climatological data, assessments, and projections to develop climate action plans, set emissions targets, negotiate international agreements, and allocate resources for climate adaptation and mitigation efforts.

1.2. HISTORY OF CLIMATOLOGY

Since the beginning of time, climate has been crucial to human understanding of various topics, including diseases, agricultural patterns, geological processes, and more. Depending on the disciplines, places, and viewpoints from which it is studied, its understanding has changed at different times throughout history.

The two parallel narratives that have shaped what we now know as climatology are the geographical and the physical. The first methods of climatology were geographic explorations, classifications, and studies. These methods defined climate as a place's local characteristic, shaped by a variety of geographical factors like latitude, proximity to an ocean, or influences from mountains. However, as time went on, physical scientists developed several instruments and specific theories, particularly in the fields of fluid dynamics and thermodynamics, that ultimately led to a more thorough understanding of the climate, which is primarily distinguished by its non-static nature and global scope.

This essay provides a general overview of the history of climatology and a brief discussion of the field's current situation before posing the question of whether there is a chance for another paradigm shift in which geography will play a major role.

1.2.1. Early Climatology

Climate, as defined by statistical weather characteristics over time, is the subject of the study of climatology (Landsberg and Oliver, 2005). Many forcing mechanisms of varying complexity influence the Earth's climate, and each one is significant. Nonetheless, the primary driver of climate activity, solar radiation, can be considered the most significant forcing.

Therefore, early studies of the distribution of sunlight across the Earth's surface would be a convenient place to start when researching the history of climatology. Eratosthenes (3rd century BCE) conducted one of the earliest experiments known to have used this feature. He measured the shadows of two vertical poles at various latitudes during the summer solstice and calculated remarkably accurate estimates of the Earth's spherical shape (Roller et al., 2010; Torretti, 2007).

Furthermore, the Greeks observed that the chilly regions to the north and the desserts to the south were really a reflection of the varying latitudes' sun incidence angles (Edwards, 2011; Miller, 2005). The significance of this conceptual model grew to such an extent that Ptolemaic maps from the second century CE labeled parallels using the day length of the specific latitudes, which are comparable to sunlight incidence angles but more comprehensible and natural. The studies of Hippocrates, who made remarkable observations of human health as a function of climates, seasonality, and weather events as early as the 4th century BCE, demonstrate that Greek philosophers were already familiar with several ideas around the effects of climate change before these early conceptions of what causes different climates (Falagas et al., 2010; Woodruff and McMichael, 2005). Climate became a central focus of their research over time. As a result, Parmenides of Elea, in the fifth century BCE, proposed the conceptual division of the Earth into horizontal climate zones. He proposed the following climate bands: extreme cold at the poles, intense winds and precipitation in the equatorial region, and two habitable zones in mid-latitudes, where Greece is located.

It is one of the well-documented tales because the Greeks' comprehension of climate had a lasting influence on the evolution of concepts in the Western world. Nonetheless, other modern societies also possessed a comprehensive comprehension of climate, and while this essay does not aim to reveal the obscure beginnings of climatology, it is necessary to at least highlight a few instances. It is known that the Chinese began to observe weather phenomena, such as typhoons, droughts, and floods, and studied how they affected harvesting and cultural customs, around the same time as the Greeks. On the other side of the globe, the Olmec culture (dating from the second century BCE) created a climatic-religious imaginary around hurricanes after realizing their impact on the development of civilization (Melgarejo Vivanco, 1975).

Numerous discussions on the history of climate claim that climatology did not advance much until the Middle Ages when Roger Bacon adopted the Greek concept of climate zones and produced more accurate climate divisions. Bacon also elaborated on the observed motion of the Sun and its effects on symmetric bodies like the Earth, as well as the optical characteristics of airborne water droplets. Bacon acknowledged that there were seven climatic regions, but he did not suggest that only those seven described the entire world, from the equator to the poles, in his writing, "Sed tamen hoc potest esse in climatibus notis, quae septem vocantur, sed non oportet quod sit in omnibus regionibus a medio mundi usque ad polum." He thus raised the possibility of more research on climates.

By the 17th century, naturalists and physicians were not the only communities interested in climate and weather; the obvious effects on agriculture and disease caused school teachers, clergy, and the general public to become interested in the subject and active observers. Heymann (2010) states that these communities approached the problem in different ways. The approaches were still very varied and, for the most part, anecdotal, despite the significance of the interest. These days, we recognize that understanding cross-border processes in the study of climate demands international collaboration. Therefore, until atmospheric observations were systematically gathered, no significant progress would be made.

1.2.2. Foundations of the Modern Climatology

James Jurin is credited with being the first to successfully organize a methodical approach to gathering meteorological data at multiple locations around 1720 (Landsberg, 1964). The

Societas Meteorologica Palatina, a German organization that gathered meteorological data from multiple stations throughout Europe, was established in 1780 (Kington, 1974). The 18th century, according to Landsberg (1964), is when climatology could be distinctly acknowledged as a science. He notes that by then, western explorers had visited almost every region of the globe and navigators had gathered a variety of atmospheric and oceanic data. After that, the development of causal relations and the systematization of analysis were the main priorities.

The popularity of the thermometer and barometer, which were created in the middle of the 17th century, led to an increase in weather observations and a shift in the public's perception of climate toward something more "objective" (Heymann, 2010). This setting allowed for additional data analysis and pattern recognition, which eventually resulted in the development of better models. The first isotherms in a map, created by von Humboldt in 1817 to depict the distribution of heat across the planet, are a prime example of this. According to Heymann (2010), the climate is defined as a collection of atmospheric variables that are detectable by human senses. Consequently, the majority of efforts in climate studies have been directed toward determining the specific distribution of these parameters for various locations and regions. Meanwhile, these systematized data also helped in the development of conceptual models.

The experiences, stories, and navigational aids were combined into a conceptual model to create theoretical explanations of the dynamics of the atmosphere. To explain the trade winds, Halley (1753) conceptualized two atmospheric cells made up of warm air masses at the equator's surface that are raised and move heat and water toward the poles. Hadley (1735) went on to discuss the Coriolis effect and the implications of air masses circulating within a rotating Earth. Additionally, Dove and Ferrel expanded the original model by including two and three circulation cells. Furthermore, Coffin's (1852) modern works, which examined a significant quantity of winds in the northern hemisphere, validated the theoretical extensions of the original two-cell circulation model.

We can begin to see a division in the field of climate research at this time. Naturalists like Alexander von Humboldt approached climatology from a geographic perspective on the one hand, while the hard sciences began to establish the scientific framework that would eventually be necessary for studying climate on the other. The main narratives of both sides during the 19th and 20th centuries are presented in the section that follows.

1.2.3. Recent History of Climate Studies

1.2.3.1. Geographical Understanding of Climate

It is easy to imagine a longer connection to any living thing when one recalls the Humboldtian understanding of climate as the changes in the atmosphere that humans could perceive. Accordingly, climatological theories of the 19th century benefited greatly from the study of botany (Landsberg, 1964). A scheme of plant cover as a function of latitude was first developed. Later, more complex climatic elements were added and even described for the first time in botanical studies. One such study is the work of Schouw (1827), which, despite focusing on Botanics, is the first to propose the relationship between sea levels and the wind's dragging force during the monsoon. Furthermore, Landsberg (1964) discusses the recent research on the relationships between temperature and tree rings as well as the addition of other significant variables to the studies of organism classification, such as the presence of mountains, seas, and continents that are thought to alter the mean latitudinal climate. This research was also motivated by the desire to understand how climate change affects agriculture (e.g., de Gasparin (1844)).

Several writers began to notice apparent causal relationships between human deforestation and climate change as a result of ongoing observations in this field. M. Becquerel, one of the main writers on the topic, produced an engaging book that expanded on this causal relationship even though he acknowledged that "so far there has been no positive evidence that wood has any real influence on the climate of a great estate or locality." Thompson (1990) discusses the early research on climatic changes in America brought on by human disturbance of forests in that context. According to Heymann (2010), conjectures about these concepts date back to the time of Christopher Columbus. These studies gave rise to various misconceptions, some of which even led to improper forest management practices (Naudts et al. 2016). The truth is that information and knowledge regarding the dynamics of the climate were still lacking.

The navigation community was another essential component in the gathering and organization of climate data. The seafarers' practical contributions to the field were the driving force behind them. The mapping of sea routes was the common goal that brought navigators together, according to Maury (1855). The physical geography of the sea, which was to include "a philosophical account of the winds and currents of the sea, of the circulation of the atmosphere and ocean, of the temperature and depth of the sea; of the wonders that lie hidden in its depth and of the phenomena that display themselves at its surface," was created as a result of the overwhelming interest in geography (Maury, 1855). The network for information exchange and data gathering that was established in this setting had a significant impact on the development of climate history (Miller, 2005).

Many questions concerning historical climates began to surface on the geological front. Huge granite blocks were observed in the Rhone Valley around 1830, and Johan von Charpentier proposed that glaciers may have carried them there (Von Charpentier, 1841). The difficulty in conceiving a cause that would have a climate suitable for the diluvian glaciers to develop to the extent that the hypothesis requires, under the latitude of the Alps, is the reason behind the most common objection, he acknowledged. Subsequently, Agassiz and Bettannier (1840) conducted a thorough investigation into the dynamics and formation of glaciers, particularly in the Alps. They concluded that the existence of ice ages was necessary because historical and geological records demonstrated enormous climatic oscillations. Subsequent research on the ice ages' consequences on historical biological diversity established a connection between botanic and climate studies (Anderson et al. 2016).

This era was marked by advancements in description and classification methods for pure climate studies. A seminal work in that field is Kendrew's "The Climates of the Continents" (1922). This is followed by additional details in terms of space, such as Geiger's (1933) descriptions of microclimates, as well as temporal measurements and the quantity of observations, such as Dove's (1857) works, which innovated on the measured elements by adding the effect of ice on the air above, diurnal and annual temperatures, and began to notice the meridional flows. According to Landsberg (1964), the majority of the dynamical explanations for modern climatology were false, even though these and other studies were crucial to it (Miller, 2005).

The abundance of false ideas was partly caused by climatology's restricted function in providing studies on environmental determinism with a classification framework. Therefore, rather than seeing advancements in our understanding of the nature of climate, we can characterize this period as one of knowledge accumulation. Concurrently, other scientists were accumulating information that would eventually guide future climate research.

1.2.4. Physical Understanding of Climate

As climatology researchers conducted exploratory and classificatory studies, another line of inquiry—studies that view the atmosphere as a fluid that exchanges mass, momentum, and energy—began to take shape. Climate science is thought to have started with the work of Jean-Baptiste Fourier in the early 19th century on terrestrial temperature. Fourier recorded the equations describing the dynamics of terrestrial heat in his most well-known work in this area, "Teorie analytique de la chaleur" (Fourier, 1822).

His additional contribution, which has been cited as the first analogy of the atmosphere to a greenhouse, stated that "the interposition of the air greatly modifies the effects of heat on the surface of the globe." Later, using the analogy of climbing a mountain, Samuel Langley calculated the temperature of the Earth's surface in the absence of air. He concluded that "when the air was altogether absent, the temperature of the Earth under direct sunshine would be excessively low" (Langley, 1884).

Further works converted the intuitive concepts into formal physics, Tyndall (1859) identified the mechanism that was previously proposed by Fourier, showing experimentally how certain gases commonly found in the atmosphere, such as water, absorbed long-wave radiation. On this train of discoveries, Svante Arrhenius (1896) calculated the influence of carbonic acid (CO_2) on the planetary energy balance. Anderson et al. (2016) point out that his close collaboration with the geologist Arvid Hogbom, who studied the carbon cycle, led them to speculate on the necessary quantity of CO_2 that had to be added or removed to get the temperature levels associated with the glaciations theorized by previous geologists.

Some writers interpreted this as a mechanism that would enable people to "be able efficaciously to regulate the future climate of the earth and consequently prevent the arrival of a new Ice Age" (Ekholm, 1901) under the auspices of this new knowledge. Many years after, Guy Callendar (1938) provided evidence that this mechanism was functioning. He gathered data on average global temperatures and estimated global fuel combustion, concluding that the atmosphere was warming at a rate of 0.033°C annually. His time-series represents a significant contribution to the field, despite the acknowledged possibility that it was influenced by internal climate variations or volcanic forcing (Hawkins et al. 2013).

In a separate story, astronomical perspectives on climate were also being investigated, which added the Earth's relative position to the Sun to the equation of climate variations. The astronomical cycle calculations were developed by Milankovitch (1920), who also calculated the Earth's energy balance as a result of them.

Additionally, the study of the atmosphere through the prism of fluid dynamics allowed for the understanding of climate dynamics and atmospheric circulation. This first occurred in many experiments involving rotating plates filled with atmosphere-like fluids. These types of experiments were the main means of simulating the atmosphere before the computational age, and they developed hand in hand with the theory of hydrodynamics (Taylor, 1921). The relationship between atmospheric and oceanic circulation—which demonstrates one of the earliest methods toward the comprehensive understanding of climatology—was a key idea developed within this type of research.

The physicist Vilhelm Bjerknes (1904) presented a system of differential equations that depicted the physical relationships between atmospheric variables at the start of the 20th century. This ground-breaking plan provided the structure for the development of contemporary meteorology and the explanation of the underlying physical processes (Gramelsberger, 2009; Thorpe et al. (2003)). The numerical approach to problem-solving was later developed after the emphasis shifted to attempting to solve those equations using non-analytical methods, which proved to be ineffective.

A globally integrated view of the climate system was demonstrated by the physical understanding of the climate. The latest theories and models established a connection between the distinct features of regional climates and a cohesive worldwide circulation. They were offering analytical tools for studying climate change rather than for classification. From the local perspective to the global one, and from a static to a dynamic climate. The physical viewpoint of a climatic science took precedence over the conventional geographic understanding of climatology.

1.2.5. Modern Climatological Studies

The war contributed to the technological and numerical revolution that accelerated scientific advancement. John von Neumann, one of the foremost figures in the field of climate at the time, began using the first computers to solve the equations governing atmospheric circulation. Subsequently, as computer power grew, atmospheric circulation models—also known as GCMs, or General Circulation Models—became more sophisticated. They advanced from solving equations in a small number of grid cells to even coupling the oceanic circulation to the atmosphere. This was followed by an increase in the number of modules that represented pertinent elements of the climatic system, including land, soils, vegetation, hydrology, and the cryosphere. The state of the art in our understanding of the climate system at the moment is represented by these models, which were formerly known as Earth System Models (Edwards, 2011).

As these models evolved, so did the international collaboration in this field. The more research teams involved, the more complex they were. In the end, the global climate research community voiced genuine concerns about climate change linked to anthropogenic global warming (Bolin et al. 2007,). The United Nations Framework Convention on Climate Change (UNFCCC), whose goal is to "stabilize greenhouse gas concentrations in the atmosphere at a level that would prevent dangerous anthropogenic interference with the climate system," was thus adopted by the UN in

1992. In this way, the UNFCCC established a set of worldwide goals and increased interest in climate research (Ramakrishna, 2000).

The Intergovernmental Panel on Climate Change (IPCC), an international organization composed of global academic experts on specific climate change topics, was established concurrently with the establishment of the UNFCCC. It is reasonable to argue that the IPCC plays a crucial role in international action and fosters communication within the scientific community, with an emphasis on the integration of politicians and society at large (Randall, 2000). Making assessment reports that convey the state of the art in climate science, particularly as it relates to climate change, accessible to the general public is one of its objectives. The IPCC's enormous and historically unprecedented body of work is what earned them the 2007 Nobel Peace Prize.

Considering that these global organisms impact most of the current climate research, it is critical to keep them in mind. We should be aware that the presence of a global authority in climate science could lead to some unintended consequences, such as uniform or biased responses to the questions posed by the field. Preventing dangerously high levels of climate change is currently the central question in climate science research. A large number of interdisciplinary research groups, each concentrating on a different aspect of the problem's solution, are required to achieve this goal. Although it is not the aim of this article to go over every single one of the many components, it is possible to draw attention to the following:

- Climate Modeling: Studies focused on generating and running the Earth System Models and General Circulation Models using specific sets of scenarios that allow its systematic analysis.
- Integrated Assessment Models: This branch of study generates models that link human systems such as agriculture, energy and economy with simplified climate models in order to find optimal socioeconomic development paths.
- Meta-Studies: Another line of research has been done by studies that link different methodologies, results and models from other major studies in order to solve climate change specific questions.

Outside this framework, there have been several critiques of the tools and research paths that the international entities have been promoting, pointing out that in most of cases, these approaches perpetuate the system that originated the initial problem. Bohm et al. (2012) state that we need to shift out of capitalism to stop climate change.

This approach has been posed within the geophysical community too. Knutti et al. (2016) question the global standards set by international negotiators as desired climate outcomes. Werner (2012) describes the human-Earth interactions within a machine-like framework fueled by the dominant global culture that generates environmental challenges as consequences, such as climate change. He proposes that the only mechanisms that could slow down the dynamics of the system are those generated out of it such as blockades, protests, and sabotages. Along this line, Naomi Klein (2017) disentangles the political and social framework in which we are immersed, trying to shed light over it and encouraging steps to move away from it (Anderson et al., 2016; Arrhenius et al., 1896).

1.3. METEOROLOGY AND CLIMATOLOGY

Meteorology and climatology are the two atmospheric sciences that are intrinsically related. The study of weather, or the general instantaneous state of the atmosphere at a specific location and time, is called meteorology. Specific atmospheric parameters, such as temperature, precipitation, humidity, wind direction, wind speed, cloud cover, and cloud type, are directly measured to describe the weather. The tangible components of the atmosphere are referred to as "weather." It might only take a glance or a stroll outside to accurately describe the weather where you are. Naturally, these observations can be contrasted with other locations' atmospheric conditions, which are typically different.

Weather discussions focus on brief time intervals because meteorology deals with precise and direct measurements of atmospheric properties. Most conversations about the weather take place over periods of a few days at most. Questions like "How's the weather today? How's it different from yesterday? What's the forecast for tomorrow or later in the week?" all entail short-term atmospheric property analysis for a specific time and location. Thus, the only periods covered by meteorology are the present, the recent past, and the near future.

Examining the forces that produce the atmospheric properties under study, however, is a far more crucial aspect of meteorology. Weather patterns vary over time due to variations in the strength or direction of these forces as well as modifications to the internal characteristics of the matter they affect. Meteorology is the only natural science where the main objective is to forecast future conditions, even though many meteorologists are not directly involved in forecasting these changes. Recent technological advancements that enable better observation, data collection, and modeling of the atmosphere, along with a better understanding of these forces, have greatly improved weather forecasting. For the majority of places, weather forecasts issued by the National Weather Service in the United States are currently accurate for up to 72 hours.

Climate, on the other hand, describes the long-term conditions of the atmosphere in a particular location. Though their areas of interest differ, climatologists and meteorologists are both interested in the same atmospheric processes. While climatologists study the processes to comprehend their long-term effects, meteorologists may study the processes for their own sake. Thus, climatology enables us to investigate atmospheric processes and their effects well beyond the current weather.

Normals, extremes, and frequencies are the three aspects of climate data that need to be taken into account. These are used to compare current atmospheric conditions to those from previous similar time periods in order to determine how the atmosphere is doing at any given point in time. The term "normals" describes the typical weather in a location.

Climatic normals provide an overview of the kinds of expected weather conditions for a location for a year and are usually calculated for 30-year periods. For instance, Crestview, Florida experiences warm, muggy summers and cool, but not frigid, winters according to climate normalcy.

Similar to how two students who both have an average of 85% in a class might not have reached that average by receiving the same grade on every graded assignment, two locations could have the same average conditions but with different ranges of those conditions. As a result, extremes are used to characterize the highest and lowest measurements of atmospheric variables that, based on a lengthy series of observations, can be anticipated to occur at a specific location and time. For instance, a temperature of 0°C (32°F) in April at Crestview would be outside of the typical range.

Lastly, frequencies describe the rate of occurrence of a specific phenomenon over an extended period at a specific location. For risk assessment, engineering, or agricultural applications, frequency data are frequently crucial. For example, a homeowner's insurance premium is affected by how frequently hailstorms occur in a city. Alternatively, if a culvert is designed by an engineer to accept 8 cm (3 in) of rain in 5 hours, but that frequency is exceeded on average twice a year; the residents impacted by the culvert may or may not find this rate of failure acceptable. A farmer might be curious to know the average number of days with more than 1.5 cm (0.6 in) of rain in October since any crop harvested in that month is negatively impacted by October rains.

Thus, we can conclude that while meteorologists and climatologists both research the same atmospheric processes, three key distinctions should be noted. First, there are differences in the time scales involved. While climatologists research the long-term patterns and trends of those short-term features, the "climate," meteorologists are primarily interested in features of the atmosphere at a specific time and place, or the "weather." Second, climatologists take into account the processes' long-term effects, whereas meteorologists are more focused on the processes themselves. Third, interactions between the atmosphere and the other "spheres" are more likely to have significant effects over longer time scales rather than shorter ones, as climatology is intrinsically more entwined with processes occurring not only in the atmosphere but also in the other "spheres." This is especially true if those processes take place over wide swaths of land, as the effects typically take longer to manifest in these situations. For example, a complete evaporation of the Great Lakes would unavoidably occur over an extended period. When comparing tomorrow's weather to todays,' there won't be much of an impact from the Great Lakes' fluctuating water level. Therefore, when predicting the weather for tomorrow, a meteorologist wouldn't really need to take this atmosphere–hydrosphere interaction into account. On the other hand, it is more likely that the Great Lakes' divergent water contents over centuries will have a discernible and significant effect on the climate during that time. Thus, when assessing climate, interactions between the atmosphere and other spheres—like in this example—must be taken into account.

Notwithstanding the distinctions between climatology and meteorology, it's critical to acknowledge that these lines are getting fuzzier with time. A proficient climatologist must possess a solid foundation in the principles of atmospheric physics and chemistry, which govern the atmosphere's instantaneous behavior. A proficient meteorologist should be aware of the significance of long-term trends, their effects, and the influence of other elements in the Earth, ocean, and atmosphere system.

Interactions between the atmosphere and social systems are included in the holistic approach to climatology. A theme that has gained popularity in climatology recently is the effect of human activity on the environment. The relationship between climate and many aspects of the human condition is becoming more widely acknowledged. This is particularly true for climatic "extremes" and "frequencies," as the conditions that surpass thresholds and "abnormal" events typically have the biggest effects on people and society.

1.4. SCALES IN CLIMATOLOGY

Climatology has the following scales (Figure 1.1):

i. Micro Scale: The smallest atmospheric scale is called the micro scale. It concerns the systems that function over a smaller region, less than 0.5 km. The study of the climate is limited to very near the ground under this scale. Take the climate around a house, for instance.

ii. Local Scale: The larger scale than the micro scale is the local scale. Systems that cover a region between 0.5 and 5 km are involved. For instance, the climate in a river valley.

iii. Meso Scale: It involves the systems that operate over an area between 5 and 100 km. For example- The climate of a basin or any identified region.

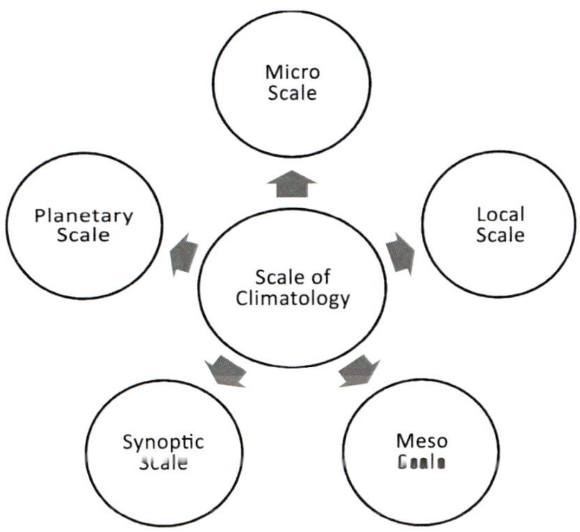

Figure 1.1. Scale of climatology.

Source: https://egyankosh.ac.in/bitstream/123456789/102663/3/Unit-1.pdf

iv. Synoptic Scale: It has to do with systems that cover a distance of 100–10,000 km. For instance, a region's climate.

v. Planetary Scale: Out of all the atmospheric scales, the planetary scale is the largest. It deals with systems that cover an area of 10,000–40,000,000 km. For instance, the climate of the entire planet or a hemisphere (Barry et al., 2013; Bjerknes et al., 1904).

1.5. SUB-FIELDS OF CLIMATOLOGY

It has been noted that, in addition to the traditional scientific disciplines working together more frequently in recent years to investigate climate-related issues, some new collaborations have also emerged. According to Critchfield, there are three fundamental subfields of climatology, and the best examples of these are bioclimatology, agro-climatology, pharma-climatology, construction climatology, and urban-climatology. However, the subfields of climatology can be broadly separated into the following subsections.

1.5.1. Physical Climatology

Meteorology and physical climatology are closely related fields. The primary focus of physical climatology is on atmospheric phenomena. The goal of physical climatology is to elucidate the variables causing both spatial and temporal variations in the atmosphere. The study of weather-related phenomena, such as insolation, temperature, air pressure, wind, evaporation, humidity, precipitation, fog visibility, etc., is the focus of this branch of climatology. Different climates were created by the combination of all or some of these elements.

In other words, this sub-field studies the systematic description and analysis of (Figure 1.2):

- Energy balance of atmosphere and earth;
- Atmospheric temperature;
- Humidity and precipitation;
- Atmospheric motions and air circulation;
- Air masses;
- Extreme atmospheric events like- atmospheric disturbances, droughts, floods, etc.

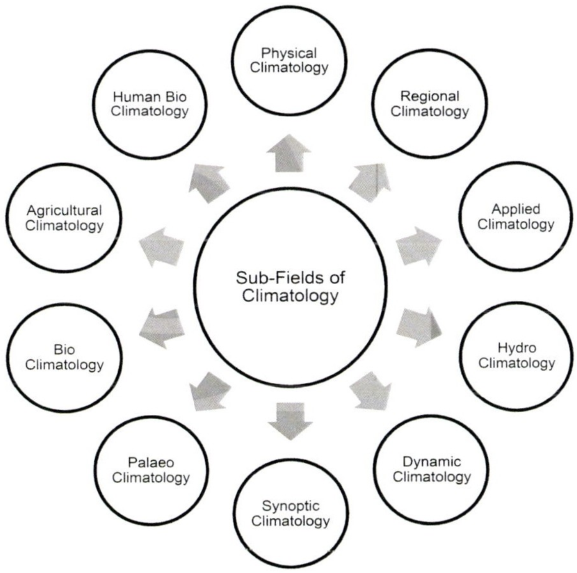

Figure 1.2. Sub- fields of climatology.

Source: https://egyankosh.ac.in/bitstream/123456789/102663/3/Unit-1.pdf.

1.5.2. Regional Climatology

Regional climatology is the study of climate patterns and variations in a specific geographical area, ranging from local to regional scales. It examines the long-term weather conditions, including temperature, precipitation, humidity, wind patterns, and other atmospheric phenomena, within a defined region.

This field integrates various disciplines such as meteorology, geography, environmental science, and climatology to analyze and understand the complex interactions between the atmosphere, land surfaces, oceans, and other factors that influence regional climate. Researchers in regional climatology aim to identify trends, variability, and anomalies in the climate system within a particular area over time.

Some key aspects of regional climatology include:

1. Climate Classification: Classifying different regions based on their climatic characteristics using systems like the Köppen climate classification.
2. Climate Change: Assessing how regional climates are changing over time due to natural variability and human activities such as greenhouse gas emissions.
3. Extreme Events: Studying the frequency and intensity of extreme weather events such as hurricanes, droughts, heatwaves, and heavy rainfall within a region.
4. Impact on Society and Environment: Understanding the impacts of regional climate variations on ecosystems, agriculture, water resources, infrastructure, human health, and socio-economic systems.

5. Modeling and Prediction: Developing models and predictive tools to forecast regional climate patterns and assess potential future scenarios under different climate change scenarios.
6. Adaptation and Mitigation Strategies: Developing strategies and policies to help societies adapt to changing regional climates and mitigate the impacts of climate change.

Regional climatology is essential for policymakers, urban planners, resource managers, and other stakeholders to make informed decisions related to climate resilience, sustainable development, and disaster risk reduction at the regional level.

1.5.3. Descriptive Climatology

Descriptive climatology is a branch of climatology that focuses on characterizing and describing the climate of various regions or locations without necessarily delving into the underlying processes or mechanisms driving those climate patterns. Instead of explaining why certain climatic conditions occur, descriptive climatology aims to provide a detailed account of the climate's observable characteristics, such as temperature, precipitation, humidity, wind patterns, and other atmospheric phenomena.

Key aspects of descriptive climatology include:

1. Climate Data Analysis: Gathering and analyzing historical climate data, including long-term records of temperature, precipitation, and other relevant variables, to describe typical climate patterns and variations within a region.
2. Spatial and Temporal Variability: Examining spatial variations in climate across different geographic areas, as well as temporal variations over various time scales, from daily and seasonal to decadal and centennial.
3. Climate Maps and Atlases: Creating maps and atlases that visually represent climatic features, patterns, and variations across different regions, providing valuable information for researchers, policymakers, and the general public.
4. Case Studies: Conducting case studies to describe and analyze specific climatic events, phenomena, or trends within a particular region, such as heatwaves, droughts, floods, or changes in climate variability.
5. Climate Normals: Calculating climate normals, which represent average climate conditions over a specific period (typically 30 years), to provide a baseline for comparing current climate conditions and assessing long-term trends.

Descriptive climatology provides valuable information for various sectors, including agriculture, forestry, water resource management, tourism, and urban planning. By understanding and describing the climate of different regions, stakeholders can make informed decisions regarding land use, infrastructure development, disaster preparedness, and climate change adaptation strategies.

1.5.4. Applied Climatology

Applied climatology is the practical application of climatological knowledge and techniques to address specific problems or challenges faced by society, industries, and various sectors. Unlike theoretical or descriptive climatology, which focuses on understanding climate patterns and characteristics, applied climatology seeks to use this understanding to solve real-world problems and improve decision-making processes. Here are some key aspects of applied climatology:

1. Climate Risk Assessment and Management: Applied climatologists assess the risks associated with climate variability and change, such as extreme weather events, droughts, and shifts in temperature and precipitation patterns. They develop strategies to mitigate these risks and adapt to changing climatic conditions, helping communities, businesses, and governments prepare for and respond to climate-related challenges.
2. Agriculture and Forestry: Applied climatology plays a crucial role in agricultural and forestry practices by providing information on suitable crops, planting dates, irrigation requirements, pest and disease management, and other factors influenced by climate conditions. Climatologists work with farmers, foresters, and agricultural extension services to optimize resource management and increase resilience to climate variability.
3. Water Resource Management: Climatologists analyze precipitation patterns, snowmelt, river flow, groundwater recharge, and other hydrological variables to support water resource planning, allocation, and management. They develop models and tools to assess water availability, predict droughts and floods, and optimize water use efficiency in various sectors, including agriculture, industry, and municipal supply.
4. Energy Planning and Infrastructure: Applied climatology informs energy planning and infrastructure development by assessing renewable energy resources (such as solar, wind, and hydroelectric power) and their variability over time. Climatologists help design and optimize energy systems, improve energy efficiency, and evaluate the potential impacts of climate change on energy supply and demand.
5. Urban Planning and Design: Climatologists collaborate with urban planners, architects, and engineers to design sustainable and climate-resilient cities. They guide mitigating urban heat islands, managing stormwater runoff, enhancing green spaces, and incorporating climate adaptation measures into urban infrastructure and building design.
6. Health and Public Safety: Applied climatology contributes to public health and safety by studying the impacts of climate on human health, including heat-related illnesses, vector-borne diseases, air quality, and natural disasters. Climatologists work with healthcare providers, emergency responders, and policymakers to develop early warning systems, heat action plans, and other interventions to protect vulnerable populations.

Overall, applied climatology bridges the gap between scientific knowledge and practical applications, helping societies and decision-makers address climate-related challenges and build resilience to climate variability and change.

1.5.5. Hydro-Climatology

Hydroclimatology is a specialized branch of climatology that focuses on the interaction between climate and the water cycle, including precipitation, evaporation, surface runoff, groundwater recharge, and water storage in various forms such as rivers, lakes, and glaciers. It explores how climate patterns and variability influence hydrological processes and how changes in the water cycle, in turn, affect climate.

Key aspects of hydro climatology include:

1. Precipitation Patterns: Hydroclimatologists analyze patterns of precipitation, including frequency, intensity, duration, and spatial distribution, to understand how climate variability and change influence regional and global precipitation regimes. This information is essential for water resource management, agriculture, and flood forecasting.
2. Evaporation and Transpiration: Hydroclimatology examines processes of evaporation from land and water surfaces and transpiration from vegetation, which collectively constitute evapotranspiration. Changes in temperature, humidity, wind, and solar radiation influence evapotranspiration rates, which in turn affect water availability and ecosystem dynamics.
3. Surface Runoff and Streamflow: Hydroclimatologists study surface runoff and streamflow processes, including infiltration, overland flow, channel routing, and river discharge. They assess how precipitation, soil properties, topography, land cover, and human activities influence the timing, magnitude, and variability of streamflow in different watersheds.
4. Groundwater Dynamics: Hydroclimatology investigates groundwater recharge, storage, and discharge processes in aquifers, which play a vital role in sustaining ecosystems and supporting human water needs. Climatic factors such as precipitation, temperature, and soil moisture influence groundwater levels and quality, with implications for water supply, irrigation, and ecosystem health.
5. Snow and Ice Hydrology: In regions with seasonal snow cover or glaciers, hydroclimatologists study snow accumulation, melting, and runoff processes, as well as glacier dynamics and mass balance. Changes in snow and ice contribute to variations in river flow, water availability, and sea level rise, with implications for water resource management and climate adaptation.
6. Climate Change Impacts: Hydroclimatology assesses the impacts of climate change on hydrological processes, including shifts in precipitation patterns, changes in snowmelt timing, alterations in river flow regimes, and increased frequency of extreme events such as floods and droughts. Understanding these impacts is crucial for adapting water management strategies to future climate conditions.

1.5.6. Dynamic Climatology

Dynamic climatology is a subfield of climatology that focuses on understanding the atmospheric processes and dynamics that drive climate variability and change. Unlike descriptive or applied climatology, which primarily examines observed climate patterns and their practical implications, dynamic climatology delves into the underlying physical mechanisms and theoretical models that govern the behavior of the Earth's atmosphere.

Key aspects of dynamic climatology include:

1. Atmospheric Circulation: Dynamic climatology studies the large-scale circulation patterns of the atmosphere, such as the jet streams, Hadley cells, and mid-latitude cyclones. These circulation features transport heat, moisture, and energy around the globe, influencing regional climate patterns and weather phenomena.
2. Climate Modeling: Dynamic climatologists develop and use numerical models of the atmosphere and oceans to simulate past, present, and future climate conditions. These models incorporate equations of motion, thermodynamics, radiation, and other physical processes to simulate the complex interactions between the atmosphere, land surface, oceans, and ice.
3. Climate Feedbacks: Dynamic climatology investigates feedback mechanisms that amplify or dampen climate changes, such as the ice-albedo feedback, water vapor feedback, and cloud feedback. These feedbacks can either reinforce or mitigate the effects of external forcings, such as greenhouse gas emissions or changes in solar radiation.
4. Climate Variability: Dynamic climatologists study natural climate variability on various time scales, from intraseasonal oscillations like the Madden-Julian Oscillation (MJO) to interannual phenomena like the El Niño-Southern Oscillation (ENSO) and decadal variations such as the Pacific Decadal Oscillation (PDO). Understanding these modes of variability is essential for predicting short-term weather and long-term climate trends.
5. Climate Change Attribution: Dynamic climatology seeks to attribute observed changes in climate to specific drivers, such as human activities, natural variability, or external forcings. By analyzing historical climate data and conducting model simulations, climatologists can quantify the relative contributions of different factors to observed changes in temperature, precipitation, and other climate variables.
6. Extreme Weather Events: Dynamic climatology investigates the role of atmospheric dynamics in shaping extreme weather events, including heatwaves, droughts, heavy rainfall, and tropical cyclones. By understanding the physical processes that lead to these events, climatologists can improve forecasting and risk assessment efforts.

1.5.7. Synoptic Climatology

Synoptic climatology is a branch of climatology that focuses on analyzing and understanding the relationship between large-scale atmospheric circulation patterns, known as synoptic weather systems, and regional climate variability. It examines how synoptic-scale weather phenomena, such as weather fronts, cyclones, anticyclones, and air masses, influence local and regional climate conditions over short to medium timescales, typically ranging from days to weeks.

Key aspects of synoptic climatology include:

1. Synoptic Weather Systems: Synoptic climatology examines the characteristics, behavior, and spatial distribution of synoptic-scale weather systems that affect a particular region. These systems include extratropical cyclones, high-pressure systems, atmospheric rivers, and other features associated with changes in temperature, precipitation, wind, and atmospheric pressure.
2. Weather Typing: Synoptic climatologists classify different types of weather patterns or circulation regimes based on synoptic-scale atmospheric conditions, such as pressure patterns, wind directions, and temperature gradients. By identifying recurring weather types, climatologists can assess their frequency, duration, and seasonal variability and analyze their impact on regional climate.
3. Teleconnections: Synoptic climatology investigates teleconnections, which are large-scale atmospheric linkages that connect weather and climate patterns across different regions of the globe. Examples of teleconnection patterns include the North Atlantic Oscillation (NAO), the Pacific-North American (PNA) pattern, and the Arctic Oscillation (AO). These teleconnections can influence regional climate variability by modulating the strength and position of atmospheric circulation features.
4. Climate Indices: Synoptic climatology uses climate indices derived from synoptic-scale atmospheric variables, such as sea level pressure, temperature, and wind, to quantify and track changes in regional climate patterns over time. These indices provide valuable information for monitoring climate variability, assessing climate trends, and understanding the drivers of climate change.
5. Extreme Weather Events: Synoptic climatology examines the synoptic-scale conditions associated with extreme weather events, including heatwaves, cold spells, heavy rainfall, droughts, and severe storms. By analyzing the synoptic patterns preceding and during extreme events, climatologists can identify the atmospheric mechanisms responsible for their occurrence and assess their frequency and intensity in a changing climate.
6. Climate Forecasting: Synoptic climatology contributes to weather and climate forecasting by providing insights into the large-scale atmospheric patterns associated with different weather conditions. Forecasters use synoptic-scale information to predict the evolution of weather systems and assess the likelihood of specific weather outcomes over short to medium-time horizons.

1.5.8. Palaeoclimatology

Paleoclimatology is the study of past climates on Earth, ranging from thousands to millions of years ago. It uses various methods to reconstruct ancient climate conditions and understand the processes and drivers of climate change over geological timescales. By examining evidence from natural archives such as ice cores, tree rings, sediment layers, corals, and fossilized remains, paleoclimatologists can reconstruct past temperature, precipitation, atmospheric composition, and other climate variables.

Key aspects of paleoclimatology include:

1. Proxy Data: Paleoclimatologists use proxy data, which are indirect indicators of past climate conditions, to reconstruct ancient climates. These proxies include isotopic ratios in ice cores and sediment layers, tree ring widths, pollen analysis, fossilized remains of plants and animals, and geochemical signatures preserved in rocks and minerals.
2. Paleotemperature Reconstruction: Paleoclimatologists use proxies such as isotopic ratios of oxygen and hydrogen in ice cores, tree ring widths, and sediment composition to reconstruct past temperature variations. By analyzing these proxies, researchers can reconstruct temperature records spanning hundreds to millions of years and identify long-term climate trends and cycles.
3. Paleoprecipitation Reconstruction: Paleoclimatologists use proxies such as sediment layers, pollen analysis, and isotopic ratios in speleothems (cave formations) to reconstruct past precipitation patterns. These reconstructions provide insights into changes in regional moisture regimes, monsoon dynamics, and drought frequency over geological timescales.
4. Paleoatmospheric Composition: Paleoclimatology investigates changes in atmospheric composition over time, including variations in greenhouse gas concentrations (e.g., carbon dioxide, methane), aerosol levels, and atmospheric circulation patterns. By analyzing air bubbles trapped in ice cores and chemical signatures in sediment layers, researchers can reconstruct past atmospheric conditions and understand the drivers of climate change.
5. Climate Forcing Mechanisms: Paleoclimatology examines natural and anthropogenic factors that influence climate change, such as variations in solar radiation, volcanic eruptions, changes in Earth's orbit and axial tilt (Milankovitch cycles), and human activities (e.g., deforestation, industrial emissions). By studying past climate changes, researchers can assess the relative importance of different forcing mechanisms and their impact on Earth's climate system.
6. Paleoclimate Modeling: Paleoclimatology integrates proxy data with climate models to simulate past climate conditions and test hypotheses about the drivers of climate change. These models help researchers understand the sensitivity of Earth's climate system to external forces and improve projections of future climate change.

1.5.9. Bioclimatology

Bioclimatology, also known as ecological climatology or climatic ecology, is a branch of climatology that focuses on studying the relationships between climate and living organisms, including plants, animals, and ecosystems. It explores how climate influences the distribution, behavior, physiology, and ecology of organisms, as well as the structure and functioning of ecosystems.

Key aspects of bioclimatology include:

1. Species Distribution: Bioclimatologists study how climate factors such as temperature, precipitation, humidity, and sunlight influence the geographical distribution of plant and animal species. They use climate data and ecological models to predict species' habitat suitability and assess the potential impacts of climate change on species ranges and biodiversity.
2. Climate Adaptations: Bioclimatology investigates how organisms have adapted to different climatic conditions over evolutionary time scales. This includes physiological adaptations, such as tolerance to temperature extremes and water stress, as well as behavioral adaptations, such as migration, hibernation, and reproductive timing.
3. Ecosystem Dynamics: Bioclimatologists examine how climate influences ecosystem structure and function, including primary productivity, species composition, and nutrient cycling. They study ecological processes such as photosynthesis, respiration, decomposition, and trophic interactions in relation to climate variability and change.
4. Climate Change Impacts: Bioclimatology assesses the impacts of climate change on biodiversity, ecosystems, and ecosystem services. Climate change can alter species distributions, disrupt ecological relationships, and lead to shifts in ecosystem boundaries and functioning. Bioclimatologists study these impacts to inform conservation strategies and ecosystem management practices.
5. Human Health: Bioclimatology explores the connections between climate and human health, including the effects of temperature, humidity, air quality, and extreme weather events on human well-being. Researchers investigate the relationships between climate variables and the incidence of vector-borne diseases, heat-related illnesses, allergies, and other health impacts.
6. Agroclimatology: Bioclimatology intersects with agroclimatology, the study of climate's influence on agriculture. Bioclimatologists work with agronomists and farmers to understand how climate variability affects crop growth, yield, and quality, and to develop climate-smart agricultural practices that enhance resilience to climate change.

1.5.10. Human Bioclimatology

Human bioclimatology is a subfield of bioclimatology that focuses specifically on studying the interactions between climate and human health, comfort, and well-being. It examines how variations in climate conditions, such as temperature, humidity, air quality, and solar radiation, affect physiological processes, behavior, and health outcomes in human populations.

Key aspects of human bioclimatology include:

1. Temperature Effects: Human bioclimatology investigates the impacts of temperature on human health, including heat-related illnesses such as heat exhaustion and heatstroke, as well as cold-related conditions like hypothermia and frostbite. Researchers study the physiological responses to temperature extremes, individual susceptibility factors, and strategies for mitigating heat and cold stress.
2. Humidity and Air Quality: Human bioclimatology explores the effects of humidity levels, air pollution, and indoor air quality on respiratory health, allergies, and cardiovascular function. High humidity can exacerbate heat-related discomfort and increase the risk of mold growth, while poor air quality can worsen respiratory conditions such as asthma and COPD.
3. Solar Radiation: Human bioclimatology examines the health impacts of solar radiation exposure, including sunburn, skin cancer, and vitamin D deficiency. Researchers investigate the role of factors such as latitude, altitude, cloud cover, and ozone depletion in modulating solar UV radiation levels and assess strategies for sun protection and vitamin D supplementation.
4. Urban Heat Islands: Human bioclimatology studies the phenomenon of urban heat islands, where cities experience higher temperatures than surrounding rural areas due to human activities and infrastructure. Urban heat islands can exacerbate heat-related health risks, particularly during heatwaves, and disproportionately affect vulnerable populations in densely populated urban areas.
5. Climate Change and Health: Human bioclimatology assesses the health impacts of climate change, including changes in temperature extremes, precipitation patterns, and the frequency and intensity of extreme weather events. Climate change can increase the risk of heat-related mortality, vector-borne diseases, food and waterborne illnesses, and mental health disorders.
6. Adaptation and Mitigation Strategies: Human bioclimatology explores adaptation and mitigation strategies to protect human health from climate-related risks. These strategies include urban planning measures to reduce urban heat islands, heatwave early warning systems, green infrastructure initiatives, and public health interventions to promote heat awareness and resilience.

1.5.11. Agricultural Climatology

Agricultural climatology is a branch of climatology that focuses on understanding the influence of climate on agricultural systems, including crop production, livestock management, and agroecosystem dynamics. It examines how climate variables such as temperature, precipitation, humidity, solar radiation, and wind affect agricultural processes, productivity, and sustainability.

Key aspects of agricultural climatology include:

1. Growing Season Conditions: Agricultural climatology assesses the climatic conditions during the growing season, including temperature regimes, frost dates, precipitation patterns, and the

distribution of growing degree days. Understanding these conditions is essential for selecting suitable crops, determining planting dates, and optimizing agricultural practices.

2. Water Availability: Agricultural climatology examines the availability and distribution of water resources, including rainfall, soil moisture, surface water, and groundwater. Researchers study the impacts of droughts, floods, and irrigation practices on crop water requirements, water use efficiency, and soil moisture dynamics.
3. Crop Suitability and Zoning: Agricultural climatology uses climate data and models to assess the suitability of different regions for specific crops and agricultural activities. This includes agroclimatic zoning, which identifies areas with favorable climate conditions for crop cultivation, as well as climate risk assessments to evaluate the potential impacts of climate variability and change on agricultural production.
4. Climate Variability and Change: Agricultural climatology investigates the impacts of climate variability and change on agricultural systems, including shifts in temperature and precipitation patterns, changes in the frequency and intensity of extreme weather events, and alterations in growing season lengths. Researchers develop adaptation strategies to mitigate the risks associated with climate change and enhance agricultural resilience.
5. Pest and Disease Dynamics: Agricultural climatology studies the influence of climate on pest and disease dynamics in agricultural ecosystems. Climatic factors such as temperature, humidity, and precipitation can affect the distribution, abundance, and behavior of pests and pathogens, as well as the development of crop diseases and pest outbreaks.
6. Agroclimatic Forecasting: Agricultural climatology provides weather and climate forecasts tailored to the needs of farmers and agricultural stakeholders. These forecasts help farmers make informed decisions about planting, irrigation, fertilization, pest management, and harvest timing, improving farm productivity and resource efficiency (Bohm et al., 2012; Bolin et al., 2007).

CASE STUDY

IMPROVING CLIMATE MODELLING ACCURACY:

Identifying climate trends frequently relies on comparing data for sensitive indicators used in predictive models that can contain variability generated by affects such as measurement altitude. Ground-based, air-borne and satellite made observations need independent scrutiny to increase confidence in data consolidation. The introduction of more rigorous measurement uncertainty derivations for all inputs to predictive models is needed to improve trend analysis and aid data comparability.

Challenge:

Observations of the Earth's atmosphere, oceans and land from space are vital for climate studies. These are built into models used by climatologists to understand and predict how weather patterns are changing. International initiatives such as the Quality Assurance Framework for Earth Observation (QA4EO) are increasing this community's understanding of the importance of quality assurance, rigorous uncertainty analysis and measurement traceability for underpinning climate trend identification.

Measurement data generated by orbiting satellites requires robust links to ground based instrument responses to validate data quality and support its use in climate modeling. However, changes in the ambient temperature and pressure between ground based and satellite borne instrumentation can introduce significant result variability. The European Space Agency, ESA, believes that there is an increasing need for climatologists to have a good understanding of all the sources of experimental data spread and the effects these introduce in climate trend analysis. Training courses that increase this community's ability to consider parameters that contribute to result variability are needed to increase their skill set and improve the quality of data that underpins reliable climate change trend identification.

Solution:

The EMRP projects European metrology for earth observation and climate and Metrology for Earth Observation and Climate generated training material suitable for analysts working with satelliteborne, air-borne and ground-based observations to evaluate the measurement uncertainties that contribute to data variability. This material was converted into an easy to assimilate course in the EMPIR project Uncertainty Analysis for Earth Observation e-Learning Training Course and is accessible via NPL's Learning Management System. All the components of the course are compatible across multiple devices, operating systems and browsers, thereby allowing users easy access remotely.

Modular in concept, the first stage provides the required skills for assimilating the concepts used in the more practically orientated subsequent stages. Module 3 highlights the APEX spectral instrument used during research flights studying natural phenomena such as vegetation cover on hills or air quality using solar reflectance measurements. By linking satellite-generated reflectance measurement data to ground-based instrument responses and understanding the variation in data this creates is achieved using worked examples for post-launch instrument calibration techniques.

Impact:

Paolo Castracane, ESA/ESRIN, evaluated the course and found it 'very well-structured and its technical content well presented with the adequate degree of repetitions and overlaps to fix concepts.' Another course student, Daniel Kükenbrink, applied its methods in his research on the radiance effects of single trees on their surroundings. These effects generate small changes and having a robust method for assessing measurement uncertainties has given the data he generated greater acceptability by the scientific community.

CLASS ACTIVITY

To understand the diverse climate patterns across different regions of the world and the factors influencing them.

- Ø Materials Needed:

1. World map
2. Marker pens
3. Internet access or reference books
4. Climate classification charts

- Ø Assessment: Assign each group a different region of the world (e.g., North America, Europe, Africa, Asia, South America, Australia). Guide students to identify the factors that contribute to these differences, such as latitude, proximity to oceans, and elevation.

SUMMARY

- Climatology is a critical branch of atmospheric science that studies the long-term patterns and changes in weather conditions over a specific region. It encompasses the analysis of atmospheric data, historical records, computer modeling, and various other techniques to understand the dynamics of climate and its relationship with the environment.
- The nature and scope of climatology is broad and interdisciplinary, encompassing many different fields and applications. It seeks to understand and predict the patterns, causes, and impacts of climate variability and change, and to develop effective solutions to the challenges it poses for sustainable development and social justice.
- The climate of any place is not decided by the weather of only a few days or few years but sometimes it is the average of many years. That is why in general terms, climatology is also called an average or general meteorology. For a common person, the weather of one day or one hour is not as important as the average weather of that place or region. That is why climatology has become more popular than meteorology.
- Physical climatology is closely related to meteorology. Physical climatology is mainly concerned with atmospheric phenomena. Physical climatology tries to explain the factors that responsible for temporal and spatial changes in atmosphere.
- Applied Climatology is concerned with the study of the effects of climate and with the application of the climatological knowledge to practical problems i.e., application of climatic principles and knowledge to solve various problems faced by human society. It also studies that, how man modifies climate by introducing various changes in the physical environment.

REVIEW QUESTIONS

1. What is Climatology? Write about the scope of climatology.
2. Describe the various spatial and temporal scales of climatology.
3. Write in detail about the development of climatology.
4. What are the various scales of climatology?
5. Write about the sub-fields of climatology.

REFERENCES

1. Anderson, T. R., Hawkins, E., & Jones, P. D. (2016). CO_2, the greenhouse effect and global warming: From the pioneering work of Arrhenius and Callendar to today's Earth system models. Endeavour, 40(3), 178–187.
2. Arrhenius, S. (1896). On the influence of carbonic acid in the air upon the temperature of the ground. The London, Edinburgh, and Dublin Philosophical Magazine and Journal of Science, 41(251), 237–276.

3. Barry, R. G. (2013). A brief history of the terms climate and climatology. International Journal of Climatology, 33(5), 1317–1320.
4. Bjerknes, V. (1904). Das Problem der Wettervorhersage: Betrachtet vom Standpunkte der Mechanik und der Physik.
5. Böhm, S., Misoczky, M. C., & Moog, S. (2012). Greening capitalism? A Marxist critique of carbon markets. Organization Studies, 33(11), 1617–1638.
6. Bolin, B., et al. (2007). A History of The Science and Politics of Climate Change. Cambridge University Press.
7. Callendar, G. S. (1938). The artificial production of carbon dioxide and its influence on temperature. Quarterly Journal of the Royal Meteorological Society, 64(275), 223–240.
8. Edwards, P. N. (2011). History of climate modeling. Wiley Interdisciplinary Reviews: Climate Change, 2(1), 128–139.
9. Ekholm, N. (1901). On the variations of the climate of the geological and historical past and their causes. Quarterly Journal of the Royal Meteorological Society, 27(117), 1–62.
10. Falagas, M. E., Bliziotis, I. A., Kosmidis, J., & Daikos, G. K. (2010). Unusual climatic conditions and infectious diseases: Observations made by Hippocrates. Enfermedades Infecciosas y Microbiología Clínica, 28(10), 716–718.
11. Ferrel, W. (1882). An essay on the winds and currents of the ocean. Popular Essays on the Movements of the Atmosphere, Professional Papers of the Signal Service, (12), 7–19.
12. Fleming, J. R. (1999). Joseph Fourier, the greenhouse effect, and the quest for a universal theory of terrestrial temperatures. Endeavour, 23(2), 72–75.
13. Fourier, J. (1822). Théorie analytique de la chaleur. Firmin Didot, Père et Fils.
14. Fourier, J. (1824). Remarques générales sur les températures du globe terrestre et des espaces planétaires. In Annales de Chimie et de Physique (Vol. 27).
15. Gramelsberger, G. (2009). Conceiving meteorology as the exact science of the atmosphere: Vilhelm Bjerknes's paper of 1904 as a milestone. Meteorologische Zeitschrift, 18(6), 669–673.
16. Hadley, G. (1735). Concerning the cause of the general trade-winds. Philosophical Transactions, 39(436–444), 58–62.
17. Halley, E. (1753). An historical account of the trade winds, and monsoons, observable in the seas between and near the tropics, with an attempt to assign the physical cause of the said winds. Philosophical Transactions (1683–1775), 16, 153–168.
18. Hawkins, E., Jones, P., et al. (2013). On increasing global temperatures: 75 years after Callendar. Quarterly Journal of the Royal Meteorological Society, 139(677), 1961–1963.
19. Landsberg, H. (1964). Roots of modern climatology. Journal of the Washington Academy of Sciences, 54(4), 130–141.

20. Landsberg, H., & Oliver, J. (2005). Climatology. In Encyclopedia of World Climatology (pp. 272–283). Springer.
21. Langley, S. P. (1884). Researches on solar heat and its absorption by the earth's atmosphere: A report of the Mount Whitney expedition. In Professional Papers of the Signal Service (Vols. 14–16). US Government Printing Office.
22. Louie, K. S., & Liu, K. B. (2003). Earliest historical records of typhoons in China. Journal of Historical Geography, 29(3), 299–316.
23. Manabe, S., Bryan, K., & Spelman, M. J. (1975). A global ocean-atmosphere climate model. Part I. The atmospheric circulation. Journal of Physical Oceanography, 5(1), 3–29.
24. Maury, M. F. (1855). The Physical Geography of The Sea. Nelson and Sons, Paternoster Row, London.
25. McMichael, A. J., & Woodruff, R. E. (2005). Climate change and human health. In Encyclopedia of World Climatology (pp. 209–213). Springer.
26. Naudts, K., Chen, Y., McGrath, M. J., Ryder, J., Valade, A., Otto, J., & Luyssaert, S. (2016). Europe's forest management did not mitigate climate warming. Science, 351(6273), 597–600.
27. Phillips, N. A. (1956). The general circulation of the atmosphere: A numerical experiment. Quarterly Journal of the Royal Meteorological Society, 82(352), 123–164.
28. Ramakrishna, K. (2000). The UNFCCC history and evolution of the climate change negotiations. In Climate Change and Development (pp. 47–62). Yale School of Forestry and Environmental Studies and UNDP.
29. Taylor, G. I. (1921). Experiments with rotating fluids. Proceedings of the Royal Society of London. Series A, Containing Papers of a Mathematical and Physical Character, 100(703), 114–121.
30. Thompson, K. (1990). Forests and climate change in America: Some early views. *Climatic Change, 3*(1), 47–64.
31. Thorpe, A. J., Volkert, H., & Ziemianski, M. (2003). The Bjerknes' circulation theorem: A historical perspective. *Bulletin of the American Meteorological Society, 84*(4), 471–480.
32. Torretti, R. (2007). Eratosthenes of Cyrene. In *The Biographical Encyclopedia of Astronomers* (pp. 341–342). Springer.
33. Tyndall, J. (1859). On the transmission of heat of different qualities through gases of different kinds. In *Proceedings of the Royal Institution* (Vol. 3, pp. 155–158).
34. Werner, B. (2012). Is Earth f**ked? Dynamical futility of global environmental management and possibilities for sustainability via direct action activism. In *AGU Fall Meeting Abstracts*, San Francisco, Calif., 3–7 Dec. Retrieved from: https://archive.org/stream/weatherpredictio00richrich (accessed on 30 July 2024).

CHAPTER 2

Atmospheric Composition and Structure

LEARNING OBJECTIVES

After studying this chapter, you will be able to:

- Describe the composition of atmosphere;
- Discuss about atmospheric structure;
- Explain vertical structure of the atmosphere: mass density, pressure, and temperature;
- Describe the height and latitude structure of the atmosphere;
- Illustrate the interpretation of atmospheric temperature structure;
- Explain atmospheric pressure and temperature variations with altitude.

KEY TERMS FROM THIS CHAPTER

Adiabatic heating	Atmosphere	Atmospheric mass
Denitrification	Exosphere	Gravitational tides
Greenhouse gas	Hygroscopic aerosols	Latitude
Mesosphere	Plasmasphere	Radiative time
Stratosphere	Thermal conduction	Thermosphere
Troposphere		

INTRODUCTION

The Earth's atmosphere is a complex and dynamic system composed of a mixture of gases that envelop the planet, playing a crucial role in sustaining life and influencing climate and weather patterns. The composition of the atmosphere is primarily nitrogen (about 78%) and oxygen (about 21%), with trace amounts of other gases such as argon, carbon dioxide, neon, helium, methane, krypton, and hydrogen. Water vapor, although variable in concentration, is also a significant component, contributing to weather processes and the greenhouse effect. The presence of aerosols, tiny solid or liquid particles, further adds to the complexity of the atmosphere, affecting air quality and climate.

Structurally, the atmosphere is divided into several layers, each characterized by distinct properties and behaviors. The troposphere is the lowest layer, extending up to about 8 to 15 kilometers above the Earth's surface. It contains most of the atmosphere's mass and is where weather phenomena occur. The temperature in the troposphere decreases with altitude. Above the troposphere lies the stratosphere, reaching up to about 50 kilometers. This layer contains the ozone layer, which absorbs and scatters ultraviolet solar radiation, leading to a temperature increase with altitude. The mesosphere extends from about 50 to 85 kilometers above the Earth, where temperatures again decrease with altitude, reaching the coldest temperatures in the atmosphere. Above this is the thermosphere, extending up to about 600 kilometers, where temperatures rise significantly due to the absorption of high-energy solar radiation. Finally, the exosphere is the outermost layer, gradually transitioning into space, where the atmosphere thins out into a sparse distribution of atoms and molecules.

Understanding the composition and structure of the atmosphere is crucial for comprehending various atmospheric processes, including weather, climate change, and the protective functions of the ozone layer. It also underpins many aspects of environmental science, meteorology, and climate studies, making it a vital area of research and education.

2.1. COMPOSITION OF ATMOSPHERE

Composition literally means "ingredients" or "constituents" of something. Stated differently, it refers to the process by which something is composed. When we use the same meaning for atmosphere, we are referring to the things or components that make up our atmosphere. Our atmosphere is a mechanical mixture of gases, water vapor, and dust particles because it is made up of many different gases and other substances. Let's talk about the elements that make up the atmosphere.

All of the planet's living things depend heavily on the atmosphere, which is a mechanical mixture of various gases and other materials. Together, the four main gases—nitrogen, oxygen, argon, and carbon dioxide—make up 99.99% of the dry air's total volume. Nitrogen has a maximum concentration of over 78%, while oxygen has a concentration of slightly less than 21% (Table 2.1).

Table 2.1. Atmospheric Gases

Groups	**Gases**	**Volume % of dry air**
Major Gases	Nitrogen	78.084
	Oxygen	20.9476
	Argon	0.934
	*C_O2	0.04
Minor Gases	*Methane	0.002
(* are also variable gases)	Neon	0.001818
	Helium	0.000524
	Krypton	0.00005
	Hydrogen	0.0000087
	Xenon	0.00006
	*Ozone	
Variable Gasses	Water vapour	Variable amount
(CO_2 methane and ozone gases are also variable)	Dust particles	Variable amount
	Aerosols	Variable amount

Source: https://dspmuranchi.ac.in/pdf/Blog/Climatology_2_Final_.pdf.

2.1.1. Nitrogen

With 78.084% of the total volume of dry gases in the atmosphere, nitrogen is the most abundant element (Figure 2.1). This is virtually chemically inert and unrelated to any chemical activity occurring in the atmosphere. Because it is not easily combined with other elements, it is referred to as a neutral substance. This gas can be found above 100 km in altitude, but its concentration is lower than 50 km above sea level. This gas is important for plant and animal growth and reproduction. A very tiny amount of atmospheric nitrogen can be converted by certain soil bacteria into nitrates, which are then fixed to the soil and water bodies where it can be ingested by plants and animals. We refer to this process as "nitrogen fixation." Denitrification is the chemical process by which bacteria transform the nitrogen fixed in the earth's surface and return it to the atmosphere.

2.1.2. Oxygen

With 20.9476% of all dry atmospheric gases, it is the second-largest gas in the atmosphere (Figure 2.1). For a great number of the planet's living things to survive, it is vital. It is an extremely active gas chemically. It can be combined with a number of other elements to create a variety of compounds. Oxygen is necessary for fuel combustion. Anything that burns consumes oxygen, which aids in the burning process. Although oxygen can be found over 100 kilometers, it is mostly present within 16 kilometers of altitude. As one rises in elevation, the concentration of oxygen rapidly falls. Because there is rarely any oxygen available for breathing on a mountainside, mountaineers are expected to carry oxygen.

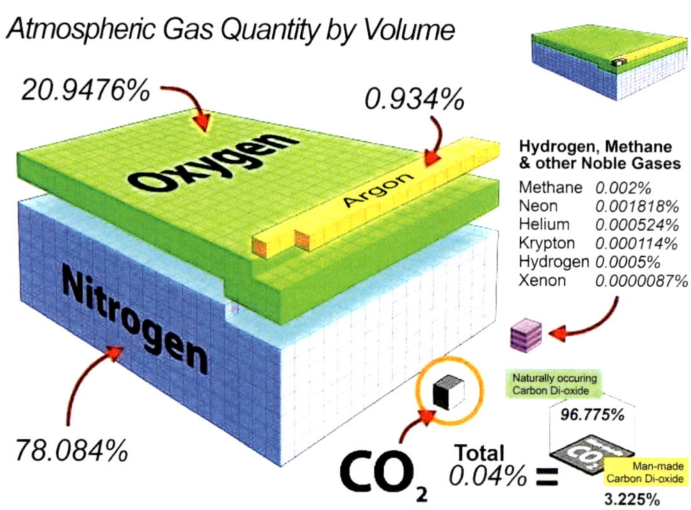

Figure 2.1. *Atmospheric gases and their proportions.*

Source: https://dspmuranchi.ac.in/pdf/Blog/Climatology_2_Final_.pdf.

2.1.3. Argon

Argon is the third-largest gas in the atmosphere in terms of percentage, making up 0.934 of the entire dry atmospheres. It is an inactive gas in terms of chemistry. It can also be found in seawater and the crust of the earth. It is a component of fluorescent and electric bulbs.

2.1.4. Carbon Dioxide (CO2)

It is the atmosphere's fourth most prevalent gas. It is the densest gas and is located at the bottom. It is primarily found in the lower strata and can be found up to a height of around 30 km. Its proportion is extremely low, i.e., 0.04%, but it is especially important for the development of the biosphere's vegetative life. Although it is transparent to incoming solar radiation, it prevents some of it from escaping. It is therefore referred to as a greenhouse gas. It is extremely important in raising the global temperature.

Because of its dependence on combustion, human activity, and the planet's vegetative cover, it is also referred to as variable gas. The use of energy, transportation, industry, agriculture, waste production, and other human activities are all contributing factors to the atmospheric concentration of carbon dioxide. In addition to these naturally occurring sources, there are also some man-made ones, such as the respiration of plants and the natural denudation process that releases carbon stored in rocks into the atmosphere. All of these are causing the atmosphere to rise. This gas is becoming more prevalent as society's economic development progresses (Figures 2.2 and 2.3). Figure 2.3 shows that economically developed nations and regions produce large volumes of carbon dioxide emissions.

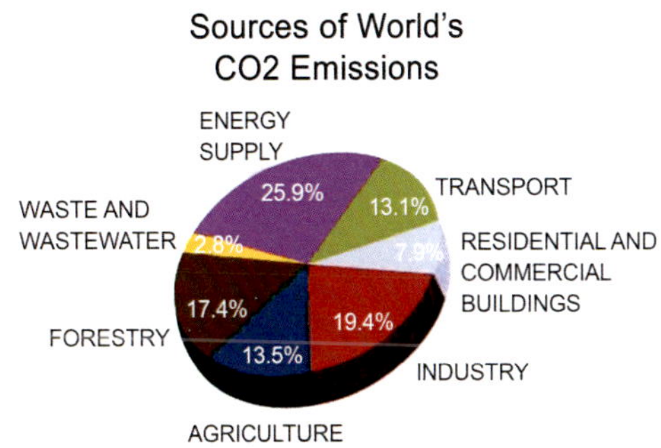

Figure 2.2. *Different sources of carbon dioxide.*

Source: https://dspmuranchi.ac.in/pdf/Blog/Climatology_2_Final_.pdf.

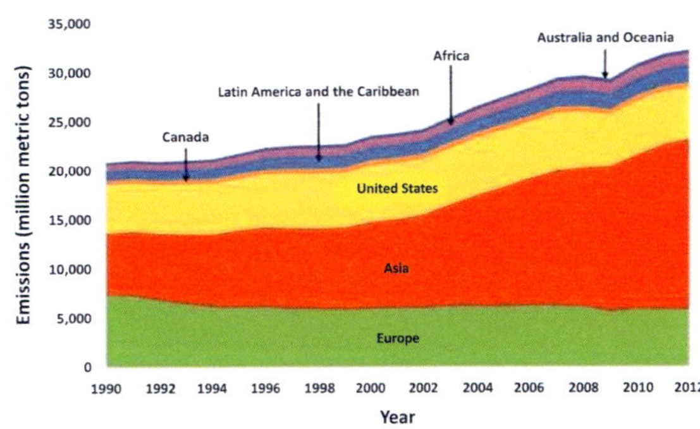

Figure 2.3. *Global carbon dioxide emission by region: 1990–2012.*

Source: https://dspmuranchi.ac.in/pdf/Blog/Climatology_2_Final_.pdf.

2.1.5. Methane

Methane is another greenhouse gas that raises air temperature by absorbing radiation. Methane is also produced in the atmosphere by paddy farming. Additionally, it is created in wetlands and soggy soils and released into the atmosphere. Methane emissions into the atmosphere can also be attributed to fossil fuels. Its atmospheric concentration varies. Figure 2.4 illustrates the various methane sources.

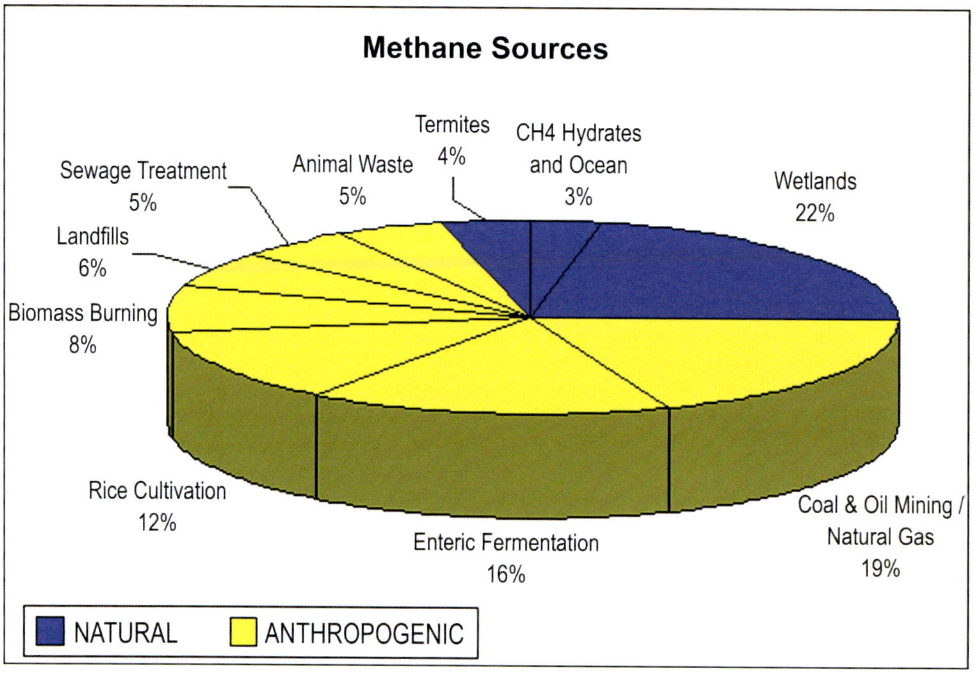

Figure 2.4. *Different sources of methane.*

Source: https://dspmuranchi.ac.in/pdf/Blog/Climatology_2_Final_.pdf.

2.1.6. Ozone

The concentration of ozone is found in a belt that spans the heights of 15 to 50 km in the atmosphere. Ozone is made up of three oxygen atoms, represented by the symbol O_3, as opposed to the usual two oxygen atoms. It is created when ultraviolet solar radiation breaks the molecules of oxygen in the atmosphere. It could even form during a thunderstorm's electrical discharge. This gas is also known as variable because a variety of activities are necessary for both its formation and disintegration. Even though ozone is extremely rare (0.00006%), it is essential to life as it shields the planet from harmful UV radiation by absorbing it. Other minor gases include helium, xenon, hydrogen, krypton, and neon. Some gases are still very rare; they are referred to as trace gases. Ammonia, carbon monoxide, sulfur dioxide, nitrogen dioxide, nitrous oxide, sulfur hexafluoride, and other substances are significant among them.

2.1.7. Water Vapor

Despite its tiny size, water vapor plays a crucial role in the distribution of life and vegetation in the atmosphere. Water vapor is always present in the atmosphere, although its concentration varies according to the time of year, the temperature, and the availability of water for evaporation and evapotranspiration. The air is rarely totally dry. Because of the high summer temperatures, the air has a larger water-holding capacity than the low winter temperatures.

When there is enough water on the surface of the earth or in a body of water, there is more vapor present; when there is less opportunity to avoid the same, there is less vapor. Examples include subtropical hot desert regions with little to no water availability and equatorial regions rich in water bodies (high vapor). Air and vapor are both gases, but while they mix and move naturally, they do not exhibit the same homogeneity.

There is never more vapor in the atmosphere than four times its total volume at any given time. Only the troposphere contains it, and the lower level is where it is most concentrated. 90% of the total vapor is found below a distance of six kilometers. About half of the population is thought to reside within two kilometers of elevation. Because water vapor has greenhouse properties, it is essential to maintaining Earth's temperature.

2.1.8. Dust Particles

There are enormous amounts of suspended dust particles in the lower atmosphere. These dust particles are made of solid materials that wind has carried with them from different sources. Higher wind velocities produce more dust particles. Not only are solid particles suspended, but they are also carried over long distances. Their sizes differ greatly. They can only be large and suspended when the wind has a high drafting ability; when the wind's carrying capacity decreases, they settle quickly. You must have seen the storms and gales for yourselves as well.

Even in perfectly still air, a vast number of microscopic dust particles are suspended. The amount of dust particles varies with the season, being more prevalent in the dry seasons when the soil is loose and easily carried by winds, and less so in the rainy seasons when the soil is compacted and settled. Around the world, it is higher in subtropical hot desert regions and less in polar and equatorial regions. While coarse-sized dust particles are prevalent close to the surface, minute dust particles can be found several kilometers above the surface. Microscopic particles are crucial because they serve as the condensation and precipitation nuclei.

2.1.9. Aerosols

Aerosols are incredibly finely ground solid particles or liquid droplets that cling to the surface of a gas for an extended period of time. When they are more focused, they can be seen; otherwise, they are invisible. Aerosols are microscopic, non-gaseous substances that are released into the atmosphere by a variety of natural and man-made sources. They could be tiny particles of dust from the earth, sea salt, pollen, carbon soot from burning fuels, dust from volcanoes, etc. Aerosols are also aided in their atmospheric entry by human activity. Their focus is primarily on urban and industrial areas. Aerosols in the air are also pumped by the burning of fossil fuels and the production of smoke. Thus, aerosols originate from both natural and human-generated sources. They are divided into two categories: non-hygroscopic (moisture non-absorbing) and hygroscopic (moisture absorbing and retaining). Hygroscopic aerosols aid in precipitation by serving as the condensation's nuclei.

It is fairly clear from the description above that our atmosphere is composed of countless tiny molecules of various gases, the topic of which was previously discussed. In addition, a number of non-gaseous substances that are an integral part of the atmosphere can also be found in the air. They are significant in and of themselves and are necessary for the earth to be a habitable planet (Figure 2.5).

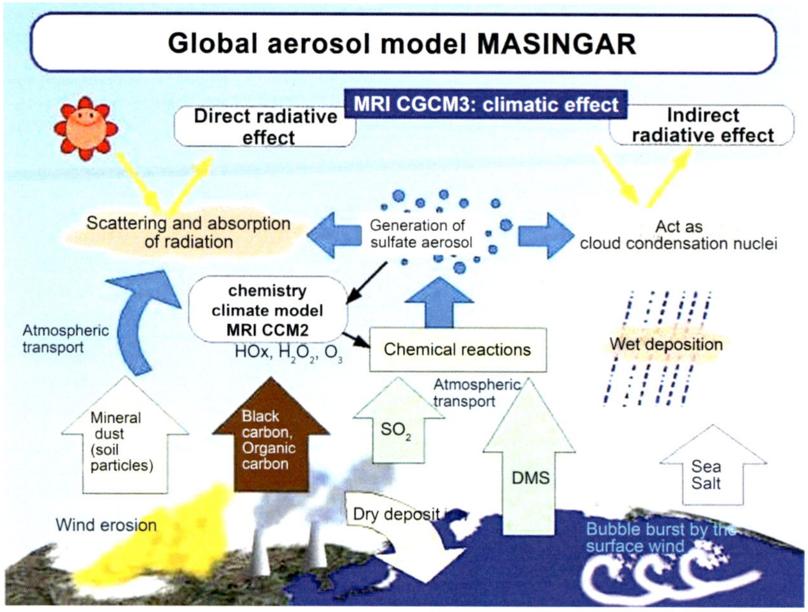

Figure 2.5. *Different sources of aerosols.*

Source: https://dspmuranchi.ac.in/pdf/Blog/Climatology_2_Final_.pdf.

2.2. ATMOSPHERIC STRUCTURE

Given this context, the study of the various components of the atmosphere and their interactions is referred to as the atmosphere's structure. The atmosphere is split vertically into various layers and sections. As a result, the study of the various layers is referred to as atmosphere structure.

There are multiple concentric strata or layers that make up the gaseous region that envelops the planet. The first 20 miles (32 km) above Earth's surface contain 99% of the entire mass of the atmosphere. An overview of the discovery of atmospheric structure in history.

Differences in temperature between the layers of the atmosphere are mainly caused by solar radiation absorption: visible light at the surface, near ultraviolet light in the middle atmosphere, and far ultraviolet light in the upper atmosphere (Figure 2.6).

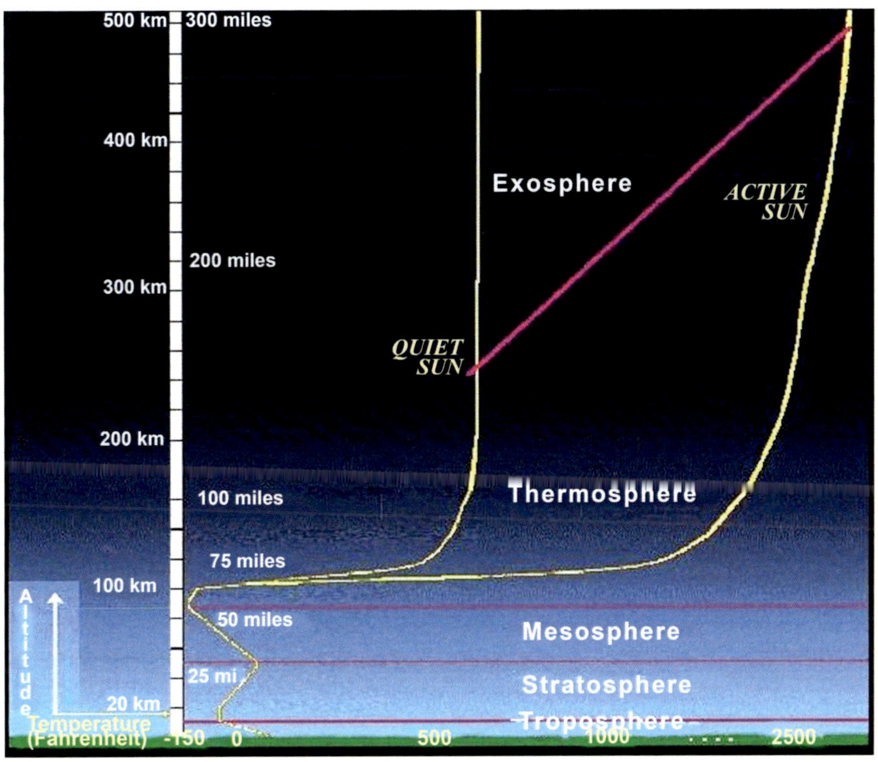

Figure 2.6. *Atmospheric layers.*

Source: https://www.albany.edu/faculty/rgk/atm101/tprofile.jpg.

2.2.1. Troposphere

Around 80% of the mass of the entire atmosphere is found in the troposphere, the atmospheric layer that is closest to the planet. As altitude increases, the troposphere's temperature and water vapor content both rapidly decrease. Since water vapor absorbs heat radiation from the planet's surface and solar energy, it has a significant impact on air temperature regulation. Ninety-nine percent of the atmospheric water vapor is found in the troposphere. The concentration of water vapor varies with latitude. They are highest above the tropics, where they can reach up to 3%, and get smaller as one moves toward the poles.

All weather phenomena take place in the troposphere, though turbulence can sometimes reach the lower stratosphere. The name "troposphere" (which translates to "region of mixing") comes from the strong convective air currents that exist within the layer.

The top border of the layer, referred to as the tropopause, varies in height from 11 miles (18 km) above the equator to 5 miles (8 km) close to the poles. Its height is also seasonal, reaching its maximum in the summer and its lowest in the winter.

2.2.2. Stratosphere

The second main layer of air in the atmosphere is called the stratosphere. It rises to a height of roughly 30 miles (50 km) above the planet's surface above the tropopause. Up to a height of fifteen miles (25 km), the temperature of the air in the stratosphere is essentially constant. After that, it progressively rises to the stratopause. Because the temperature of the air in the stratosphere rises with altitude, convection is prevented and the atmospheric conditions in the area are stabilized. Because there is very little water vapor in the stratosphere, ozone plays a major role in controlling the temperature regime of the layer. The concentration of ozone causes a rise in temperature. The conversion of solar energy to kinetic energy occurs when ultraviolet radiation is absorbed by ozone molecules, leading to the heating of the stratosphere.

The ozone layer is centered between ten and twenty miles (15 and 25 kilometers) above Earth. The stratosphere is home to about 90% of the ozone in the atmosphere. The ozone concentration in this area is roughly 10 parts per million by volume (ppmv), while the troposphere's concentration is roughly 0.04 ppmv.

The majority of solar UV-B radiation, or wavelengths between 290 and 320 nm, is absorbed by ozone. These wavelengths can be absorbed by cells' nucleic acid, which makes them hazardous to life. Increased UV radiation penetration to the planet's surface would be detrimental to the environment and harm plant life. Significantly high solar UV radiation would have a wide range of biological effects, including a sharp rise in cancer cases.

2.2.3. Mesosphere

Lower temperatures are a feature of the mesosphere, a layer that lies between 30 and 50 miles (50 and 85 km) above the surface. The mesopause, the top layer of the atmosphere, has the lowest temperatures on Earth, particularly in the summertime close to the pole. Because it has likely been the least studied of the atmospheric layers, the mesosphere has occasionally been jokingly referred to as the "ignorosphere." The middle atmosphere is sometimes used to refer to the stratosphere and mesosphere combined.

2.2.4. Thermosphere

Above the mesosphere is the thermosphere. Depending on solar activity, the thermosphere's temperature typically rises with altitude, reaching 600–3000 F (600–2000 K). The limited amount of molecular oxygen that is still present is what is causing this temperature increase. It is the result of intense solar radiation absorption. The gas molecules are widely separated at this very high altitude. The chemical makeup of the air changes significantly with altitude above 60 miles (100 km) above Earth's surface, and lighter gases (atomic oxygen, helium, and hydrogen) are added to the atmosphere. Earth's atmosphere becomes too thin to sustain aircraft at elevations of 60 miles (100 km), so vehicles must reach orbital velocities in order to stay in the air. The Karman Line is the dividing line between astronautics and aeronautics. Atomic oxygen becomes the predominant atmospheric component above a height of roughly 100 miles (160 km). Because of gravitational separation, the residual gases start to stratify according to molecular mass at very high altitudes.

2.2.5. Exosphere

The atmosphere's furthest point from Earth's surface is known as the exosphere. A upward-traveling molecule in the exosphere has little chance of colliding with other molecules and can either escape to space if it is moving quickly enough or be drawn back to Earth by gravity if it isn't. Depending on solar activity, the thermopause, also known as the exobase, is located at a height of between 150 and 300 miles (250 and 500 km). The theoretical upper bound can be found at the altitude (roughly 120,000 miles, or half the distance to the Moon) where the gravitational pull of Earth surpasses the effect of solar radiation pressure on atomic hydrogen velocities. At least 60,000 miles are seen to separate the Earth's surface from the exosphere, which is visible from space as the geocorona. A region that lies between interplanetary space and Earth's atmosphere is called the exosphere.

The behavior and quantity of free electrons and other charged particles are another factor used to categorize the upper atmosphere into different regions.

2.2.6. Ionosphere

Because of the presence and fluctuation of free electron concentration in the atmosphere, the ionosphere is defined by atmospheric effects on radiowave propagation.

- *D-region* is about 35 to 55 miles (60 – 90 km) in altitude but disappears at night.
- *E-region* is about 55 to 90 miles (90 – 140 km) in altitude.
- *F-region* is above 90 miles (140 km) in altitude.

It is divided into two regions during the day: the F_1-region, which is located between 90 and 115 miles (140 and 180 km) in altitude, and the F_2-region, which is located between 150 and 300 miles (about 250 and 500 km) in altitude, where the electron concentration peaks. The most recent Height of Maximum (hmF2) map. The term "Topside Ionosphere" typically refers to the ionosphere that is located above the peak electron concentration (Figure 2.7).

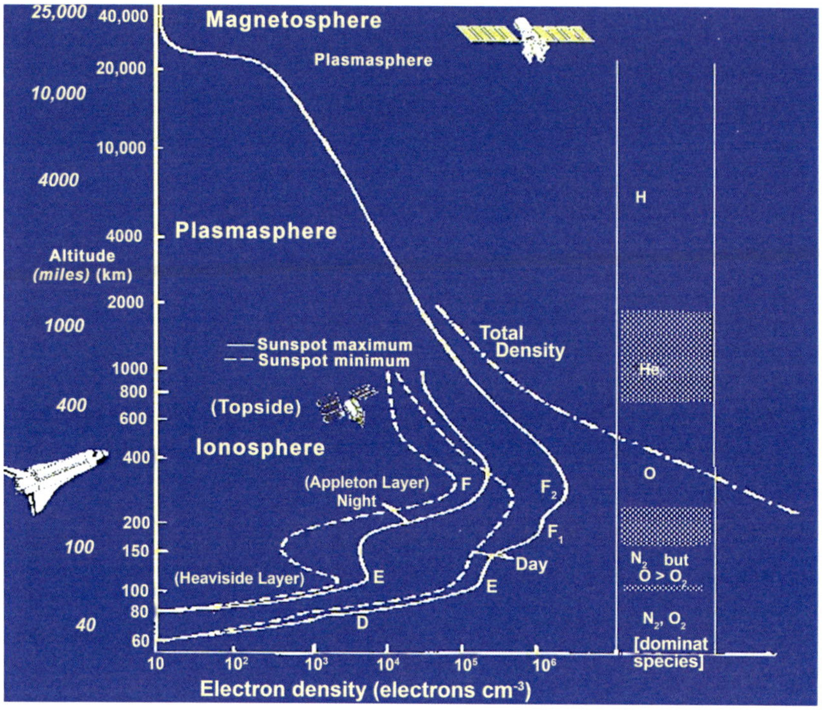

Figure 2.7. *Ionosphere.*

Source: https://www.albany.edu/faculty/rgk/atm101/eprofile.jpg.

2.2.7. Plasmasphere

The hole in the plasmasphere is aligned with Earth's magnetic axis, making it resemble a torus rather than a spherical region. [The suffix -sphere is used here more figuratively to refer to a "sphere of influence." The fourth state of matter, plasma, makes up the Earth's plasmasphere. (Use the Matter Sorter to test your ability to sort the states of matter. The primary constituents of this plasma are electrons and hydrogen ions, or protons. It has a very sharp edge called the plasmapause. The outer edge of this doughnut over the equator is typically 12,000–20,000 miles (19,000–32,000 km) above the surface, or 4–6 Earth radii from the planet's center. In essence, the plasmasphere is the ionosphere's extension. Geomagnetic field lines rotate with the Earth inside the plasmapause. The altitude at which protons overtake oxygen as the dominant species in the ionospheric plasma is considered the inner edge of the plasmasphere; this typically happens at an altitude of roughly 600 miles (1000 km). Another way to think of the plasmasphere is as a structure inside the magnetosphere (Figure 2.8).

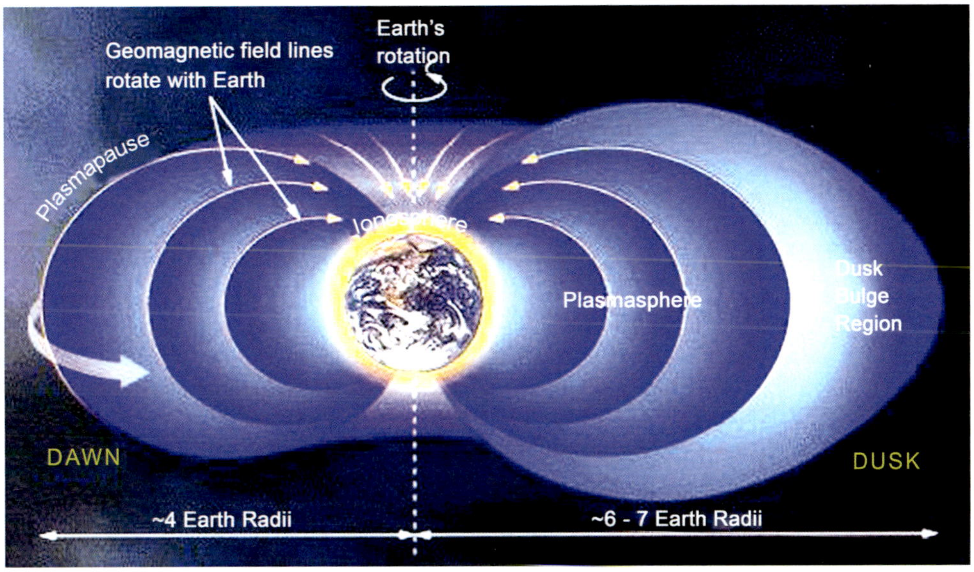

Figure 2.8. *Plasmasphere.*

Source: https://www.albany.edu/faculty/rgk/atm101/plasmasf.jpg.

2.2.8. Magnetosphere

Magnetic field lines outside the plasmapause are strongly influenced by electric fields of solar wind origin, which prevents them from corotating. The Earth's magnetic field is limited by the solar wind and interplanetary magnetic field (IMF) in the magnetosphere, a cavity that is likewise not spherical.

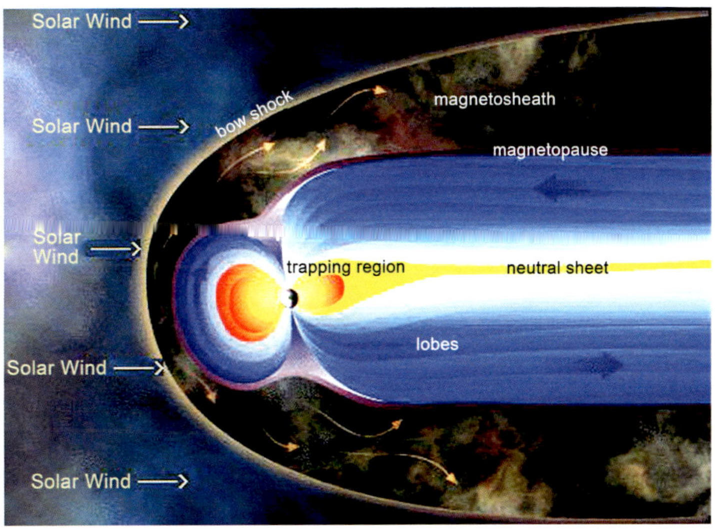

Figure 2.9. *Magnetosphere.*

Source: https://www.albany.edu/faculty/rgk/atm101/magneto.gif.

The magnetopause is the outermost point of the magnetosphere. The magnetosphere has a tail that points away from the sun and is shaped like an elongated teardrop, much like an ornament for a Christmas tree. The magnetopause is normally found on the day side of the Earth at roughly 10 Earth radii, or about 35,000 miles (or 56,000 km) above the surface. On the night side of the Earth, it extends into a long tail called the magnetotail that is a few million miles long (or about 1000 Earth radii), well past the Moon's orbit (at about 60 Earth radii). The Moon, however, is typically outside of the magnetosphere, except a few days during the Full Moon. The magnetosheath and bow shock, parts of the solar wind disrupted by Earth's presence and magnetic field, are located beyond the magnetopause (Atreya et al., 1978; Bacmeister et al., 1995) (Figure 2.9).

2.3. VERTICAL STRUCTURE OF THE ATMOSPHERE: MASS DENSITY, PRESSURE, AND TEMPERATURE

A few fundamental principles of solar system dynamics must be remembered in order to comprehend the composition of Titan's atmosphere. First, seasonal effects are significant because of Saturn and Titan's 26.73° axial tilt. Second, Saturn has an orbital eccentricity of 0.05415, which results in a variation of about 20% in the solar flux and slightly more than 1 AU in the planet's distance from the Sun.

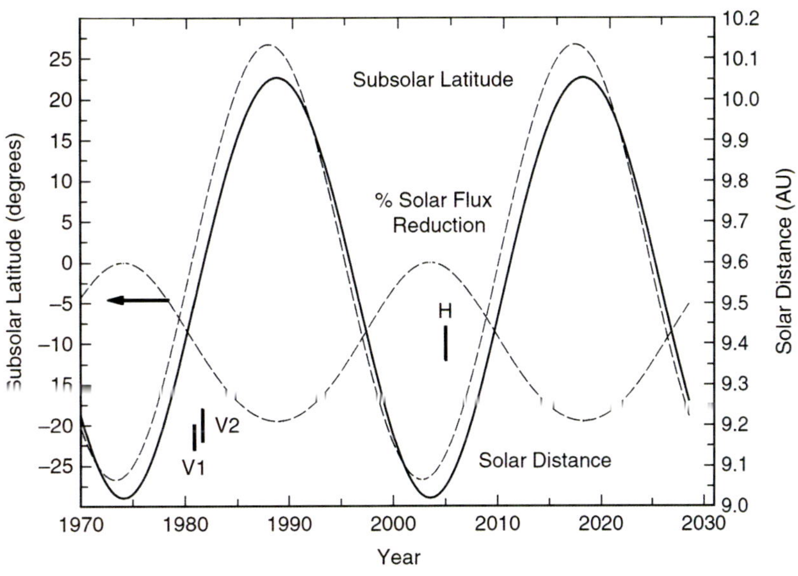

Figure 2.10. *The latitude of the subsolar point (dashed line), the distance from the Sun in AU (solid line), and the percentage variation in solar flux relative to perihelion as a function of time in years (dashed line, read ordinate as %). The last perihelion was 2003.56; the last summer solstice in the southern hemisphere was 2002.76. The voyager flybys (V1, V2) and the huygens probe entry (H) are indicated.*

Source: https://websites.umich.edu/~atreya/Chapters/Strobel-Atreya%20etal%20Titan%20fr%20C-H.pdf.

In Titan's southern hemisphere, the summer solstice took place in 2002.76, and perihelion took place last in 2003.56. While the much larger solar cycle variations in UV and EUV solar radiation will overwhelm the smaller eccentricity effects in the thermosphere, the variation in solar distance and hence solar flux will be most significant in the troposphere and stratosphere. While the Cassini spacecraft's orbit insertion in July 2004 occurred during the descending phase from peak activity, which occurred in April 2000, to solar minimum conditions in the August–November 2007 timeframe, the Voyager spacecraft flybys occurred during high solar activity. As a result, the Cassini-Huygens Mission operated primarily at low solar activity (Figure 2.10).

The vertical stratification caused by gravity identified by an e-folding pressure scale height H and varies from 15 to 100 km on Titan, is the most notable spatial variation in a planetary atmosphere. It follows that the distribution of the three thermodynamic variables—mass (r) or number (n) density, pressure (p), and temperature (T)—that characterize planetary atmospheres can be written as e. G. the latter as

$$T(\phi,\theta,z,t) = \langle T \rangle(z) + \overline{T}(\theta,z,t) + T'(\phi,\theta,z,t).$$

The first term is the vertical temperature profile with T averaged over the globe on each height or pressure surface if $z = -\int H(d\ln p)$, depending on the altitude variable. The second term is temperature departure from the first term averaged over longitude, ϕ, on a latitudinal circle at latitude θ and height z. The last term is the longitudinally varying departure from the first two terms and characterizes waves in the atmosphere. Seasonal variations are presumed to be present in the latter two quantities and absent from the first term in the troposphere and lower stratosphere, because of the long radiative time constants. Since most pronounced seasonal time variations are expected at high latitudes, 1D vertical profiles obtained at equatorial latitudes should be approximately equal to rigorously calculated <T>(z) from global data.

Within Different 1D profiles obtained at equatorial latitudes are compared with Figure 2.11. Regarding the radio occultation data obtained by Voyager at 6.2°N and 8.5°S, Lellouch et al. The 1989 analysis is used. With regard to the UVS solar occultation data collected at 4°N and 16°S, Vervack et al. (2004) profile is presented, which, if isothermal, would indicate an atmosphere at T ~ 155 K. These data sets were used by Yelle et al. An engineering model for the Cassini-Huygens mission was created in 1997. Keep in mind that this model and the Huygens Atmospheric Structure Instrument (HASI, Fulchignoni et al.) agree. Below 600 km, the 2005) density profile was outstanding. HASI measured the entry deceleration, from which density was directly derived up to approximately 175 km in altitude; hydrostatic equilibrium was used to infer pressure and temperature. Following the aeroshell's ejection and parachute's deployment at z ~ 150–175 km, the temperature and pressure were measured separately, and the density was calculated using the equation of state for an actual gas.

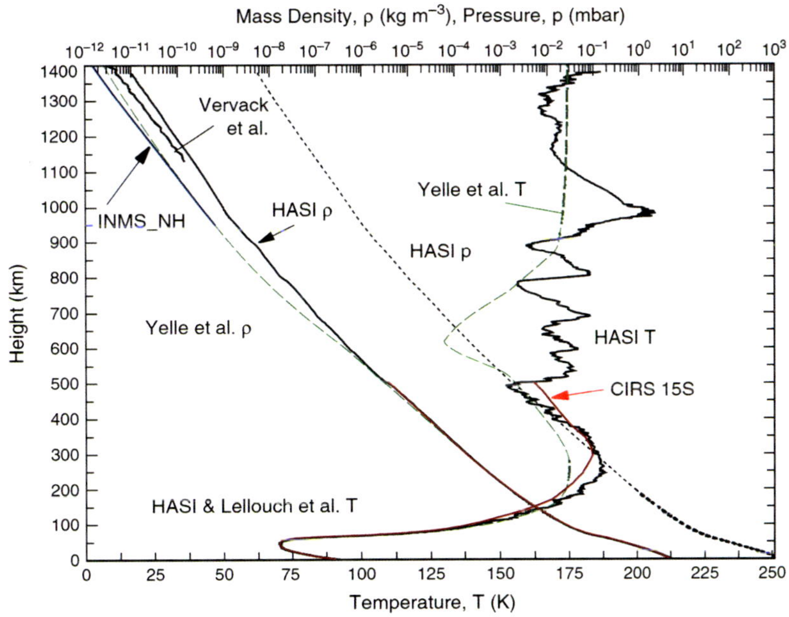

Figure 2.11. *Comparison of Cassini–Huygens HASI density, pressure, and temperature profiles (black lines, Fulchignoni et al., 2005), CIRS density and temperature profiles at 15 S (red lines, Vinatier et al., 2007a), average INMS Northern Hemisphere density profile (blue line, Müller-Wodarg et al., 2008) with the Voyager Lellouch et al. (1989); and Vervack et al. (2004) results, and the Yelle et al. (1997) engineering model (green lines), all as functions of altitude above surface.*

Source: https://websites.umich.edu/~atreya/Chapters/Strobel-Atreya%20etal%20Titan%20fr%20C-H.pdf.

Compared to a Saturnian year (= 29.5 years), the extremely long radiative time constants (~300 years at 10–20 km, ~60 years at 50 km) readily explain the absence of any change in Titan's lower atmosphere between the Voyager flybys and Huygens Probe. The mass density in the thermosphere determined by HASI differs from Vervack et al.'s calculations by up to a factor of 2. 2004 based on solar occultation measurements made by Voyager. While the HASI data were collected between medium and minimum solar activity, the Voyager data were collected during solar maximum activity. Therefore, it was expected that Titan's upper atmosphere would have shrunk in comparison to the Voyager era. Note that the Vervack et al. (2004) estimated temperature of ~155 K, which is roughly 20 K colder than HASI. Therefore, their densities would merge with the HASI densities at ~1000 km when extrapolated to lower altitudes. Titan was, however, at 9.04 AU at the time of Huygens Probe entry and 9.44 AU during the Voyager 1 flyby because of Saturn's elliptical orbit around the Sun. While the solar EUV and UV fluxes at solar maximum activity for Voyager flybys ratioed to solar fluxes pertinent to the Probe descent would be ~2, more than enough to offset the (9.44/9.04) 2 ~1.09 factor due to Titan being closer to the Sun during the Probe descent, this could explain the expanded atmosphere but not the colder temperatures in the thermosphere. For now, there is no explanation for Vervack et al.'s temperature difference, and HASI.

Moreover displayed in figure. The averaged in situ data from the Ion Neutral Mass Spectrometer (INMS) for the northern hemisphere is 2.11 (Müller-Wodarg et al. 2008), which is roughly a factor of 2.4 lower than the HASI densities at 10.3°S. The Cassini spacecraft's drag and torque allow for the determination of densities, which are consistently higher than the measured INMS densities by a factor of roughly 2.6 during each pass through the upper atmosphere when INMS is measuring the neutral atmosphere. The values derived by the Cassini spacecraft are support determined drag-determined densities obtained from HASI.

Considering Figure 2.11 Titan has an equatorial tropopause that is clearly defined, with a temperature of T = 70.43 ± 0.25 K at 44 km. It is statistically stable in the troposphere, according to the mean temperature profile. It is necessary to treat the atmosphere as a real gas rather than an ideal gas due to the low temperatures and high densities found in the troposphere. Titan's lower stratosphere has a high static stability due to strong solar heating caused by its haze absorbing visible and UV light as well as near-IR CH_4 absorption. Its dT/dz is approximately 1 K km^{-1}. Based on HASI data, the stratopause is situated at 260 km, or T = 187 K. However, the CIRS stratospheric temperature retrievals, which place T = 183 K at ~312 km, disagree with the HASI location (Vinatier et al. 2007a). The CIRS temperature at 265 km is 181 K, which is more significant than the HASI temperature of 184 K at this altitude. The IR radiance computed using the HASI-derived atmosphere does not match the observed CIRS CH_4 7.7 μm band radiance. Despite these temperature variations, there is a notable consensus in mass densities up to 500 km, which is the upper limit for CIRS retrievals.

Fulchignoni and associates. (2005) designate the Titan mesopause as the lowest temperature, measured at 494 km and 153 K. The Yelle & Co. To satisfy Voyager data collected in the thermosphere and lower/middle atmosphere, the (1997) engineering model needed a well-defined mesopause somewhere, but its minimum temperature and location were less restricted. An obvious temperature inversion with a peak temperature of 169 K and a temperature gradient of about 3.5 K km^{-1} is located just above this HASI minimum. This feature was previously identified in stellar occultation measurements, (e.g., Sicardy and associates. 2006) and is where a detached haze layer can be found (Porco et al., 2005). To balance the implied significant heat loss by thermal conduction, only a very small optical depth would be necessary to absorb enough solar radiation (Lavvas et al. 2009).

Waves propagating in gas with exponentially decreasing density are consistent with the oscillations in the HASI temperature profile, whose amplitudes are growing above 200 km. This possibility was predicted by Strobel and Sicardy (1997) in their engineering study. Titan's atmosphere may be interpreted as essentially isothermal (~170 K) from 500–1100 km, with large amplitude thermal waves (10 K) superimposed on an isothermal basic state, according to one interpretation of the HASI temperature profile. The wave temperature lapse rate (dT/dz) is calculated at several altitudes throughout the atmosphere and produces values that are greater than the adiabatic lapse rate (= g/cp), which varies from 1.24 at the surface to 0.56 K km^{-1} at 1400 km. This leads to the conclusion that the waves reach breaking/saturation amplitudes (Fulchignoni et al. 2005). Strong horizontal winds and/or high phase speeds for wave propagation are implied by the large vertical wavelengths corresponding to the largest amplitudes. The mean temperature profile above 1000 km does not have to be isothermal. Strobel (2008) has theorized that the large escape rates deduced from Cassini INMS data for CH_4 and H_2 (Yelle et al. 2008; Cui et al. 2008) can be explained by slow hydrodynamic escape, with no requirement for large non-thermal escape. The bulk outflow produces adiabatic cooling and approximately T ⊠ r^{-1}. The superposition of waves on this slowly decreasing mean temperature profile could also account for the derived HASI temperature (and density) data above ~1000 km. Potential wave sources are gravitational tides, large wind shear creating Kelvin–Helmholtz instability, and gravity waves forced by topography.

2.4. THE HEIGHT AND LATITUDE STRUCTURE OF THE ATMOSPHERE

With a focus on the variations in density, pressure, and temperature at different latitudes and heights, the pertinent Cassini observations from the nominal mission—namely, the RSS radio occultations, CIRS remote sensing, UVIS solar and stellar occultations, and INMS in situ measurements—are discussed below.

2.4.1. Radio Occultations

The Voyager spacecraft (Vgr 1) only experienced one Titan earth occultation, with an entry at 6°N and an exit at 9°S. In the lowest 3.4–4 km, the retrieved temperatures were almost identical, with a lapse rate that was almost identical to the dry adiabat. With ingress and egress soundings at 74°S, 69°S, 53°S, 34°S, 33°S, 32°S, 53°N, and 74°N, Cassini experienced four occultations by Titan. The occultations support the Voyager IRIS observations in pointing to a slight meridional temperature variation in the troposphere, despite the fact that the data have not yet been thoroughly examined. Temperatures recovered from the occultations at the tropopause differ by 4–5 K. They vary by no more than 3 K close to the surface. Higher northern latitudes experience the lowest temperatures. Above the surface, temperature profiles close to 30°S are almost adiabatic. In winter, northern latitudes experience isothermal or inverted temperatures above the surface. As latitudes rise, they become more stably stratified above the surface. Because it is extremely difficult to accurately retrieve data from CIRS far-IR limb soundings and because CIRS mid-IR soundings of the polar lower stratosphere are too cold, information specific to the lower northern polar stratosphere is of particular interest.

2.4.2. Remote Sensing

Figure 2.12 based on CIRS data collected between July 2004 and September 2006, shows the temperature of Titan's stratosphere and mesosphere as a function of latitude (75°S to 75°N) and pressure (0.001 to 10 mbar). 2008). Concerning Figure 2.10, the majority of this time is solstitial. Although there isn't a scale of altitude displayed, Figure 2.11 and apply the CIRS 15S profile's altitude scale, which ought to be appropriate for latitudes between 65°S and 30°N. Surprisingly, the equator experiences the warmest temperatures in the stratosphere, and the winter (now northern) hemisphere experiences the largest meridional temperature gradients instead of the summer (southern) hemisphere. The other notable feature is the very cold lower polar stratosphere in the northern winter hemisphere, where temperatures on constant pressure surfaces between 1 and 10 mbar drop by about 25 to 30 K from 30 to 75° north latitude. From the summer pole to roughly 45°N, the stratopause stays at a fairly constant height of 300 km and pressure of 0 mbar, respectively. At the winter north pole, it rises quickly to a height of about 400 km. The winter, northern polar "stratopause" is actually the warmest place in the entire atmosphere, measuring ~0.01 mbar/400 km and 20 K warmer than the equatorial and summer, southern regions. Since the polar night only reaches a distance of 315 km, it is plausible that the origin of this warm region is diabetic. On the other hand, increased nitrile and hydrocarbon mixing ratios in the winter at northern polar latitudes indicate that adiabatic heating and subsidence are undoubtedly happening (Figure 2.12).

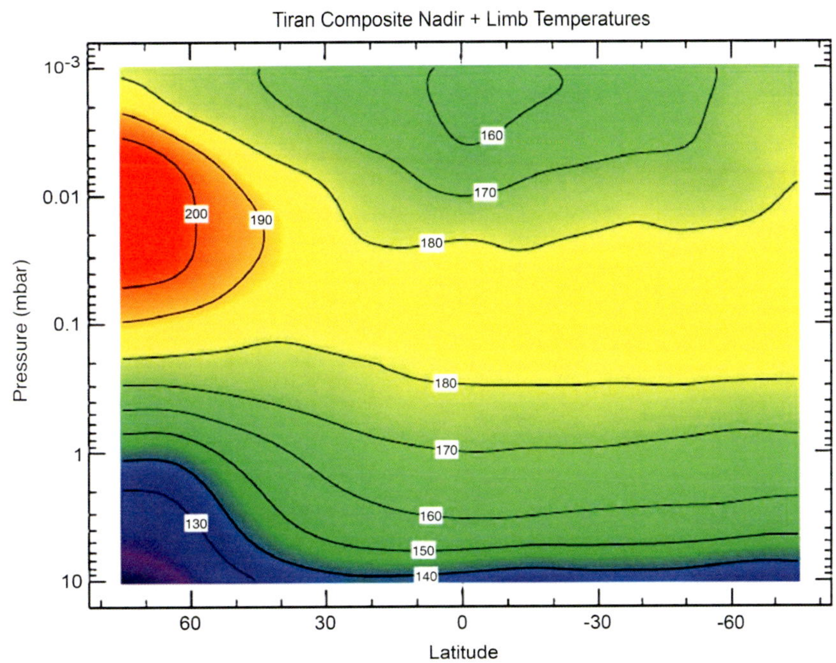

Figure 2.12. *CIRS zonal mean temperatures (K) from limb and nadir spectra recorded between July 2004 and September 2006. Retrieved temperatures were averaged over 5° latitudinal intervals and smoothed three times with a 10° boxcar function (after Achterberg et al., 2008).*

Source: https://websites.umich.edu/~atreya/Chapters/Strobel-Atreya%20etal%20Titan%20fr%20C-H.pdf.

Because of the low atmospheric opacity at 19 μm and the ability to deduce the latitudinal variation of surface temperature from brightness temperatures, CIRS can detect thermal radiation from the surface. Jennings along with others. (2009) discovered that the equatorial surface temperature of 93.7 ± 0.6 K, which is equal to the HASI value at 10°S, is 3 K lower at the north pole and 2 K lower at the south pole.

Wave structure in the atmosphere has been detected by HASI and refractive stellar occultations (discussed below), whereas CIRS has not detected wave structure to date and can only constrain wave temperature amplitudes to ~1 K. However, one must remember that CIRS can only explore the atmosphere in the limb mode and up to about 500 km. The extent of the atmosphere's total data coverage to date is constrained, and waves might eventually be picked up by CIRS. Nevertheless, CIRS spectra must be averaged, which results in a loss of spatial resolution.

2.4.3. Solar and Stellar Occultations

Equatorial Voyager UVS solar occultations yielded the remarkable measurement that at least 1000 km of well-mixed CH_4 was present (Vervack et al. Since confirmed by the INMS (Waite et al., 2004). 2005; Yelle and others. 2008). During the nominal mission, the Cassini Ultraviolet Imaging Spectrometer (UVIS) conducted two solar occultations; however, there are no reports on the results that have been published. Shemansky along with others published a report on the initial UVIS star occultations in 2005. Their estimated temperature profile above 400 km consists of two parts: (1) an adiabatic temperature gradient in a convective region around 450 km, and (2) a mesopause with T = 114 K at 615 km. In contrast, that location's HASI temperature was about 175 K.

The N_2 density profile at 1200 km is approximately eight and twenty times lower than that measured by INMS and HASI, respectively, when the UVIS retrieved temperature profile is combined with the hydrostatic equation and lower boundary condition at 400 km from CIRS and HASI data. Additionally, the calculated N_2 density combined with the inferred CH_4 densities from UVIS at 1200 km suggests that the CH_4/N_2 ratio is approximately 0.5, which is also different from the INMS values of approximately 0.3 at 1200 km. Absolute values of individual densities should not be nearly as accurate as the INMS mixing ratios for major species. Keep in mind that below 91.1 nm, interstellar hydrogen ionizes and shuts off the stellar UV flux. N_2 UV absorption only happens in pre dissociated electronic bands above this wavelength, which makes them extremely challenging to understand at UVIS spectral resolution.

A few refractive stellar occultations have also occurred, mainly at visible wavelengths, in addition to absorptive occultations (July 1989; 21 August 1995; 20 December 2001; two on November 14, 2003). While the occultations that were previously discussed were measured by Cassini instruments onboard the spacecraft, these were measured using ground-based telescopes. In their capacity as representatives, Sicardy et al. (2006) investigated Titan's atmosphere in the 0.1–250 μbar region. Nonetheless, the 0.1–1 μbar region is susceptible to initial conditions in their onion peeling analysis process, and in addition to the typical refraction effects, the stellar fluxes are attenuated by haze absorption at pressures higher than 12 μbar (400 km).

The temperature gradients that surpass the adiabatic lapse rate and the abundance of wave signatures in the retrieved temperature profile are in essential agreement with the HASI temperature profile. A prominent thermal inversion layer is present at 1.5 μbar (~507–515 km), as was already mentioned. Sicardy et al. extracted latitudinal information from the central flash analysis. (2006) discovered an isopycnic surface that is roughly equal to the 250 μbar surface, which rises to a height of around 50 km above a point chosen as the North Pole reference height at 45°N and stays mostly stable to the South Pole. This is in line with CIRS's findings (Achterberg et al.) that very strong stratospheric zonal winds occur in the northern hemisphere during the winter and much weaker winds occur in the summer. 2008).

2.4.4. In-Situ Measurements

During the first targeted Titan flyby (TA) on October 26, 2004, the Cassini INMS conducted the first *in-situ* measurements of Titan's upper atmosphere. With the help of later targeted flybys, of which 19 provided the INMS with suitable observing conditions and generated data of sufficient quality, this original dataset has been significantly increased. During 13 flybys, mostly in the northern hemisphere, and the remaining 6 flybys in the southern hemisphere, the INMS measured gas densities. The closest approach altitudes were between 950 km (T16) and 1175 km (TA). To obtain neutral or ion densities, respectively, the INMS can measure in the Open Source Ion (OSI) or Closed Source Neutral (CSN) modes. Neutral densities were measured throughout all 19 flybys, but not always on both the inbound and outbound portions. Ion composition was only measured during 9 of the 19 flybys.

Müller-Wodarg et al. used the measurements of N_2 and CH_4 neutral gas densities in the northern hemisphere. (2008) used flybys TA-T32 to build a basic empirical model of northern hemisphere gas densities and temperatures. Titan's solar declination angle changed from –23 to –11 during these flybys, indicating that the season was halfway between summer and equinox conditions in the southern hemisphere. The Müller-Wodarg and colleagues model. Therefore, (2008) can be thought of as an average depiction of Titan's winter hemisphere. The vertical range spans altitudes that the INMS sufficiently sampled, from 1000 to 1600 km. Müller-Wodarg et al. conducted a statistical analysis of the INMS N_2 and CH_4 densities. (2008) discovered no consistent patterns with longitude or local time; however, this could be partially explained by a biased sampling that favored measurements taken from polar nightside or low latitude dayside locations, mostly in Titan's longitude sector facing Saturn (Cui et al., 2008). The INMS densities at fixed height levels above approximately 1100 km showed a distinct latitudinal trend, with densities increasing by about 70% in the direction of the equator. The thermospheric temperatures in figure were inferred using the density model. 2.13 decreases with latitude below 1200 km with values near 1000 km from 164 ± 6 K at 20°N to 131 ± 6 K near 80°N, and with height from 149 ± 10 K at 1000 km to 140 ± 13 K near 1600 km.

Even though the INMS has made nine passes of *in-situ* observations in the southern hemisphere, there has been too much statistical variance between them to identify a consistent latitudinal trend. As would be expected from solar heating, a preliminary study of the southern hemisphere's structure indicates that average temperatures there are ~10–15 K higher than those in the northern hemisphere.

Even though the INMS has produced the largest collection of thermosphere observations to date, there is still one question regarding the precise calibration of densities. When INMS densities at equatorial latitudes close to 1000 km latitude are compared to those from HASI observations, the results reveal that INMS values are 2.4 times lower than the HASI values. Every time the spacecraft enters Titan's upper atmosphere during a flyby, the Cassini Attitude and Articulation Control Subsystem (AACS) measures torques on the craft, offering a second, independent estimate of total density. When the densities from the INMS are compared to those derived from the AACS at 1000 km, the INMS densities are again smaller than those from the AACS by an average of 2.6, which is quite similar to the difference between the INMS and HASI at that altitude. The determined temperatures are unaffected by this unresolved discrepancy, though (Baines et al., 2006; Bézard et al., 2009) (Figure 2.13).

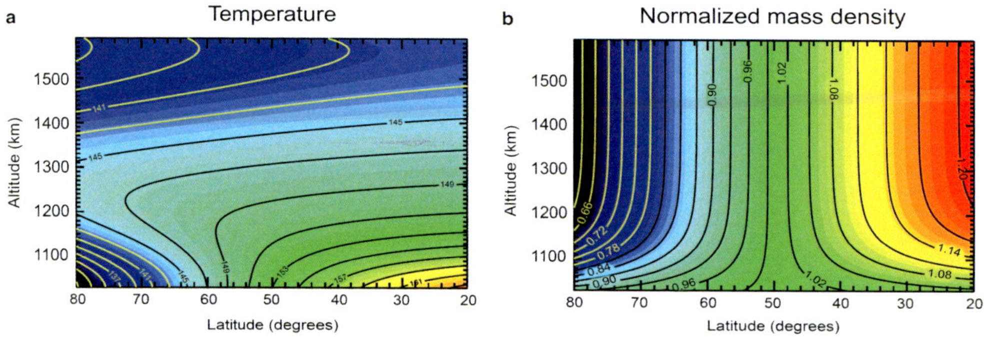

Figure 2.13. *(a) Temperatures in Titan's thermosphere are inferred from an empirical model of atmosphere densities measured by INMS in the northern hemisphere. While the atmosphere is nearly isothermal above around 1,200 km altitude, temperatures below 1,100 km increase towards the equator. (b) Normalized mass densities at each height to the value of 50N from the empirical model. Note the bulge at the equator (after Müller-Wodarg et al., 2008).*

Source: https://websites.umich.edu/~atreya/Chapters/Strobel-Atreya%20etal%20Titan%20fr%20C-H.pdf.

2.5. INTERPRETATION OF ATMOSPHERIC TEMPERATURE STRUCTURE

2.5.1. Radiative Budget of Troposphere and Stratosphere

Titan's atmosphere absorbs solar radiation both at the surface and through the absorption of methane and haze. The Huygens probe's Descent Imager/Spectral Radiometer (DISR) measured sunlight fluxes both upward and downward at wavelengths between 350 and 1600 nm and altitudes between 150 and 0 km. The vertical distribution and optical characteristics of haze aerosols, the methane absorption coefficients under Titan's conditions, and the solar energy deposition profile at the Huygens landing latitude (10°S) have all been determined by analyzing these data (Tomasko et al. 2008). At the site and time of the Huygens landing, roughly 78% of the incoming sunlight is absorbed overall over a Titan day: roughly 11% at the surface, 48% in the atmosphere below 150 kilometers, and 19% above. The Huygens haze model was used to calculate the disk-averaged solar flux profile. It is remarkably consistent with the previous McKay et al. estimate. (1989) using Voyager and ground-based albedo data.

Through thermal emission by haze particles and gases beyond ~7 μm (molecular bands and collision-induced opacity), the atmosphere is cooled radiatively. Tomasko and associates. (2008) used Cassini/CIRS nadir and limb measurements to calculate the radiative cooling rate profile at latitudes and times near the Huygens landing. Below 80 km, collision-induced opacity is the primary cause of atmospheric cooling; above that altitude, molecular band, and haze opacities predominate and are equally significant (Figure 2.14).

At altitudes between 250 and 400 km, the cooling rate reaches 10–20 K per Titan day and rapidly drops below that. A radiative time constant can be obtained by dividing the temperature by the cooling rate. In the mesosphere and upper stratosphere, it is brief (~0.5 Earth year, i.e., At 300–400 km, the 0.015 Titan year approaches a Titan year (or 29.5 Earth years) at about 80 km, and at 20 km, it is roughly equal to 10 Titan years. The cooling rate profile resembles that which Bézard et al. found. (1995) at 53°S derived at the northern spring equinox from Voyager data. The two profiles agree within 20% in the region best constrained by the Voyager/IRIS measurements (140–250 km).

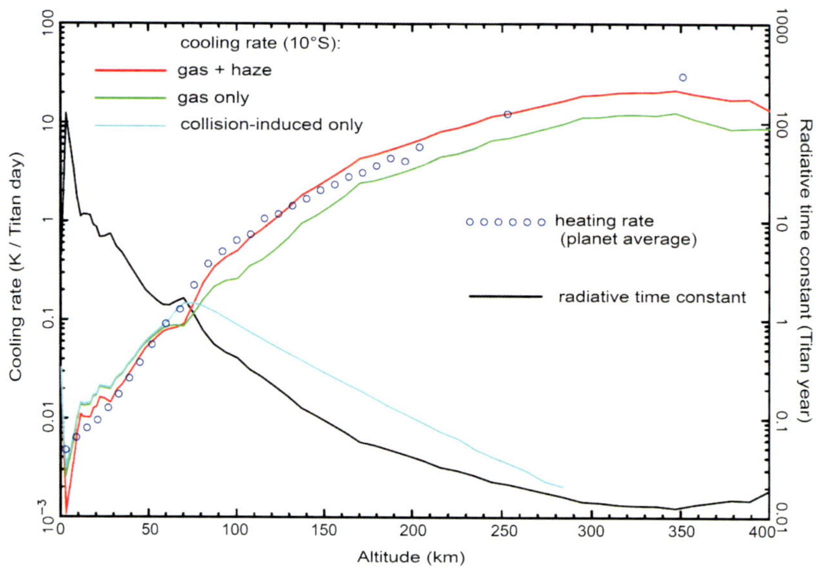

Figure 2.14. *Cooling rate profiles calculated around 10°S from Cassini/CIRS and Huygens data are compared with the solar heating rate profile, averaged over the planet, calculated from Huygens/DISR measurements (Tomasko et al., 2008). Cooling rates are best constrained in the ranges 10–65 km and 90–400 km and heating rates below 200 km. The radiative time constant (defined here as the temperature divided by the cooling rate) is also shown (right y-scale).*

Source: https://websites.umich.edu/~atreya/Chapters/Strobel-Atreya%20etal%20Titan%20fr%20C-H.pdf.

The disk-averaged heating rate profile based on the Huygens haze model and the cooling rate profile derived by CIRS around 10°S are extremely similar. From 350 to 10 km, both decrease by three orders of magnitude, and they agree at all levels within 50%. Conversely, at all altitudes below 160 km, the solar heating rate at 10°S outpaces the cooling rate, with a maximum net heating of approximately 0.5 K per Titan day at a distance of about 120 km. This extra heat is probably transferred to higher latitudes by the general circulation.

Bézard and colleagues. (1995) have demonstrated that the latitudinal compositional variations have a significant impact on the stratospheric radiative forcing. The higher concentrations of infrared emitters (gas and particles) at 50°N resulted in cooling rates derived from Voyager in the range 120–270 km (5–0.15 mbar), which were about 20–40% larger than at 53°S despite the colder temperatures (7 to 15 K). However, the radiative balance was still roughly attained at this location within error bars because the heating rate was also higher. Lebonnois et al. used a general circulation model. (2003b) also noted how higher concentrations of certain gases, such as C_2H_6, C_2H_2, and HCN, strongly enhance radiative cooling at high winter latitudes. According to research by Rannou et al., this thermal effect is comparable to that caused by the coupling between aerosols and the general circulation. 2002) in the mesosphere, but only about 1 mbar (190 km) in the stratosphere.

2.5.2. Radiative Processes in the Upper Atmosphere

Above 400 km (~0.03 μbar) non-LTE effects become important for hydrocarbon cooling rates as shown by Yelle (1991). Above 700 km (~0.1 μbar) he showed that HCN LTE rotational line cooling dominates radiative cooling to the exobase. The dominant heating processes above 400 km are near-IR CH_4 heating by absorption of solar radiation in the 1.7, 2.3, and 3.3 μm bands. Above ~650 km solar UV CH_4 heating emerges as the principal heat source and above 900 km, the thermosphere is, to zeroth order, in radiative equilibrium where solar EUV/UV heating is balanced by HCN rotational line cooling (Yelle 1991). Without HCN cooling calculated thermospheric temperatures would far exceed the observed values.

While HASI data indicated the mesopause is at 494 km, Yelle's models consistently predicted a prominent mesopause at 600 km (~0.5 m bar) with a ~25 K temperature drop from the 300 km temperature (Fulchignoni et al. 2005) to compare with ~185 K at 300 km, and a minimum temperature of 153 K. Within Figure 2.11: Yelle and associates engineering model is comparable to the HASI profi les and is representative of the models created by Yelle (1991). As mentioned above, Sicardy et al. deduced that the HASI temperature profi le above 494 km had a steep temperature gradient. From stellar occultations, (2006) requires solar heating of the Porco et al.-detected detached haze layer 2005. Liang et al., above this hazy layer. (2007b) provide proof that the aerosols can travel up to 1000 kilometers. The Lavvas et al. model included heating linked to discrete haze layers and the wide dispersion of aerosols above Titan's visible disk, which was not included in the Yelle (1991) model. (2009), who further disagreed with the HASI inferred mesopause at 494 km, discovered that the mesopause was raised by about 50 km without a change in temperature using the Yelle computer code.

2.6. ATMOSPHERIC PRESSURE AND TEMPERATURE VARIATIONS WITH ALTITUDE

The pressure within the Earth's atmosphere at any given time is known as atmospheric pressure. Most of the time, the hydrostatic pressure created by the weight of the air above the measurement point closely approximates atmospheric pressure. There is less atmospheric mass above low pressure areas and more atmospheric mass above high pressure areas. In a similar vein, pressure falls with increasing elevation because there is less atmospheric mass above. From sea level to the top of the atmosphere, a column of air with a cross section of one square inch would weigh about 14.7 pounds. A column of air that is 1 m^2 (11 sq ft) in size would weigh roughly 100 kilonewtons, or 10.2 tons at the surface.

2.6.1. Standard Atmospheric Pressure

The definition of the standard atmosphere (symbol: atm), a unit of pressure, is precisely equal to 101.325 kPa. The equivalent non-standard units are 29.92 inHg or 760 mmHg (torr). For pneumatic fluid power (ISO R554), as well as in the aerospace (ISO 2533) and petroleum (ISO 5024) industries, one standard atmosphere is the standard pressure.

The International Union of Pure and Applied Chemistry (IUPAC) recommended in 1999 that, instead of the 101.325 kPa value of "one standard atmosphere," "the standard pressure" should be precisely defined as 100 kPa (⊠75053 inHg to define the physical properties of substances. According to ISO 2787, this number serves as the industry standard pressure for compressors and pneumatic tools. The equivalent amount of moisture at standard temperature and pressure is referred to as the "standard," and compressed air flow in the US is frequently expressed in "standard cubic yards" per unit of time. A slightly different definition of this standard atmosphere is given by the following factors: temperature = 20°C (68°F), relative humidity = 20%, altitude = sea level, and air density = 1.225 kg/m^3 (0.0765 lb/cu ft). Instead, temperature = 0°C

(32°F) is frequently the standard in the air conditioning industry. The petroleum industry uses a standard temperature of 15°C (60°F) and pressure of 101°56 kPa (14 psi) for natural gas.

2.6.2. Mean Sea Level Pressure

The pressure at sea level, also known as mean sea level pressure (MSLP or QFF), is equal to the station pressure reduced to sea level (assuming an isothermal layer at the station temperature) when measured at a specific elevation on land.

This is the pressure typically provided in weather reports found online, in newspapers, and on radio and television. When household barometers are calibrated to coincide with local weather reports, they register pressure at sea level rather than the true atmospheric pressure in the area. Refer to the Altimeter (barometer versus. absolute).

When sea level is reached, everyone experiences pressure variations within the same typical range. Geographical location has no bearing on the pressure levels that are deemed high or low. As a result, isobars on a weather map become valuable and practical tools.

The altimeter setting in aviation, set either QNH or QFE, is another atmospheric pressure reduced to sea level, but the method of making this reduction differs slightly. See altimeter.

- **QNH:** The barometric altimeter setting which will cause the altimeter to read airfield elevation when on the airfield. In ISA temperature conditions the altimeter will read altitude above mean sea level in the vicinity of the airfield
- **QFE**: The barometric altimeter setting which will cause an altimeter to read zero when at the reference datum of a particular airfield (generally a runway threshold). In ISA temperature conditions the altimeter will read height above the datum in the vicinity of the airfield.

Pilots frequently use the mnemonics "Nautical Height" (for QNH) and "Field Elevation" (for QFE) to distinguish between QNH and QFE, even though they are arbitrary Q codes rather than acronyms.

101.325 kPa (1013.25 mbar), or 29.921 inches of mercury (inHg) or 760 millimeters (mmHg), is the average sea-level pressure. QNH is transmitted globally in millibars or hectopascals (1 millibar = 1 hectopascal) for aviation weather reports (METAR), except the United States and Canada, where it is reported in inches (or hundredths of inches) of mercury. (The sea level pressure (SLP), which is also reported by the United States and Canada and is determined by a different method, is reported in hectopascals or millibars in the remarks section of the code, which is not a part that is transmitted internationally. While Environment Canada uses the same standard unit of pressure, sea level pressure is reported in kilopascals in public weather reports in Canada. For example, 1013.2 mbar or 101.32 kPa is transmitted as 132; 1000.0 mbar or 100.00 kPa is transmitted as 000; 998.7 mbar or 99.87 kPa is transmitted as 987; etc. Only three digits are required in the weather code; decimal points and the one or two most significant digits are omitted. The Siberian High, which frequently reaches a sea-level pressure above 1032.0 mbar, is home to the highest sea-level pressure on Earth. The centers of hurricanes (typhoons, Baguio) have the lowest detectable sea-level pressure.

2.6.3. Altitude Atmospheric Pressure Variation

From the surface of the earth to the uppermost layer of the mesosphere, pressure varies smoothly. NASA has averaged the conditions for every region of the earth year-round, although the pressure varies with the weather. The air pressures (as a percentage of one atmosphere) and their corresponding average altitudes are listed below. The air pressure at different altitudes is roughly indicated in the table:

Fraction of 1 atm	**Average Altitude**	
	(m)	**(ft)**
1	0	0
1/2	5,486	18,000
1/3	8,376	27,480
1/10	16,132	52,926
1/100	30,901	101,381
1/1000	48,467	159,013
1/10000	69,464	227,899
1/100000	96,282	283,076

Source: https://www.chemeurope.com/en/encyclopedia/Atmospheric_pressure.html.

2.6.4. Calculating Variation with Altitude

Below 86 km (or 278,400 ft), there are two distinct equations for calculating pressure at different height regimes. When the standard temperature lapse rate is less than or equal to zero, one uses equation 1, and when it is equal to zero, one uses equation 2.

$$P = P_b \cdot \left[\frac{T_b}{T_b + L_b \cdot (h - h_b)}\right]^{\frac{g_0 \cdot M}{R^* \cdot L_b}} \quad (1)$$

$$P = P_b \cdot \exp\left[\frac{-g_0 \cdot M \cdot (h - h_b)}{R^* \cdot T_b}\right] \quad (2)$$

where; P is the static pressure (pascals); T is the standard temperature (kelvins); L is the standard temperature lapse rate (kelvins per m); h is the height above sea level (meters); R^* is the universal gas constant: 8.31432×10^3 N·m / (kmol·K); g_0 is the gravitational constant (9.80665 m/s^2); M is the molar mass of Earth's air (28.9644 g/mol).

Or converted to English units:

where; P is the static pressure (inches of mercury); T is the standard temperature (kelvins); L is the standard temperature lapse rate (kelvins per ft); h is the height above sea level (feet); R^* is the universal gas constant (using feet and kelvins and gram moles: 8.9494596×10^4 kg·sq ft·s^{-2}·K^{-1}·kmol^{-1}); g_0 is the gravitational constant (32.17405 ft/s^2); and M is the molar mass of Earth's air (28.9644 g/mol).

The value of subscript b ranges from 0 to 6 by each of the seven successive layers of the atmosphere shown in the table below. In these equations, g_0, M, and R^* are each single-valued constants, while P, L, T, and h are multivalued constants by the table below. It should be noted that the values used for M, g_0, and R^* with the U.S. Standard Atmosphere, 1976, and that the value for R^* in particular does not agree with standard values for this constant. The reference value for P_b for $b = 0$ is the defined sea level value, $P_0 = 101325$ pascals or 29.92126 inHg. Values of P_b of $b = 1$ through $b = 6$ are obtained from the application of the appropriate member of the pair equations 1 and 2 for the case when $h = h_{b+1}$

Subscript b	Height Above Sea Level		Static Pressure		Standard Temperature (K)	Temperature Lapse Rate	
	(m)	(ft)	(pascals)	(inHg)		(K/m)	(K/ft)
0	0	0	101325	29.92126	288.15	-0.0065	-0.0019812
1	11,000	36,089	22632.1	6.683245	216.65	0.0	0.0
2	20,000	65,617	5474.89	1.616734	216.65	0.001	0.0003048
3	32,000	104,987	868.019	0.2563258	228.65	0.0028	0.00085344
4	47,000	154,199	110.906	0.0327506	270.65	0.0	0.0
5	51,000	167,323	66.9389	0.01976704	270.65	-0.0028	-0.00085344
6	71,000	232,940	3.95642	0.00116833	214.65	-0.002	-0.0006097

Source: *https://www.chemeurope.com/en/encyclopedia/Atmospheric_pressure.html*

2.6.5. Sample Calculation

Find the pressure at 30,000 meters. First, take note that 30,000 meters is in the range of subscript b=2 in the above chart since it is above 20,000 but below 32,000. It should be noted that equation 1 is valid because the temperature lapse rate in that area is not equal to zero.

$$P = P_2 \cdot \left[\frac{T_2}{T_2 + L_2 \cdot (h - h_2)}\right]^{\frac{g_0 \cdot M}{R^* \cdot L_2}}$$

Or

$$P = 5474.89 \cdot \left[\frac{216.65}{216.65 + 0.001 \cdot (30,000 - 20,000)}\right]^{\frac{9.80665 \cdot 28.9644}{8314.32 \cdot 0.001}}$$

$$P = 5474.89 \cdot \left[\frac{216.65}{226.65}\right]^{34.163195}$$

$$P = 5474.89 \cdot 0.214044$$

$$P = 1171.867 \text{ Pascals at 30,000 meters}$$

2.6.6. Local Atmospheric Pressure Variation

Earth's atmospheric pressure varies greatly, and understanding these variations is crucial to understanding weather and climate. For information on how changes in air pressure affect the weather, see pressure system.

The highest known atmospheric pressure was recorded on December 19, 2001, at Tosontsengel in the Khövsgöl Province of Mongolia, at 108.6 kPa (1,086 mbar, or 32.06 inches of mercury). (Mechtly, E. A. 1973).

During Typhoon Tip on October 12, 1979, in the Western Pacific, the lowest recorded non-tornadic atmospheric pressure of 87.0 kPa (870 mbar, or 25.69 inches of mercury) was recorded. (Mechtly, E. A. 1973) During Hurricane Wilma on October 19, 2005, the Atlantic Ocean record was 88.2 kPa (882 mbar, or 26.04 inches of mercury).

Global atmospheric tides are the cause of the diurnal (twice-daily) cycle in atmospheric pressure. With an amplitude of a few millibars, this effect is greatest in tropical regions and nearly nonexistent in Polar Regions. Northern Europe's rhythmic variations are depicted in the graph at the top of this page. The two superimposed cycles in these variations are semi-circadian (12 h) and circadian (24 h).

2.6.7. Atmospheric Pressure Based on Height of Water

A mercury barometer is frequently used to measure atmospheric pressure, and a height of about 760 mm (30 inches) of mercury is frequently used in teaching, visualization, and measurement. But since humans don't often come into contact with mercury, water frequently offers a more natural way to think about how much pressure is in one atmosphere.

A pressure of one atmosphere, equivalent to 101.325 kPa or 14.7 lbf/sq in, is sufficient to lift water by roughly 10.3 m (33.9 ft). As a result, a diver in a freshwater lake at a depth of 10.3 meters feels pressure equivalent to roughly two atmospheres (1 atm for the water and 1 atm for the air). This is also the highest point at which suction can be used to draw up a column of water.

Aneroid barometers or strain gauges are the most common types of non-professional barometers. Barometers are described in Pressure measurement.

2.6.8. Atmospheric Pressure's Relation to Water's Boiling Point

While it is commonly believed that water boils at 100°C (212°F), in reality, water evaporates when the atmospheric pressure surrounding it equalizes the vapor pressure. As a result, water's boiling point rises at higher pressures and falls at lower pressures (Burt 2004). This is the reason specific baking instructions are needed when baking at elevations higher than 3,500 feet above sea level (Cordier et al., 2008; Courtin et al., 1995).

CASE STUDY

HOW DO CARBON MONOXIDE AND AEROSOL CONCENTRATIONS AFFECT EARTH'S ATMOSPHERE?

Earth's Dynamic Atmosphere:

Anyone who watches the weather knows that Earth's atmosphere is a complex, dynamic system—it is constantly changing. The important questions about our thin, life-sustaining blanket of gases in the 21st century are:

- How is it changing?
- What factors and events are driving these changes?
- What do these changes mean for life on Earth?

What drives atmospheric changes, and why should we care? That's a fair question. After all, humans have evolved, and indeed thrived, in this unpredictable and highly dynamic environment. The answer is this: We should care because human activity appears to be altering Earth's atmosphere at a very high rate. Indeed, recent findings indicate that important components of the atmosphere are changing more rapidly now than at any other time in human history. Because the composition of the atmosphere is so important in regulating conditions on Earth, and we are so totally dependent on those conditions, we need to develop an accurate understanding of how the atmosphere is changing, and why.

Carbon Monoxide, CO

Carbon monoxide is an invisible, odorless, highly toxic gas that is produced through incomplete combustion. Its name is often abbreviated to its chemical formula, CO. Global background concentrations of carbon monoxide range from about 0.05 to 12 parts per million (ppm). The major *natural* sources of this gas are brush fires, forest fires, and volcanoes. Burning of fossil fuels such as oil, gasoline, natural gas, and coal are the major *anthropogenic* sources. According to the U.S. Environmental Protection Agency, 60% of the total global emissions of CO are from human activities. Internal combustion engines in automobiles are the largest single producers of this poisonous gas.

Although carbon monoxide has always been a component of Earth's atmosphere, its concentrations have increased since the Industrial Revolution when humans began using technologies that dramatically heightened the amount of burning that occurs. CO is an important part of the atmosphere because it reacts with other atmospheric compounds to form pollutants that harm human health and can potentially influence the global climate system. In sunny urban areas where large amounts of CO are produced, it reacts with other pollutants to form ozone and "smog," both of which are responsible for serious human health problems. At larger scales, CO can impact the global climate system by slowing the rate at which carbon dioxide is removed from the atmosphere. In addition to this reduced removal rate, CO ultimately forms additional carbon dioxide over time, when it reacts with oxygen (O_2) in the atmosphere.

Aerosols, Often Called "Particulates"

Aerosols are very different from CO but also play a role in human health and environmental quality. Aerosols, unlike gases, are *solid particles* that are very small—so small that they become airborne. Once they are in the air, they can travel thousands of miles before settling out in the oceans or on land. Unlike CO, aerosols are not invisible, even though they may be difficult or impossible to see at low concentrations. However, at high concentrations, they are easily visible. In certain places and at certain times, such as in the desert during a strong wind storm, aerosols from the dry, dusty soils may become so concentrated as to block out the sun. At lower concentrations, they absorb and scatter less sunlight, but may still be visible.

There are many reasons scientists are interested in aerosols. These include:

- Monitoring environmental "disturbances" such as forest fires and volcanic eruptions;
- Monitoring air quality for public health;
- Studying the Earth's radiation budget;
- Studying climate change.

CLASS ACTIVITY

Using graph paper and the descriptions of the atmosphere from the worksheet, students should draw the layers of the atmosphere proportionally on a graph. (The exosphere may not fit on a single sheet of graph paper without making the other layers too small.)

SUMMARY

- The Earth's atmosphere is a complex and dynamic system composed of a mixture of gases that envelop the planet, playing a crucial role in sustaining life and influencing climate and weather patterns.
- Nitrogen is the most abundant found in the atmosphere constituting 78.084% of the total volume of dry gases.
- In terms of percentage, argon is the third largest gas in the atmosphere constituting 0.934% of a total dry atmosphere.
- Methane is also a greenhouse gas that absorbs radiation and causes higher temperatures in the air.
- Water vapor is small in amount but it is one of the most important part of atmosphere with respect to the distribution of vegetation and life.
- Aerosols are extremely fine-sized solid particles or liquid droplets that continue to be in suspended form in gas for a very long time.
- The troposphere is the atmospheric layer closest to the planet and contains the largest percentage (around 80%) of the mass of the total atmosphere.
- The stratosphere is the second major strata of air in the atmosphere. It extends above the tropopause to an altitude of about 30 miles (50 km) above the planet's surface.
- The thermosphere is located above the mesosphere. The temperature in the thermosphere generally increases with altitude reaching 600 to 3000 F (600–2000 K) depending on solar activity.
- The ionosphere is defined by atmospheric effects on radiowave propagation because of the presence and variation in concentration of free electrons in the atmosphere.
- Atmospheric pressure is the pressure at any point in the Earth's atmosphere. In most circumstances, atmospheric pressure is closely approximated by the hydrostatic pressure caused by the weight of air above the measurement point.

REVIEW QUESTIONS

1. What is the composition of the atmosphere?
2. What do you understand by atmospheric structure?
3. What is the height and latitude structure of the atmosphere?
4. Explain the radiative processes in the upper atmosphere.
5. Describe altitude atmospheric pressure variation.

REFERENCES

1. Achterberg, R. K., Conrath, B. J., Gierasch, P. J., Flasar, F. M., & Nixon, C. A. (2008). Titan's middle-atmospheric temperatures and dynamics observed by the Cassini composite infrared spectrometer. *Icarus, 194*, 263–277. https://doi.org/10.1016/j.icarus.2007.09.
2. Atreya, S. K., Adams, E. Y., Niemann, H. B., Demick-Montelara, J. E., Owen, T. C., Fulchignoni, M., Ferri, F., & Wilson, E. H. (2006). Titan's methane cycle. *Planetary and Space Science, 54*, 1177–1187. https://doi.org/10.1016/j.pss.2006.05.028.
3. Atreya, S. K., Donahue, T. M., & Kuhn, W. R. (1978). Evolution of a nitrogen atmosphere on Titan. *Science, 201*, 611–613.
4. Bacmeister, J. T., Schoeberl, M. R., Summers, M. E., Rosenfield, J. E., & Zhu, X. (1995). Descent of long-lived trace gases in the winter polar vortex. *Journal of Geophysical Research, 100*, 11669–11684.
5. Baines, K. H., Drossart, P., Lopez-Valverde, M. A., Atreya, S. K., Sotin, C., Momary, T. W., Brown, R. H., Buratti, B. J., Clark, R. N., & Nicholson, P. D. (2006). On the discovery of CO nighttime emissions on Titan by Cassini/VIMS: Derived stratospheric abundances and geological implications. *Planetary and Space Science, 54*, 1552–1562. https://doi.org/10.1016/j.pss.2006.06.020.
6. Bézard, B. (2009). Composition and chemistry of Titan's stratosphere. *Philosophical Transactions of the Royal Society A, 367*, 683–695. https://doi.org/10.1098/rsta.2008.0186.
7. Bézard, B., Coustenis, A., & McKay, C. P. (1995). Titan's atmospheric temperature asymmetry: A radiative origin? *Icarus, 111*, 267–276.
8. Bézard, B., Nixon, C. A., Kleiner, I., & Jennings, D. E. (2007). Detection of 13CH_3D on Titan. *Icarus, 191*, 397–400.
9. Bjoraker, G., Achterberg, R., Anderson, C., Samuelson, R., Carlson, R., & Jennings, D. (2008). Cassini/CIRS observations of water vapor in Titan's stratosphere. *Bulletin of the American Astronomical Society, 40*(31.12), 446.
10. Broadfoot, A. L., et al. (1981). Extreme ultraviolet observations from Voyager 1 encounter with Saturn. *Science, 212*, 206.
11. Burt, C. C. (2004). *Extreme Weather: A Guide & Record Book*. W. W. Norton & Company. ISBN 0-393-32658-6
12. Cordier, D., Mousis, O., Lunine, J. I., Moudens, A., & Vuitton, V. (2008). Photochemical enrichment of deuterium in Titan's atmosphere: New insights from Cassini–Huygens. *Astrophysical Journal Letters, 689*, L61–L64.
13. Courtin, R. D., Sim, C., Kim, S., Gautier, D., & Jennings, D. E. (2008). Latitudinal variations of tropospheric H_2on Titan from the Cassini CIRS investigation. *Bulletin of the American Astronomical Society, 40*(31.01), 446.
14. Courtin, R., Gautier, D., & McKay, C. P. (1995). Titan's thermal emission spectrum: Reanalysis of the Voyager infrared measurements. *Icarus, 114*, 114–162.
15. Coustenis, A., & Bézard, B. (1995). Titan's atmosphere from Voyager infrared

observations. IV. Latitudinal variations of temperature and composition. *Icarus, 115*, 126–140.

16. Coustenis, A., Achterberg, R. K., Conrath, B. J., Jennings, D. E., Marten, A., Gautier, D., Nixon, C. A., Flasar, F. M., Teanby, N. A., Bézard, B., Samuelson, R. E., Carlson, R. C., Lellouch, E., Bjoraker, G. L., Romani, P. N., Taylor, F. W., Irwin, P. G. J., Fouchet, T., Hubert, A., Orton, G. S., Kunde, V. G., Vinatier, S., Mondellini, J., Abbas, M. M., & Courtin, R. (2007). The composition of Titan's stratosphere from Cassini/CIRS mid-infrared spectra. *Icarus, 189*, 35–62. https://doi.org/10.1016/j.icarus.2006.12.022.
17. Coustenis, A., Jennings, D. E., Jolly, A., Bénilan, Y., Nixon, C. A., Vinatieri, S., Gautier, D., Bjoraker, G., Romani, P., Carlson, R., & Flasar, F. M. (2008). Detection of C_2HD and the D/H ratio on Titan. *Icarus, 197*, 539–548. https://doi.org/10.1016/j.icarus.2008.06.003.
18. Coustenis, A., Salama, A., Lellouch, E., Encrenaz, T., Bjoraker, G. L., Samuelson, R. E., de Graauw, T., Feuchtgruber, H., & Kessler, M. F. (1998). Evidence for water vapor in Titan's atmosphere from ISO/SWS data. *Astronomy & Astrophysics, 336*, L85–L89.
19. Coustenis, A., Salama, A., Schulz, B., Ott, S., Lellouch, E., Encrenaz, T. H., Gautier, D., & Feuchtgruber, H. (2003). Titan's atmosphere from ISO mid-infrared spectroscopy. *Icarus, 161*, 383–403. https://doi.org/10.1016/S0019-1035(02)00028-3.
20. Crespin, A., Lebonnois, S., Vinatieri, S., Bézard, B., Coustenis, A., & Teanby, N. A. (2008). Diagnostics of Titan's stratospheric dynamics using Cassini/CIRS data and the 2-dimensional IPSL Circulation Model. *Icarus, 197*, 556–557. https://doi.org/10.1016/j.icarus.2008.05.010.
21. Cui, J., Yelle, R. V., & Volk, K. (2008). Distribution and escape of molecular hydrogen in Titan's thermosphere and exosphere. *Journal of Geophysical Research, 113*, E10004. https://doi.org/10.1029/2007JE003032.
22. Danielson, R. E., Caldwell, J., & Larach, D. R. (1973). An inversion in the atmosphere of Titan. *Icarus, 20*, 437–443.
23. de Kok, R., Irwin, P. G. J., Teanby, N. A., Lellouch, E., Bézard, B., Vinatieri, S., Nixon, C. A., Fletcher, L., Howett, C., Calcutt, S. B., Bowles, N. E., Flasar, F. M., & Taylor, F. W. (2007). Oxygen compounds in Titan's stratosphere as observed by Cassini CIRS. *Icarus, 186*, 354–363. https://doi.org/10.1016/j.icarus.2006.09.016.
24. Dire, J. R. (2000). Seasonal photochemical and meridional transport model for the stratosphere of Titan. *Icarus, 145*, 428–444.
25. Dubouloz, N., Raulin, F., Lellouch, E., & Gautier, D. (1989). Titan's hypothesized ocean properties – The influence of surface temperature and atmospheric composition uncertainties. *Icarus, 82*, 81–96. https://doi.org/10.1016/0019-1035(89)90025-0.
26. Flasar, F. M., et al. (2005). Titan's atmospheric temperatures, winds, and composition. *Science, 308*, 975–978.
27. Fulchignoni, M., Ferri, F., Angrilli, F., et al. (2005). Titan's physical characteristics measured by the Huygens Atmospheric Structure Instrument (HASI). *Nature, 438*, 785–791. https://doi.org/10.1038/nature04314.

28. Gillett, F. C., Forrest, W. J., & Merrill, K. M. (1973). 8–13-micron observations of Titan. *Astrophysical Journal, 187*, L37–L38.

29. Gurwell, M. A. (2004). Submillimeter observations of Titan: Global measures of stratospheric temperature, CO, HCN, HC3N, and the isotopic $^{12}C/^{13}C$ and $^{14}N/^{15}N$. *Astrophysical Journal, 616*, L7–L10.

30. Hanel, R. A., et al. (1981). Infrared observations of the Saturnian system from Voyager 1. *Science, 212*, 192–200.

31. Hartle, R. E., Sittler, E. C., Neubauer, F. M., Johnson, R. E., Smith, H. T., Crary, F., McComas, D. J., Young, D. T., Coates, A. J., Simpson, D., Bolton, S., Reisenfeld, D., Szego, K., Berthelier, J. J., Rymer, A., Vilppola, J., Steinberg, J. T., & Andre, N. (2006). Preliminary interpretation of Titan plasma interaction as observed by the Cassini Plasma Spectrometer: Comparisons with Voyager 1. *Geophysical Research Letters, 33*, 8201. https://doi.org/10.1029/2005GL024817

32. Holton, J. R., & Schoeberl, M. R. (1988). The role of gravity wave generated advection and diffusion in the transport of tracers in the mesosphere. *Journal of Geophysical Research, 93*, 11075–11082.

33. Hörst, S., Vuitton, V., & Yelle, R. V. (2008). The origin of oxygen species in Titan's atmosphere. *Journal of Geophysical Research, 113*, E10006. https://doi.org/10.1029/2008JE003135.

34. Hunten, D. M. (1974). *The Atmosphere of Titan*. NASA SP-340.

35. Hunten, D. M. (1978). A Titan atmosphere with a surface temperature of 200 K. In D. M. Hunten & D. Morrison (Eds.), *The Saturn System* (NASA Conf. Publ. 2068, pp. 127–140). Washington, DC.

36. Jacquemart, D., Lellouch, E., Bézard, B., de Bergh, C., Coustenis, A., Lacome, N., Schmitt, B., & Tomasko, M. (2008). New laboratory measurements of CH_4 in Titan's conditions and a reanalysis of the DISR near-surface spectra at the Huygens landing site. *Planetary and Space Science, 56*, 613–623.

37. Jennings, D. E., Flasar, F. M., Kunde, V. G., Samuelson, R. E., Pearl, J. C., Nixon, C. A., & Carlson, R. C. (2009). Titan's surface brightness temperatures. *Astrophysical Journal, 691*, L103–L105.

38. Jennings, D. E., Nixon, C. A., Jolly, A., Bézard, B., Coustenis, A., Vinatieri, S., Irwin, P. G. J., Teanby, N., Romani, P. N., Achterberg, R. K., & Flasar, F. M. (2008). Isotopic ratios in Titan's atmosphere from Cassini CIRS limb sounding: HC_3N in the north. *Astrophysical Journal Letters, 681*, L109–L111.

39. Kuiper, G. P. (1944). Titan: A satellite with an atmosphere. *Astrophysical Journal, 100*, 378–383.

40. Lara, L. (1998). Correction to "Vertical distribution of Titan's atmospheric neutral constituents" by L. M. Lara et al. *Journal of Geophysical Research, 103*(E11), 25775.

41. Lara, L. M., Lellouch, E., López-Moreno, J. J., & Rodrigo, R. (1996). Vertical distribution of Titan's atmospheric neutral constituents. *Journal of Geophysical Research, 101*, 23261–23283. https://doi.org/10.1029/96JE02036.

42. Lavvas, P. P., Coustenis, A., & Vardavas, I. M. (2008). Coupling photochemistry

with haze formation in Titan's atmosphere, Part II: Results and validation with Cassini/Huygens data. *Planetary and Space Science, 56*, 67–99.

43. Lavvas, P. P., Yelle, R. V., & Vuitton, V. (2009). The detached haze layer in Titan's mesosphere. *Icarus*. (In press)
44. Lebonnois, S., Bakes, E. L. O., & McKay, C. P. (2003a). Atomic and molecular hydrogen budget in Titan's atmosphere. *Icarus, 161*, 474–485. https://doi.org/10.1016/S0019-1035(02)00039-8.
45. Lebonnois, S., Hourdin, F., Rannou, P., Luz, D., & Toublanc, D. (2003b). Impact of the seasonal variations of composition on the temperature field of Titan's stratosphere. *Icarus, 163*, 164–174.
46. Lellouch, E. (1990). Atmospheric models of Titan and Triton. *Annales Geophysicae, 8*, 653–660.
47. Lellouch, E., Coustenis, A., Gauthier, D., Raulin, F., Dubouloz, N., & Frere, C. (1989). Titan's atmosphere and hypothesized ocean: A reanalysis of the Voyager 1 radio-occultation and IRIS 7.7 μm data. *Icarus, 79*, 328–349.
48. Lellouch, E., Coustenis, A., Sebag, B., Cuby, J. G., López-Valverde, M., Schmitt, B., Fouchet, T., & Crovisier, J. (2003). Titan's 5-μm window observations with the Very Large Telescope. *Icarus, 162*, 125–142. https://doi.org/10.1016/S0019-1035(02)00079-9.
49. Lewis, J. S. (1971). Satellites of the outer planets: Their physical and chemical nature. *Icarus, 15*, 174–185.
50. Liang, M. C., Heays, A. N., Lewis, B. R., Gibson, S. T., & Yung, Y. L. (2007a). Source of nitrogen isotope anomaly in HCN in the atmosphere of Titan. *Astrophysical Journal, 664*, L115–L118.
51. Liang, M. C., Yung, Y. L., & Shemansky, D. E. (2007b). Photolytically generated aerosols in the mesosphere and thermosphere of Titan. *Astrophysical Journal, 661*, L199–L202.
52. Lindal, G. F., Wood, G. E., Hotz, H. B., & Sweetnam, D. N. (1983). The atmosphere of Titan: An analysis of the Voyager radio occultation measurements. *Icarus, 53*, 348–363.
53. Lodders, K. (2003). Solar system abundances and condensation temperatures of the elements. *Astrophysical Journal, 591*, 220–1247.
54. Lunine, J. I., Yung, Y. L., & Lorenz, R. D. (1999). On the volatile inventory of Titan from isotopic abundances in nitrogen and methane. *Planetary and Space Science, 47*, 1291–1303.
55. Lutz, B. L., de Bergh, C., & Owen, T. (1983). Titan–Discovery of carbon monoxide in its atmosphere. *Science, 220*, 1374–1376.
56. Marten, A., Hidayat, T., Biraud, Y., & Moreno, R. (2002). New millimeter heterodyne observations of Titan: Vertical distributions of nitriles HCN, HC_3N, CH_3CN, and the isotopic ratio $^{15}N/^{14}N$in its atmosphere. *Icarus, 158*, 532–544.
57. McKay, C. P., Coustenis, A., Samuelson, R. E., Lemmon, M. T., Lorenz, R. D., Cabane, M., Rannou, P., & Drossart, P. (2001). Physical properties of the organic aerosols and clouds on Titan. *Planetary and Space Science, 49*, 79–99.

58. McKay, C. P., Pollack, J. B., & Courtin, R. (1989). The thermal structure of Titan's atmosphere. *Icarus, 80*, 23–53.

59. Mechtly, E. A. (1973). *The International System of Units, Physical Constants and Conversion Factors*. NASA SP-7012, Second Revision, National Aeronautics and Space Administration.

60. Mousis, O., Gautier, D., & Coustenis, A. (2002). The D/H ratio in methane in Titan: Origin and history. *Icarus, 159*, 156–165.

61. Müller-Wodarg, I. C. F., Yelle, R. V., Cui, J., & Waite, H. J. Jr. (2008). Horizontal structure and dynamics of Titan's thermosphere. *Journal of Geophysical Research, 113*, E10005. https://doi.org/10.1029/2007JE003033.

62. Niemann, H. B., et al. (2005). The abundances of constituents of Titan's atmosphere from the GCMS instrument on the Huygens probe. *Nature, 438*, 779–784. https://doi.org/10.1038/nature.04122.

63. Nixon, C. A., Achterberg, R. K., Vinatier, S., Bézard, B., Coustenis, A., Irwin, P. G. J., Teanby, N. A., de Kok, R., Romani, P. N., Jennings, D. E., Bjoraker, G. L., & Flasar, F. M. (2008a). The $^{12}C/^{13}C$ isotopic ratio in Titan hydrocarbons from Cassini/CIRS infrared spectra. *Icarus, 195*, 778–791. https://doi.org/10.1016/j.icarus.2008.01.012.

64. Nixon, C. A., Jennings, D. E., Bézard, B., Teanby, N. A., Achterberg, R. K., Coustenis, A., Vinatier, S., Irwin, P. G. J., Romani, P. N., Hewagama, T., & Flasar, F. M. (2008b). Isotopic ratios in Titan's atmosphere from Cassini CIRS limb sounding: CO_2 in the equator and south. *Astrophysical Journal Letters, 681*, L101–L103.

65. Owen, T., Biver, N., Marten, A., Matthews, H., & Meier, R. (1999). Saturn VI (Titan). *IAU Circular, 7306*.

66. Owen, T., Mahaffy, P. R., Niemann, H. B., Atreya, S., & Wong, M. (2001). Protosolar Nitrogen. *Astrophysical Journal, 553*, L77–L79.

67. Penteado, P. F., Griffith, C. A., Greathouse, K., & de Bergh, C. (2005). Measurements of CH3D and CH4 in Titan from infrared spectroscopy. *Astrophysical Journal, 629*, L53–L56.

68. Pinto, J. P., Lunine, J. I., Kim, S. J., & Yung, Y. L. (1986). The D to H ratio and the origin and evolution of Titan's atmosphere. *Nature, 319*, 388–390.

69. Porco, C. C., et al. (2005). Imaging of Titan from the Cassini spacecraft. *Nature, 434*, 159–168.

70. Rannou, P., Hourdin, F., & McKay, C. P. (2002). A wind origin for Titan's haze structure. *Nature, 418*, 853–856.

71. Roe, H. G., de Pater, I., & McKay, C. P. (2004). Seasonal variation of Titan's stratospheric ethylene (C2H4) observed. *Icarus, 169*, 440–461.

72. Samuelson, R. E., Maguire, W. C., Hanel, R. A., Kunde, V. G., Jennings, D. E., Yung, Y. L., & Aikin, A. C. (1983). CO_2 on Titan. *Journal of Geophysical Research, 88*, 8709–8715.

73. Samuelson, R. E., Nath, N. R., & Borysow, A. (1997). Gaseous abundances and methane supersaturation in Titan's troposphere. *Planetary and Space Science, 45*, 959–980.

74. Sekine, Y., Imanaka, H., Matsui, T., Khare, B. N., Bakes, E. L. O., McKay, C. P., & Sugita, S. (2008a). The role of organic haze in Titan's atmospheric chemistry: II. Effect of heterogeneous reaction to the hydrogen budget and chemical composition of the atmosphere. *Icarus, 194*, 201–211. https://doi.org/10.1016/j.icarus.2007.08.030.

75. Sekine, Y., Imanaka, H., Matsui, T., Khare, B. N., Bakes, E. L. O., McKay, C. P., & Sugita, S. (2008b). The role of organic haze in Titan's atmospheric chemistry: I. Laboratory investigation on heterogeneous reaction of atomic hydrogen with Titan tholin. *Icarus, 194*, 186–200. https://doi.org/10.1016/j.icarus.2007.08.031

76. Shemansky, D. E., Stewart, A. I. F., West, R. A., Esposito, L. W., Hallett, J. T., & Liu, X. (2005). The Cassini UVIS Stellar Probe of the Titan Atmosphere. *Science, 308*, 978–982. https://doi.org/10.1126/science.1111790.

77. Sicardy, B., Colas, F., Widemann, T., et al. (2006). The two stellar occultations of November 14, 2003. *Journal of Geophysical Research, 111*. https://doi.org/10.1029/2005JE002624.

78. Strobel, D. F. (1974). The photochemistry of hydrocarbons in the atmosphere of Titan. *Icarus, 21*, 466–470.

79. Strobel, D. F. (2008). Titan's hydrodynamically escaping atmosphere. *Icarus, 193*, 588–594. https://doi.org/10.1016/j.icarus.2007.08.014.

80. Strobel, D. F. (2009). Titan's hydrodynamically escaping atmosphere: Escape rates and the structure of the exobase region. *Icarus, 202*, 632–641. https://doi.org/10.1016/j.icarus.2009.03.007

81. Strobel, D. F., & Shemansky, D. E. (1982). EUV emission from Titan's upper atmosphere: Voyager 1 encounter. *Journal of Geophysical Research, 87*, 1361.

82. Strobel, D. F., & Sicardy, B. (1997). Gravity wave and wind shear models. In J. P. Lebreton (Ed.), *Huygens Science, Payload and Mission* (ESA SP, vol. 1177, pp. 299–311).

83. Strobel, D. F., Hall, D. T., Zhu, X., & Summers, M. E. (1993). Upper limit on Titan's atmospheric argon abundance. *Icarus, 103*, 333–336.

84. Teanby, N. A., et al. (2006). Latitudinal variations of HCN, HC_3N, and C_2N_2 in Titan's stratosphere derived from Cassini CIRS data. *Icarus, 181*, 243–255.

85. Teanby, N. A., et al. (2007). Vertical profiles of HCN, HC_3N, and C_2H_2 in Titan's atmosphere derived from Cassini/CIRS data. *Icarus, 186*, 364–384.

86. Teanby, N. A., et al. (2008). Titan's polar vortex structure revealed by chemical tracers. *Journal of Geophysical Research, 113*, E12003. https://doi.org/10.1029/2008JE003218.

87. Teanby, N. A., et al. (2009). Titan's stratospheric C_2N_2, C_3H_4, and C_4H_2 abundance from Cassini/CIRS far-infrared spectra. *Icarus, 202*, 620–631. https://doi.org/10.1016/j.icarus.2009.03.022.

88. Teanby, N. A., Irwin, P. G. J., de Kok, R., & Nixon, C. A. (2009). Dynamical implications of seasonal and spatial variations in Titan's stratospheric composition. *Philosophical Transactions of the Royal Society A, 367*, 697–711. https://doi.org/10.1098/rsta.2008.0164.

89. Tomasko, M., Bézard, B., Doose, L., Engel, S., Karkoschka, E., & Vinatier, S. (2008). Heat balance in Titan's atmosphere. *Planetary and Space Science, 56*, 648–659. https://doi.org/10.1016/j.pss.2007.10.012.
90. Toublanc, D., Parisot, J. P., Brillet, J., Gautier, D., Raulin, F., & McKay, C. P. (1995). Photochemical modeling of Titan's atmosphere. *Icarus, 113*, 2–26. https://doi.org/10.1006/icar.1995.1002
91. Trafton, L. M. (1972). On the possible detection of H_2 in Titan's atmosphere. *Astrophysical Journal, 175*, 285–293.
92. Vervack, R. J. Jr., Sandel, B. R., & Strobel, D. F. (2004). New perspectives on Titan's upper atmosphere from a reanalysis of the Voyager 1 UVS solar occultations. *Icarus, 170*, 91–112.
93. Vinatier, S. (2007). *Analyse Des Spectres Infrarouges Thermiques Émis Par L'atmosphère De Titan Enregistrés Par L'instrument Cassini/CIRS*. Ph.D. Thesis, University of Denis Diderot, Paris VII.
94. Vinatier, S., Bézard, B., & Nixon, C. A. (2007b). The Titan $^{14}N/^{15}N$ and $^{12}C/^{13}C$ isotopic ratios in HCN from Cassini/CIRS. *Icarus, 191*, 712–721. https://doi.org/10.1016/j.icarus.2007.06.001.
95. Vinatier, S., et al. (2007a). Vertical abundance profiles of hydrocarbons in Titan's atmosphere at 15°S and 80°N retrieved from Cassini/CIRS spectra. *Icarus, 188*, 120–138.
96. Vuitton, V., Yelle, R. V., & Anicich, V. G. (2006). The nitrogen chemistry of Titan's upper atmosphere revealed. *Astrophysical Journal, 647*, L175–L178.
97. Vuitton, V., Yelle, R. V., & Cui, J. (2008). Formation and distribution of benzene on Titan. *Journal of Geophysical Research, 113*, E05007. https://doi.org/10.1029/2007JE002997.
98. Vuitton, V., Yelle, R. V., & McEwan, M. J. (2007). Ion-chemistry and N-containing molecules in Titan's upper atmosphere. *Icarus, 191*, 722–742. https://doi.org/10.1016/j.icarus.2007.06.023.
99. Waite, J. H. Jr., et al. (2005). Ion neutral mass spectrometer results from the first flyby of Titan. *Science, 308*, 982–986.
100. Waite, J. H., Young, D. T., Cravens, T. E., Coates, A. J., Crary, F. J., Magee, B., & Westlake, J. (2007). The process of tholin formation in Titan's upper atmosphere. *Science, 316*, 870–875. https://doi.org/10.1126/science.1139727.
101. Wilson, E. H., & Atreya, S. K. (2004). Current state of modeling the photochemistry of Titan's mutually dependent atmosphere and ionosphere. *Journal of Geophysical Research, 109*, E06002. https://doi.org/10.1029/2003JE002181.
102. Wong, A. S., Morgan, C. G., Yung, Y. L., & Owen, T. (2002). Evolution of CO on Titan. *Icarus, 155*, 382–392. https://doi.org/10.1006/icar.2001.6720.
103. Yelle, R. V. (1991). Non-LTE models of Titan's upper atmosphere. *Astrophysical Journal, 383*, 380–400.
104. Yelle, R. V., Cui, J., & Müller-Wodarg, I. C. F. (2008). Methane escape from Titan's atmosphere. *Journal of Geophysical Research, 113*, E10003. https://doi.org/10.1029/2007JE003031.

105. Yelle, R. V., Strobel, D. F., Lellouch, E., & Gautier, D. (1997). Engineering models for Titan's atmosphere. In J. P. Lebreton (Ed.), *Huygens Science, Payload and Mission* (ESA SP, vol. 1177, pp. 243–256).

106. Yung, Y. L., Allen, M., & Pinto, J. P. (1984). Photochemistry of the atmosphere of Titan – Comparison between model and observations. *Astrophysical Journal, 55*, 465–506. https://doi.org/10.1086/190963.

CHAPTER 3

Solar Radiation and Energy Balance

LEARNING OBJECTIVES

After studying this chapter, you will be able to:

- Discuss the solar radiation and the earth's energy balance;
- Explain the concepts of insolation and albedo;
- Learn about Earth's heat budget;
- Describe the mechanisms of heat transfer.

KEY TERMS FROM THIS CHAPTER

Albedo	Carbohydrates	Carbon dioxide
Chlorofluorocarbons	Earth surface	Greenhouse effect
Greenhouse gases	Halocarbons	Hydrofluorocarbons
Infrared radiation	Insolation	Methane
Nitrous oxide	Ozone	Ozone absorption
Perfluorocarbons	Photosynthesis	Rayleigh scattering

INTRODUCTION

There is a layer of atmospheric gases surrounding the Earth. The majority of these gases, including carbon dioxide, are created naturally, but they are also produced by burning fossil fuels, among other things. Numerous gases, like CFCs, are only produced by human activity. Certain gases possess the ability to retain infrared radiation in the atmosphere instead of losing it to space by absorbing it when it is reflected from the earth's surface. The Earth has been warming gradually but steadily due to a greenhouse effect as a result of the concentration of these gases growing, changing the global climate.

3.1. SOLAR RADIATION AND THE EARTH'S ENERGY BALANCE

Solar radiation is the primary energy source that powers the climate system. High energy particles and electromagnetic waves are continuously emitted by the sun. The sun is the primary energy source that the earth has access to. A tiny quantity of energy may come from the radioactive elements found in the crust of the Earth. Factors like solar output, the sun's altitude, the earth's distance from the sun, and the duration of the day all have an impact on how much solar energy reaches the top of the atmosphere. About 7% of solar radiation falls into the ultraviolet (0.2–0.4 µm) wavelength range, 41% is visible light (0.4–0.7 µm), and 52% is near-infrared (>0.7 µm) wavelength range. Here, it is appropriate to introduce Wien's displacement law.

The wavelength of maximum emission varies inversely with the radiating body's absolute temperature, according to Wien's law. The terrestrial radiation emitted from the earth's surface has a peak intensity of approximately 10 µm and a range from about 4 to 100 µm. The mean temperature of the earth's surface is approximately 288 K (15°C). In terms of solar radiation, they are very strong, with a peak intensity of 0.5 µm for shortwave radiation. Moreover, the solar constant is defined as "energy received at the top of the atmosphere on a surface perpendicular to the solar beam for mean solar distance." Due to sunspot activity, the solar constant shows tiny periodic variations of roughly 1 Wm–2. Put differently, the number of sunspots affects the solar output. The sunspot cycles are said to have wavelengths of 88 years (Gleissberg), 22 years (Hale) magnetic cycle, much less significant 37.2 years (18.6 years – the luni–solar oscillation), and an average of 11 years (Schwabe cycle, which varies between 8 and 13 years). Additionally, it states that "assuming earth behaves like a black body, a persistent anomaly of 1% in the solar constant could change the effective mean temperature of the earth's surface by as much as 0.6°C."

Furthermore, the amount of solar energy that the Earth receives varies according to its changing distance from the sun. The Earth's surface receives different amounts of solar energy at perihelion and aphelion due to the eccentricity of the Earth's orbit around the sun. However, the continentality and atmospheric circulation obscure this uneven absorption of solar radiation. Moreover, the aphelion took place during the northern hemisphere winter some 11,000 years ago. Currently, the summertime in the northern hemisphere is when the aphelion occurs.

The sun's altitude has an impact on the quantity of solar radiation that reaches Earth's surface. The latitude of the location, the time of day, and the season are the variables that affect the sun's altitude. Higher sun altitudes result in greater radiation intensity per unit area at the earth's surface as well as shorter radiation paths for solar radiation to travel before reaching the surface. In terms of the relationship between the duration of the day and the amount of solar radiation that reaches the surface of the earth, it is clear that the equator experiences nearly a 12-hour day, and as a result, more solar radiation eventually reaches the earth's surface there. In actuality, some variables, including solar output, the sun's altitude, the Earth's distance from the sun, and the duration of the day, all work together to affect how much solar radiation reaches the Earth's surface.

Solar radiation falls into the short wavelength range, or i.e., less than 4 μm. "Approximately 18% of the insolation is absorbed directly by ozone and water vapor," according to reports. Since ozone absorption is concentrated in spectral bands like 0.20–0.31 μm, 0.31–0.35 μm, and 0.45–0.85 μm, it can be said that the ozonosphere is actually in charge of absorbing the harmful UV radiation. Only 70% of the insolation is used to heat the Earth and its atmosphere because the remaining 30% is reflected into space by the atmosphere, clouds, and the Earth's surface. The albedo is defined as the "proportion of incident radiation that is reflected." Another name for the albedo is the reflection coefficient. Albedo is also influenced by the type of cloud. For example, albedo values for cumulonimbus are reported to be 90%, while albedo values for cirrostratus range from 44% to 50%. Nearly half of the energy available at the top of the atmosphere is reportedly absorbed by the earth's surface and reradiated as terrestrial long-wave infrared radiation with a wavelength greater than 3 μm. In addition, a large portion of the radiation reaching Earth is absorbed by gases found in the atmosphere, including carbon dioxide, ozone, and water vapor. The atmospheric gases' ability to absorb terrestrial radiation is crucial to the continued existence of life on Earth. However, the amount of solar radiation that reaches different regions of the earth varies. The variations in solar radiation received depending on the season, day length, altitude, and continentality are the causes of this. The length of the day and the path taken by solar radiation through the atmosphere are both determined by latitude. In addition, the aspect and elevation of the land affect how much solar radiation reaches the surface. It is obvious that, in clear skies, high elevations receive more solar radiation than either lower elevations or areas close to mean sea level.

The average annual solar radiation received by each square meter of the Earth's spherical surface outside of the atmosphere is 342 Watts per square meter (Wm^{-2}). 31% of this is instantly reflected into space by the Earth's surface, atmosphere, and clouds. The majority (168 Wm^{-2}) warms the Earth's surface (the land and the ocean), with the remaining 235 Wm^{-2} being partially absorbed by the atmosphere.

The heat is reflected back into space by the earth's surface in the form of water vapor, which releases heat when it condenses into clouds higher in the atmosphere, sensible heat, and infrared radiation. Under current circumstances, the global mean temperature close to the earth's surface is maintained at 15°C by this energy exchange between the surface and atmosphere. At the top of the troposphere, the mean temperature drops off quickly as altitude increases to –58°C. A balance between the amount of radiation from the sun and the amount of radiation leaving the climate system is necessary for a stable climate. Consequently, the climate system itself needs to radiate back into space at a rate of 235 Wm^{-2} on average.

Figure 3.1, which depicts the details of this energy balance, shows how the atmosphere emits infrared radiation out of it on the right side and what happens to the incoming solar radiation on the left.

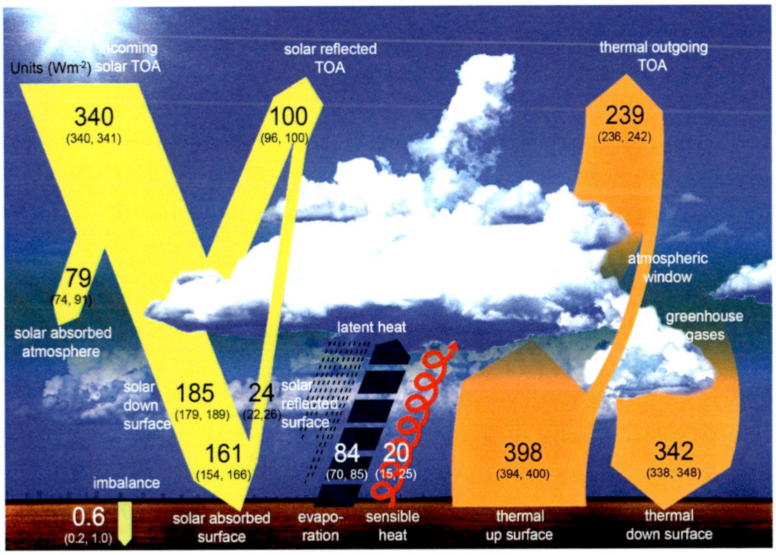

Figure 3.1. *Global heat budget.*

Source: *https://egyankosh.ac.in/bitstream/123456789/53800/1/Unit-2.pdf.*

Every physical object emits energy to a certain extent and at wavelengths that are characteristic of its temperature. Shorter wavelengths radiate more energy at higher temperatures. The Earth must radiate at an effective emission temperature of −19°C and typical wavelengths in the infrared region of the spectrum to emit 235 Wm^{-2}. This is 34°C colder than the surface of the Earth's average temperature of 15°C.

3.1.1. Solar Radiation Basics

The term "solar radiation," which can also refer to sunlight or solar resources, refers to the electromagnetic radiation that the sun emits. Many technologies can be used to capture solar radiation and convert it into useful forms of energy, like heat and electricity. However, the solar resource that is available at a given location determines the technological viability and cost-effective operation of these technologies.

The common term for the electromagnetic radiation that the sun emits is solar radiation. Solar radiation, which makes up the majority of the radiant energy that Earth receives, comes from the sun. Without this radiant energy input, the planet Earth would be lifeless, cold, and dark. The oceanic and atmospheric circulation systems are propelled by solar radiation that is absorbed by the Earth's surface and, to a lesser degree, the surrounding atmosphere, in conjunction with the planet's rotation. Plants use energy from the spectrum of solar radiation to change carbon dioxide and water into carbohydrates through a process called photosynthesis. Scattered solar radiation indeed makes up the sky's light.

1. **Basic Principles:** Every location on Earth receives sunlight for at least part of the year. The amount of solar radiation that reaches any one spot on the Earth's surface varies according to:
   - Geographic location;
   - Time of day;
   - Season;
   - Local landscape;
   - Local weather.

The sun strikes the surface of the round Earth at varying angles, from 0° (just above the horizon) to 90° (directly overhead). The Earth's surface receives maximum energy when the sun's rays are directed vertically. The sun's rays become more dispersed and diffuse as they pass through the atmosphere with increasing slant. The icy Polar Regions never see a high sun because the Earth is round, and for a portion of the year, they receive no sun at all due to the tilted axis of rotation.

During a portion of the year, the Earth is closer to the sun due to its elliptical orbit around it. The Earth's surface receives slightly more solar energy when the sun is closer to the planet. In the summertime in the southern hemisphere and the wintertime in the northern hemisphere, Earth is closer to the sun. The warmer summers and colder winters that one would anticipate in the southern hemisphere are, nevertheless, moderated by the presence of large oceans.

A more important factor in determining how much sunlight strikes the Earth at a given location is the 23.5° tilt in the Earth's axis of rotation. Longer days are the result of tilting, with longer days in the southern hemisphere for the remaining six months and longer days in the northern hemisphere from the spring (vernal) to the fall (autumnal) equinox. On the equinoxes, which fall on or around March 23 and September 22 of each year, days and nights are both precisely twelve hours long.

Middle-latitude nations like the United States benefit from increased summer solar radiation due to longer days and the sun's close approach to overhead. During the shorter days of the winter, the sun's rays are much more slanted. In June, cities close to 40° latitude, like Denver, Colorado, receive almost three times as much solar energy as they do in December.

The hourly variations in sunlight are also caused by the Earth's rotation. The sun is low in the sky in the morning and late in the afternoon. In comparison to noon, when the sun is at its highest point, its rays travel farther through the atmosphere. Around solar noon on a clear day, a solar collector receives the maximum amount of solar energy.

2. **Diffuse and Direct Solar Radiation:** As sunlight passes through the atmosphere, some of it is absorbed, scattered, and reflected by:
   - Air molecules;
   - Water vapor;
   - Clouds;
   - Dust;
   - Pollutants;
   - Forest fires;
   - Volcanoes.

We refer to this as diffuse solar radiation. Direct beam solar radiation is the portion of solar radiation that reaches the Earth's surface undiluted. Global solar radiation is the total of both direct and diffuse solar radiation. On clear, dry days, atmospheric conditions can reduce direct beam radiation by 10%, and on cloudy, thick days, they can reduce it by 100%.

3. **Measurement:** Scientists track how much light reaches particular spots at different times of the year. Next, they calculate how much sunlight reaches areas with comparable climates at the same latitude. Total radiation on a horizontal surface or total radiation on a surface tracking the sun are the common ways that solar energy measurements are expressed.

The unit of measurement for radiation for solar electric (photovoltaic) systems is frequently kWh/m^2 or kilowatt-hours per square meter. Watts per square meter (W/m^2) is another way to express direct estimates of solar energy.

The standard unit of measurement for radiation data in space heating and solar water heating systems is British thermal units per square foot or Btu/ft^2.

4. **Distribution:** Because photovoltaic (PV) systems use both direct and scattered sunlight, the solar resource in the United States is abundant. Other technologies might have more restrictions. However, the amount of solar energy that reaches a given location determines how much power is produced there by any solar technology. As a result, solar technologies perform best in the United States southwest, which gets the most solar radiation.

3.1.2. Solar Output

Based on the total solar emission, the emission temperature of the solar disk, which is roughly 5777K, determines the quantity and spectral type of this radiant energy (LeNoble, 1993). Shorter wavelengths change more than longer wavelengths when it comes to the spectral (wavelength) characteristics of the radiant flux, although small variations in the emission temperature do not significantly affect the total amount of energy produced. At wavelengths shorter than 250 nm, less than 0.2% of the sun's energy output is released, and at wavelengths longer than roughly 6 μm, a comparable amount is radiated. The visible spectrum, between 400 and 700 nm, is the wavelength region with the highest emission, with the peak of the irradiance distribution occurring at 465 nm (Figure 3.2). The energy flux that Earth receives from the sun is known as total solar irradiance (TSI), or solar constant. Its average value is 1367 Wm^{-2}. The TSI variation as determined by multiple satellite instruments between 1979 and 2002 is shown in Figure 3.3. An approximate 11-year cycle connects solar activities like sunspots, auroras, and geomagnetic disturbances to variations in the sun's emittance (see bottom panel of Figure 3.2). Irradiance rises in response to increased solar activity (Hoyt and Schatten, 1997).

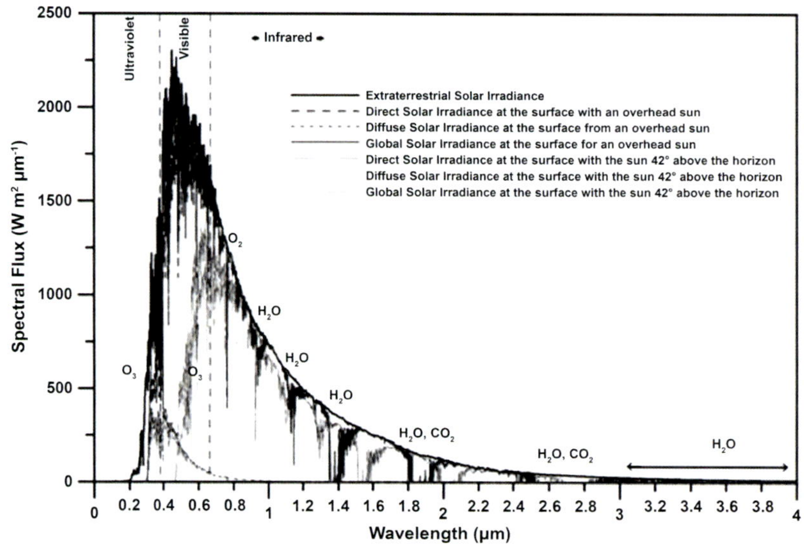

Figure 3.2. *Extraterrestrial spectrum, direct beam spectrum, and diffuse spectrum for an overhead sun (solar elevation 90) and when the sun is at 42 elevations.*

Source: John E. Oliver. Encyclopedia of World Climatology. Springer. https://doi.org/10.1007/0–387-30749-4_161.

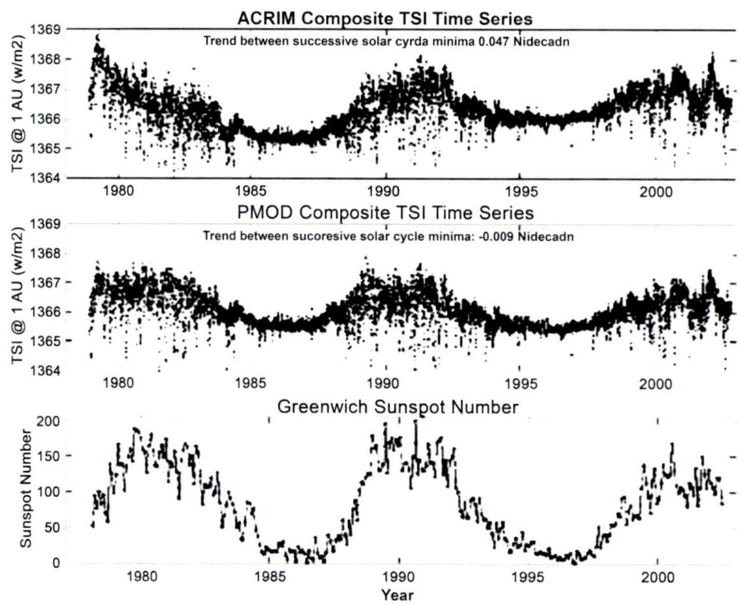

Figure 3.3. *Variation in the total solar irradiance as published by two laboratories. The lower panel indicates the sunspot cycle (Wilson & Mordvinov, 2003).*

Source: John E. Oliver. Encyclopedia of World Climatology. Springer. https://doi.org/10.1007/0-387-30749-4_161.

3.1.3. Solar Geometry

The geometry of the sun and Earth as well as the absorption and scattering characteristics of the planet's atmosphere determine how much solar radiation actually reaches the Earth's surface. The average distance between Earth and the sun is 1.496×10^8 km, or one astronomical unit (AU). The Earth's orbit is not concentric, ranging between 1.47×10^8 km and 1.52×10^8 km. An orbit diagram of the Earth around the sun is shown in Figure 3.4, which indicates that the Earth approaches the sun closest on January 3 and the greatest distance between the two bodies is on July 4. The TSI fluctuates in proportion to the square root of the distance from the sun. The eccentricity correction factor is the name given to this variation.

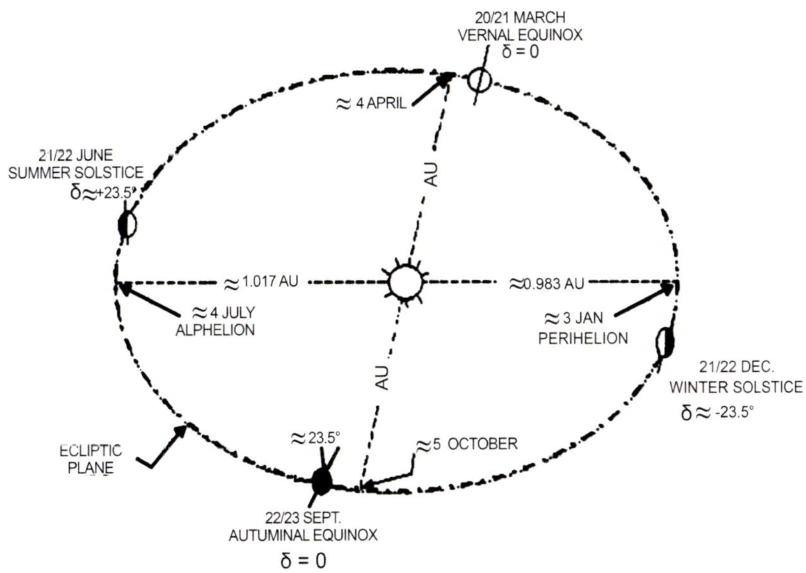

Figure 3.4. *Orbital characteristics of the Earth about the sun (adapted from Iqbal, 1983).*

Source: John E. Oliver. Encyclopedia of World Climatology. Springer. https://doi.org/10.1007/0-387-30749-4_161.

The tilt of the Earth's axis concerning the plane around which it orbits the sun (the ecliptic plane) also affects the amount of energy at the top of the atmosphere at any given location above the planet. Because of the shifting angle between the sun and the Earth, the sun appears to move north and south of the equator as the Earth revolves around it. The angular measure of this variation is called solar declination, and it varies from –23°27' (south) at the winter solstice in the Northern Hemisphere to 23°27' (north) at the summer solstice. The solar declination is zero at the equinox when the sun is directly overhead at the equator. Based on changes in the sun-Earth distance and the Earth-sun angle, Figure 3.5 illustrates how much solar radiation would reach the Earth's surface.

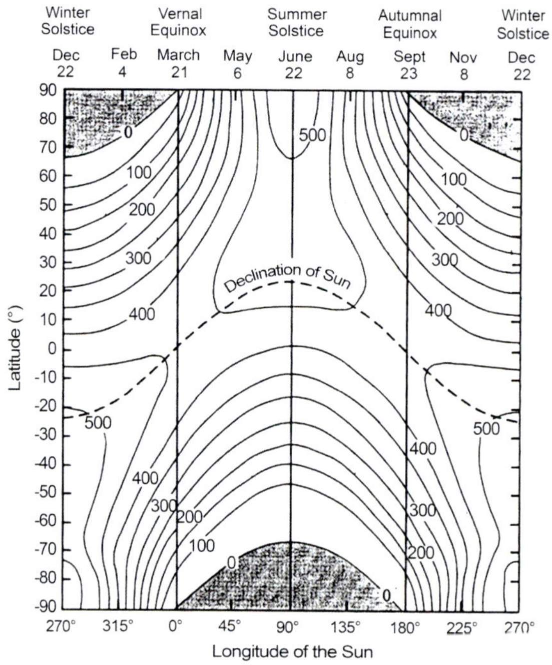

Figure 3.5. *Geographical and temporal distribution of solar radiation (W m^{-2}) at the surface of the Earth.*

Source: John E. Oliver. Encyclopedia of World Climatology. Springer. https://doi.org/10.1007/0-387-30749-4_161.

The daily eastward rotation of the Earth on its axis is the last geometric factor influencing the amount of solar radiation received at a given location. Between sunrise and sunset, the angle of elevation between the solar disk and the surface varies to zero, and when the sun is equatorward of the location, it reaches a maximum of [90 ⊠ |latitude ⊠ solar declination|]. When the sun reaches its highest point in the sky, known as solar noon, the amount of solar radiation that a flat surface receives increases with solar elevation.

3.1.4. The Atmosphere

Once it moves through the atmosphere, this solar radiation changes in both quantity and spectral type. While air molecules scatter radiation and particulates (aerosols) both scatter and absorb it, gases like ozone, oxygen, and water vapor absorb radiation and warm the atmosphere in the process. The total spectrum flux is divided into two parts by radiation scattering: (1) the direct beam spectral flux, or the spectrum that reaches the surface unabsorbed or scattered, and (2) the diffuse spectral radiation, or the spectrum that has been scattered in any way. Global spectral radiation, also known as global spectral irradiance, is the total amount of solar radiation that reaches the surface. Global, direct, and diffuse irradiance are the terms used to describe these integrated fluxes when they are calculated across the whole wavelength range that the sun emits.

Gases absorb at different levels of the atmosphere and in different spectral regions. The majority of the UV radiation from the sun is absorbed by ozone when it travels through the stratosphere, which contains the highest concentration of O_3. The troposphere has a higher concentration of water vapor, which absorbs solar radiation with longer infrared wavelengths. In the atmosphere, gases like carbon dioxide and oxygen are more evenly distributed. In the UV and visible wavelength ranges of the solar spectrum, solar radiation is also absorbed by two gases linked to human activity: sulfur dioxide and nitrogen oxides.

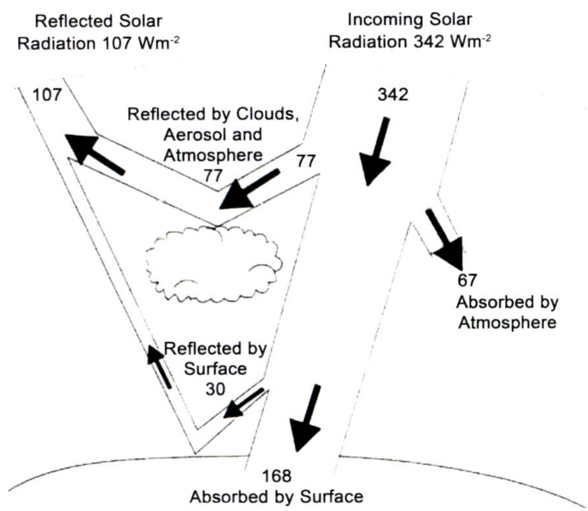

Figure 3.6. *The global average shortwave radiation budget showing the distribution of solar radiation in the atmosphere and at the surface (adapted from IPCC, 2001).*

Source: *John E. Oliver. Encyclopedia of World Climatology. Springer. https://doi.org/10.1007/0-387-30749-4_161.*

Air molecules are the main source of solar radiation scattering in skies free of clouds. This type of light scattering, called Rayleigh scattering, occurs when particles with a diameter substantially smaller than the wavelength of the radiation being scattered scatter it. The blue hue of the sky on clear days is caused by scattering, which is roughly proportional to the inverse of the wavelength to the fourth power (λ^{-4}). Figure 3.2 illustrates how Rayleigh scattering disappears at wavelengths longer than roughly 600 nm. Mie scattering is the term used to describe scattering by particles larger than the wavelength of radiation, such as smoke, dust, water droplets, or volcanic aerosols. Compared to Rayleigh scatter, this is more directed and less reliant on the solar radiation's wavelength. Mie scatter is the cause of the sky's diminished color contrast on cloudy or dirty days. The most significant scatterer of solar radiation is clouds, which can reduce the flux to almost zero when thick and increase the amount of solar radiation that reaches the Earth's surface instantaneously to values above the TSI. The efficiency of clouds in scattering and intercepting radiation is demonstrated in Figure 3.6, a schematic of the global shortwave radiation budget. Because clouds are so effective in this process, variations in the amount of cloud cover on Earth can have a big impact on climate change (IPCC, 2001a). The impact of a single cloud passing between the sun and the surface is depicted in the 5 March plot in Figure 3.7. It first shows a sharp decline in the quantity of solar radiation that reaches the surface, followed by an increase significantly higher than the radiation received from a clear sky when the direct beam irradiance is least attenuated.

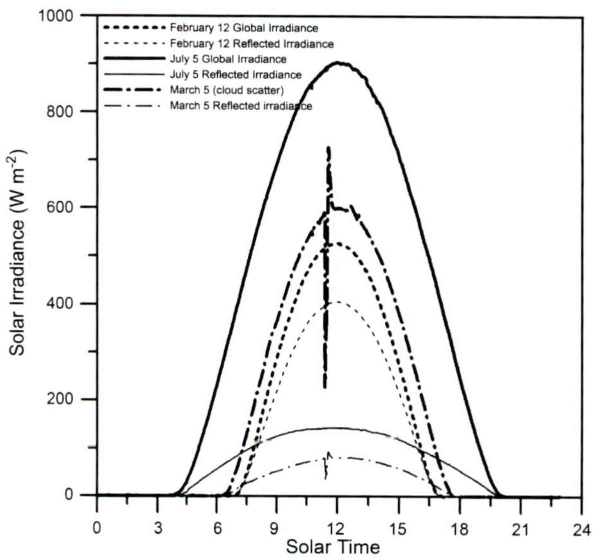

Figure 3.7. *Global and reflected irradiance observations for three nearly clear-sky days from the Bratt's lake observatory (104.7°W, 50.28°N). The blue curve illustrates the effect of clouds passing between the sun and the surface. Note the increase above the normal irradiance levels caused by the scattering of radiation from the side of the cloud after reducing the surface flux by more than half. This can also be observed in the reflected irradiance. The 12 February and 5 July days show the changes in the radiation received at the surface, primarily because of solar geometry. The difference in the reflected irradiance between the two days is a function of the surface cover.*

Source: John E. Oliver. Encyclopedia of World Climatology. Springer. https://doi.org/10.1007/0-387-30749-4_161.

3.1.5. Solar Radiation on the Surface

Geometric and atmospheric factors determine how much solar radiation reaches the Earth's surface. On the other hand, the characteristics of the surface affect how much of this radiation is absorbed there. The Bratt's Lake Observatory, which is part of a global surface radiation budget monitoring network, displays changes in the incoming and reflected global solar fluxes in Figure 3.7 (Ohmura et al., 1998). The graph's plots illustrate how variations in the surface's ability to absorb solar radiation and the amount of radiation received at the surface throughout the summer and winter are caused by variations in solar geometry. The variations in the land cover between summer (wheat) and winter (snow) are the cause of the reflected fluxes in the figure. The reflected flux also shows how the cloud passage affected the solar irradiance curve on March 5.

The energy that is not reflected is absorbed and put to use in some processes, such as photosynthesis, evaporating water, and heating the surface. Albedo is the ratio of incident solar radiation on a surface to that reflected by the surface. While most plants have an albedo between 0.16 and 0.25, snow has an albedo between 0.4 and 0.95. Man-made urban surfaces have an albedo that ranges from 0.05 for asphalt to 0.4 for light-colored brick (Oke, 1987).

Satellite instrumentation is used to gain a better understanding of the global distribution of the amount of solar radiation available at the surface, even though precise observations of the incoming and reflected solar radiation can be obtained at the Earth's surface. The surface solar radiation distribution worldwide, as determined by NASA/AtSR ERBE observations. It is evident how solar geometry and cloud cover relate to one another. Near the North Pole, there is no radiation reaching the surface in the winter, but there is maximum radiation flux in the South Polar Region. The June distribution shows a correlation between low cloud cover and high solar radiation values over the Sahara region. In a similar vein, little solar radiation reaches the surface over the southern ocean region, which is primarily cloud-covered year-round. Currently, some nations are generating near-real-time satellite distributions of solar radiation. The USA and Australia are two of these.

3.1.6. Solar Radiation and Humans

Despite the abundance and free nature of solar radiation, most of the Western world has struggled to fully utilize this resource due to several factors, including the relatively high costs of products that harness solar radiation, the relatively cheap costs of energy from fossil fuels and nuclear power, the high costs of research and development when switching from conventional methods, and the lack of incentives to change (Interlaboratory Working Group, 2000). The use of solar radiation is predicted to rise as the price of conventional energy rises and more people worldwide become conscious of the possibility of climate change brought on by the burning of fossil fuels. Solar energy is primarily used for water heating, building heating and, to a lesser extent, cooling, photovoltaic off-grid electricity generation, thermal heating for electricity production, and cooking in rural developing countries using home parabolic collectors (Duffie and Beckman, 1991). According to the IPCC (2001b), between 1986 and 1998, the use of photovoltaics grew from 37 MW to 150 MW, with growth predicted to reach 1000 MW by 2005. Between 15 and 20% of photovoltaic modules are sold worldwide each year, and this growth rate is predicted to continue through 2020 (Inter-laboratory Working Group, 2000). According to IPCC (2001b) estimates, Europe could have up to 5 million square meters of solar thermal panels installed for domestic hot water heating by 2005. By directing solar radiation onto a central superheat boiler, concentrating collectors have been used to generate power for commercial purposes. The steam produced by the boiler is then directed through a turbine. The California Solar One generating plant is an example of this kind of system; it was initially operational at 10 MW (Duffie and Beckman, 1991), but it has since been upgraded to operate at 30 MW, with the ultimate goal of producing 100 MW of electricity (IPCC, 2001b).

Compared to developed countries, developing nations and those without easy access to alternative energy sources have embraced solar energy use more enthusiastically. Large solar energy projects are being planned or carried out in Brazil, India, Japan, South Africa, Uganda, and Vietnam. The World Bank and the United Nations Environment Program are both actively working to increase the use of solar energy globally, especially in developing nations. Globally, there is thought to be between 1575 and 49 837 EJ of solar radiation that is available for use each year (IPCC, 2001b). By contrast, the current global energy consumption amounts to about 425 EJ annually (IEA, 1998).

3.1.7. The Greenhouse Effect

The majority of the radiation coming from the sun is short-wave, but the radiation coming from the earth's surface is long-wave, or infra-red, radiation. The atmosphere is absorbent to long-wave and infrared radiation due to the presence of gases like carbon dioxide, water vapor, and other trace gases, except the "atmospheric window," which is located between approximately 8.5 and 13.0 μm. "The greenhouse effect" refers to how opaque the atmosphere is to infrared radiation compared to how transparent it is to short-wave radiation. Stated differently, "the infrared radiative effect of all infrared-absorbing constituents in the atmosphere" is what is meant to be understood by the term "greenhouse effect." The Earth's surface and other atmospheric regions emit terrestrial radiation, which is absorbed by greenhouse gases, clouds, and (to a lesser degree) aerosols. The magnitude of this effect increases as greenhouse gas concentrations rise. In the spectrum of terrestrial radiation emitted by the Earth's surface, the atmosphere, and clouds, greenhouse gases are defined as "those gaseous constituents of the atmosphere, both natural and anthropogenic, that absorb and emit radiation at specific wavelengths." The main greenhouse gases in the Earth's atmosphere are water vapor (H_2O), carbon dioxide (CO_2), nitrous oxide (N_2O), methane (CH_4), and ozone (O_3). In addition, man-made greenhouse gases like sulfur hexafluoride (SF6), halocarbons, and other compounds containing chlorine and bromine are significant radiatively active gases to be aware of. Instantaneous radiative forcing is caused by greenhouse gas emissions caused by humans. "In response to this forcing, the troposphere and surface temperature warm, progressively reestablishing the radiative balance at the upper atmosphere."

The "net infra-red absorption capacity of water vapor, carbon dioxide, and other trace gases – methane (CH_4), nitrous oxide (N_2O), and tropospheric ozone (O_3)" is essentially what causes the "total greenhouse effect." However, the "natural greenhouse gases' contribution to the mean 'effective' planetary temperature of 255 K is about 33 K." Of this, 21 K comes from water vapor, 7 K from carbon dioxide, 2 K from ozone, and roughly 3 K from other trace gases like methane and nitrous oxide. It is discovered that water vapor significantly absorbs infrared light in the vicinity of the spectral bands 2.4–3.1 μm, 4.5–6.5 μm, and above 16 μm. Because it absorbs both incoming harmful UV radiation and outgoing terrestrial long-wave infra-red radiation, the thermal role that ozone plays is extremely complex. Thus, the rise in the stratosphere's ozone concentration i.e., would benefit from being above 30 km since it absorbs short-wave radiation. Conversely, a rise in ozone concentration below 25 km would absorb more solar radiation from the Earth, raising the temperature of the surface air.

It is important to remember that the infrared radiation-absorbing qualities of natural atmospheric gas constituents like carbon dioxide, methane, and water vapor are what cause the natural greenhouse effect. The Earth is a livable planet because of natural greenhouse gases, as evidenced by the planet's surface air temperature of 15°C. Put another way, without these gases, the earth would have been a frozen, uninhabitable planet with an average temperature that is about 34°C lower than it is today. However, the increased emissions of greenhouse gases brought on by human activity are what cause the enhanced greenhouse effect (Figure 3.8).

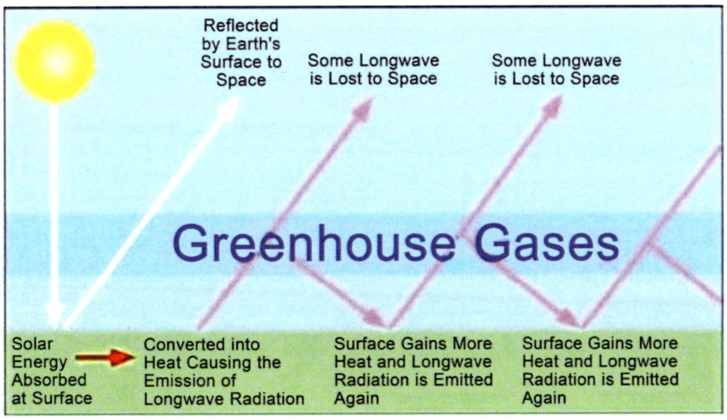

Figure 3.8. *Greenhouse effect.*

Source: https://egyankosh.ac.in/bitstream/123456789/53800/1/Unit-2.pdf.

3.1.8. Greenhouse Gases

Many refer to gases that trap heat in the atmosphere as greenhouse gases. Methane, nitrous oxide, ozone, carbon dioxide, and water vapor are the main greenhouse gases found in the Earth's atmosphere. A gas's contribution to the greenhouse effect is determined by both its properties and its abundance. Water vapor, carbon dioxide, methane, and ozone are the gases that contribute most to the greenhouse effect when these gases are ranked.

1. **Water Vapor:** Over half of the greenhouse effect is caused by water vapor. It is created spontaneously when water evaporation occurs continuously. But when the temperature rises, more evaporation takes place, which causes the atmospheric concentration of water vapor to increase. Condensation and precipitation eliminate it.
2. **Carbon Dioxide:** It is a "naturally occurring gas that is also a by-product of burning biomass, changing land use, burning fossil fuels from fossil carbon deposits, such as oil, gas, and coal, and industrial processes (e.g., cement manufacturing). It is the main greenhouse gas produced by humans that has an impact on the radiative balance of the Earth. Plants require it to carry out the process of photosynthesis. Plants and animals release carbon dioxide into the atmosphere as a result of physiological processes like respiration. The burning of fossil fuels like coal, oil, and gas in power plants, cars, factories, and other sources is the main cause of CO_2 emissions worldwide. With a Global Warming

Potential of 1, it serves as the benchmark gas for measuring other greenhouse gases.

3. **Methane:** It is produced spontaneously in wetland areas by termites and the decomposition of organic matter. In addition, a variety of man-made activities, such as the burning of biomass, the combustion of fossil fuels, the cultivation of rice, and the landfilling of household waste, have increased the average atmospheric concentration of CH_4. One of the six greenhouse gases that must be reduced under the Kyoto Protocol, methane is a major component of natural gas and is linked to all forms of hydrocarbon fuels, agriculture, and animal husbandry.
4. **Nitrous Oxide (N_2O):** One of the six greenhouse gases that must be reduced per the Kyoto Protocol is nitrous oxide. The primary human-caused sources of nitrous oxide are agricultural practices (manure management of the soil and animals), but sewage treatment, fossil fuel combustion, and chemical industrial processes also make significant contributions. Additionally, nitrous oxide is naturally generated in soil and water by a wide range of biological processes, especially microbial activity in moist tropical forests.
5. **Chlorofluorocarbons (CFCs):** These are a class of synthetic compounds that consist of carbon, fluorine, and chlorine. To facilitate refrigeration, CFCs were first produced in the 1930s. They have been widely used ever since as air conditioners, blowing agents in the production of foam, aerosol propellants, and in numerous other applications. The lower atmosphere does not contain any CFC sinks. Thus, they are carried to the stratosphere (10 to 50 km altitude), where UV radiation breaks them down, releasing free chlorine atoms that significantly reduce the amount of ozone in the atmosphere. Because of these compounds' comparatively long lifetimes, the concentration of CFCs in the atmosphere will continue to be significant well into the next century.
6. **Ozone:** One type of oxygen is ozone. It is made up of three oxygen atoms that are bonded together (O_3). The majority of the ozone in the atmosphere is created naturally in the stratosphere, which is located between 19 and 30 km above the surface of the earth. The "ozone layer" refers to this region of air that is rich in ozone. Harmful sunlight is blocked by stratospheric ozone. On the other hand, ground-level ozone is just a pollutant that absorbs solar radiation. It is created when sunlight reacts with nitrogen oxides (NOx) and volatile organic compounds (VOCs), some of which are created by human activity like driving an automobile. One element of urban smog that can be hazardous to human health is ground-level ozone.
7. **Fluorinated Gases:** Strong greenhouse gases produced by some industrial processes include sulfur hexafluoride, perfluorocarbons, and hydrofluorocarbons. Fluorinated gases are occasionally utilized to replace substances that deplete the atmosphere (i.e., halons, HCFCs, and CFCs). These gases are strong greenhouse gases even though they are normally released in smaller amounts.

3.1.9. Global Warming Potential

The amount that a specific mass of any greenhouse gas is expected to contribute to the risks associated with global warming is measured by its global warming potential or GWP. It is evaluated to an identical mass of carbon dioxide. It is dependent upon the molecule's lifetime in the atmosphere as well as its ability to absorb infrared light. The term "global warming potential" (GWP) refers to an index that measures the radiative forcing of a unit mass of a particular greenhouse gas pulsed out in the current atmosphere over a selected time horizon, relative to that of carbon dioxide. It is based on the radiative properties of greenhouse gases. The combined effect of these gases' varying atmospheric half-lives and relative potencies to produce radiative forcing is represented by the Greenhouse Gas Potential (GWP). The rate at which the unit mass of injected GHG decays over time, the cumulative radiative forcing that the unit addition of GHG will have over a certain amount of time into the future, and the radiative forcing associated with the addition of a unit mass of each GHG into the atmosphere are some of the factors that affect the global warming potential (GWP) (Table 3.1).

Table 3.1. *Global Warming Potential of GHGs Relative to Carbon Dioxide*

Chemical Species	Lifetime (Years)	GWP 20-Years	GWP 100-Years
Carbon dioxide	50–200	1	1
Methane	12	84	28
Nitrous oxide	114	264	265
CFC-11	45	6900	4660
CFC-12	100	10800	10200
CFC-113	85	6490	5820
CFC-115	1700	5860	7670
HCFC-22	12	5280	1760
HFC-23	270	10800	12400
Sulphur hexafluoride	3200	17500	23500
Carbon tetrachloride	35	3480	1730
Methyl bromide	1	9	2
Halon-1301	65	7800	6290

Source: https://egyankosh.ac.in/bitstream/123456789/53800/1/Unit-2.pdf.

3.1.10. Trends in Greenhouse Gases Emissions

Most greenhouse gases have been present in higher concentrations since the start of the industrial revolution. The concentration is rising faster, according to recent data. Naturally, human activity is associated with the burning of fossil fuels (e.g., for the production of electricity), land-use changes (primarily deforestation), and agricultural activities (Table 3.2).

Table 3.2. *The Pre-and Post-Industrial Concentrations of GHGs*

Greenhouse	Concentration 1755	Concentration 2011	Natural and Anthropogenic Sources
Carbon dioxide	276.7 ppm	390.5 ppm	Organic decay; forest fires; volcanoes; fossil fuels burning; deforestation; land-use change.
Methane	723 ppb	1803 ppb	Wetlands; organic decay; termites; natural gas and oil extraction; biomass burning; rice cultivation; cattle; refuse landfills.
Nitrous oxide	272.8 ppb	324 ppb	Forests; grasslands; oceans; soils; fertilizers; biomass burning; burning of fossil fuels.
Sulphur hexafluoride	0	7.3 ppt	Used as electrical insulators.
HFC-23	0	24 ppt	Refrigerators; aerosol spray propellants; cleaning solvents
HFC-32	0	4.9 ppt	
HFC-125	0	9.6 ppt	
HFC-134a	0	63 ppt	
HFC-143a	0	12 ppt	
CFC-11	0	238 ppt	
CFC-12	0	528 ppt	
CFC-113	0	74 ppt	
CFC-114	0	15.8 ppt	
CCl_4	0	86 ppt	
Ozone	Unknown	Varies with latitude and altitude in the atmosphere	Ozone is formed by photochemical reactions involving short-lived precursor gases like NOx, VOC, CO.

Source: https://egyankosh.ac.in/bitstream/123456789/53800/1/Unit-2.pdf.

The agriculture and related industries also cause greenhouse gas emissions. Since pre-industrial times, there has been an increase in greenhouse gas (GHG) emissions worldwide; between 1970 and 2004, there was a 70% increase. The main way human activity intensifies the greenhouse effect is through carbon dioxide emission. However, the impact of humans on the emissions of other greenhouse gases is also significant. The atmospheric concentrations of several significant variable greenhouse gases, including carbon dioxide (CO_2), methane (CH_4), nitrous oxide (N_2O), chlorofluorocarbons (CFCs), and troposphere ozone (O_3), can be affected by human activity. Between 2000 and 2010, there was an approximate increase of 10 $GtCO_2$ – eq in total annual anthropogenic GHG emissions. The energy (47%), industrial (30%), transportation (11%), and building (3%) sectors were directly responsible for this increase. Furthermore, according to the report, "the energy sector released 35% of GHG emissions in 2010, AFOLU released 24% (net emissions), industry released 21%, transportation released 14%, and the building sector released 6.4%." Without a doubt, as economic development and population growth are fueled by the intensive use of energy resources, this results in an increase in atmospheric GHG emissions.

Three greenhouse gases (GHGs) in particular, viz. Since before the industrial era, the atmosphere's concentrations of carbon dioxide (CO_2), methane (CH_4), and nitrous oxide (N_2O) have significantly increased. Indeed, the main factor causing climate change is the concentration of greenhouse gases rising. Furthermore, it is stated that the combined radiative forcing from these three greenhouse gases is responsible for 80% of the total radiative forcing from well-mixed greenhouse gases. The burning of fossil fuels, changes in land use, and agricultural practices all contribute to an increase in the concentration of these three GHGs.

Regarding the atmospheric carbon dioxide concentration, it varied from 180 parts per million during glacial (cold) periods to 300 parts per million during interglacial (warm) periods over the "past 800,000 years before 1750." Furthermore, the concentration of carbon dioxide in the atmosphere fluctuated between 260 and 280 parts per million for the last 7000 years before 1750. However, compared to pre-industrial times, "between 1750 and 2011, CO_2 concentration increased by 40%, from 278 ppm to 390 ppm." N_2O increased by 20% in 2011 from 271 ppb to 324 ppb, while the concentration of CH_4 increased by 150% from 722 ppb to 1803 ppb. The main causes of anthropogenic carbon dioxide emissions are the burning of fossil fuels, the manufacture of cement, and modifications to land use. Over the past ten years, there has been an annual increase in the atmospheric concentration of carbon dioxide of 2.0 ± 0.1 parts per million. According to reports, intensive agricultural practices will probably cause methane emissions into the atmosphere to rise in the future. Rice farming, domestic grazing animals, termites, landfills, coal mining, and oil and gas extraction are the main sources of methane emissions. Of total methane emissions, 50–65% are caused by human activity. The methane budget for the decade of 2000–2009 (bottom-up estimates), according to the IPCC (2013), is 177–284 $Tg(CH_4)$/year for emissions from natural wetlands, 187–224 $Tg(CH_4)$/year for emissions from agriculture and waste (rice, animals, and waste), 85–105 $Tg(CH_4)$/year for emissions related to fossil fuels, 61–200 $Tg(CH_4)$/year for other natural emissions, including geological, termite, and freshwater emissions, and 32–39 $Tg(CH_4)$/year for burning biomass and biofuel. Furthermore, the use of synthetic nitrogenous fertilizers, which raises N_2O emissions from soils, may be necessary for intensive agricultural practices to meet the world's rising food demand. In actuality, the amount of nitrous oxide in the atmosphere has grown at a rate of 0.73 ± 0.03 ppb/year during the past three decades. The processes of "denitrification" and "nitrification" are what cause the atmospheric emissions of nitrous oxide. Additionally, emissions of nitrous oxide were noted from "biomass burning" and "fossil fuel use and industrial processes."

It is incorrect to refer to ozone as a "well-mixed" greenhouse gas. Its concentration varies globally depending on the location and time. However, when it is present in the troposphere, it is a very significant GHG. The photochemical reactions involving short-lived precursor gases like NOx, CO, and volatile organic compounds (VOC) create the ozone in the troposphere. Furthermore, the northern hemisphere has a higher concentration of tropospheric ozone than the southern hemisphere. Summertime is said to have a higher tropospheric ozone concentration.

3.2. CONCEPTS OF INSOLATION AND ALBEDO

Insolation and Albedo are both concepts related to the sun's energy and how it affects the Earth's surface:

1. **Insolation**: The amount of solar energy that hits a given area over some time. It can be measured in kilowatt-hours per square meter per day (kWh/m^2/day) for the average amount of energy hitting an area each day, or in watts per square meter (W/m^2) for the average amount of power hitting an area over a year. The amount of insolation a given area receives varies by time and location.
2. **Albedo**: The amount of sunlight a surface reflects into space. It's expressed as a percentage of reflected insolation to incoming insolation, with 0% being total absorption and 100% being total reflection. Surfaces with light colors, like fresh snow, have high albedo, while dark surfaces, like pavement, have low albedo.

3.2.1. Insolation

The majority of the energy that the surface of the earth receives is in short wavelengths. We refer to the energy that the earth receives from the sun as incoming solar radiation or insolation. Because the earth is a geoid and only absorbs a small portion of the sun's energy, the sun's rays fall obliquely at the top of the atmosphere. The planet receives, on average, 1.94 calories per square centimeter per minute at the top of its atmosphere.

There are obvious differences in the climate and temperatures of different parts of the world, such as those found in Florida and Alaska. Florida may experience a pleasant, sunny day in the eighties, while Alaskans deal with below-freezing temperatures and biting surface winds that make it feel even colder. Through a concept called insolation, each location's latitude—that is, its placement north or south of the Equator—plays a direct role in this varied climate disparity.

To put it simply, insolation is the solar radiation that arrives on Earth in the form of brief waves from the sun. The Earth receives a comparatively small amount of insolation due to its small size and distance from the sun.

Merely two billionths of a percent of the sun's total energy, or two energy units out of 1,000,000,000 energy units emitted, make it to Earth's surface. Nevertheless, this tiny portion of solar radiation is crucial because it serves as the main energy source for the majority of physical and biological processes on Earth. Additionally, ozone, water vapor, dust particles, and other gases in the atmosphere all absorb a tiny amount of insolation.

The average amount of insolation received at the earth's surface decreases from the equator to the poles, but the amount of solar irradiance received at different latitudes at different times of the year varies over time.

- Insolation is defined as the solar energy (radiation) received by Earth.
- It is the solar radiation reaching any particular area.
- It is the source of energy for the entire Earth.

The amount of solar radiation that reaches the outer edge of our atmosphere is significantly higher at different latitudes than the amount of insolation that is received at the surface. This pattern shows that cloud cover, earth's atmosphere sediment (scattering), ability to reflect, and solubility (through ozone) cause a large amount of incoming solar radiation to be lost as it passes through the atmosphere. Maximum insolation occurs at the outer edge of the atmosphere at the North Pole during the summer solstice, while on June 21st, due to the presence of an anticyclonic and subtropical high-pressure belt, maximum insolation is received at the ground surface between latitudes 30° and 40°N.

Based on the amount of insolation received over a year, the globe is split into 3 zones.

1. **Low Latitude Zone:** Between the Tropics of Cancer and Capricorn, there is a low latitude or tropical zone. Due to the sun's northward and southerly motions, every location experience overhead sun (vertical solar rays) twice a year. As a result, every location experiences its highest and lowest insolation levels twice a year. The region experiences the maximum insolation of any other zone, and seasonal change is minimal.
2. **Middle Latitude Zone:** In both hemispheres, this zone is extended between 23.5° and 66° latitudes.
3. **Polar Zone:** In both hemispheres, the polar zone covers the area between latitudes 66° and 900 (poles). Every place experience its maximum and minimum insolation once per year, however, occasionally insolation reaches zero due to the lack of direct sunlight.

3.2.1.1. Major Characteristics of Insolation

Having understood what insolation is, let's know more about its fundamental characteristics:

- Only 51% of the sun's insolation reaches the earth's surface.
- The atmosphere reflects 35% of the remaining 49% into space.

3.2.1.2. Factors Affecting the Distribution of Insolation

The amount and intensity of insolation vary during the day and night and with seasons as well. The factors which cause these variations are:

1. **The Angle of Inclinations of the Sun's Rays:**
    - The area that receives direct sun rays or makes less angle receives more radiation than the areas that make a greater angle.
    - The amount of insolation that will be received at a given location is mostly determined by the angle formed by the sun's rays and the tangent to

the earth's surface.

- At the equator, the sun's beams are roughly vertical (the maximum angle is 90° between the sun's rays as well as the tangent towards the earth's surface) and get increasingly oblique as you move poleward. In other terms, poleward is where the sun's rays are at a smaller angle.
- The amount of energy received per unit area rises when vertical rays are dispersed over the smallest feasible portion of the earth's surface and heat it. The quantity of energy received per unit area drops, however, when oblique rays cover a bigger portion of the earth's surface.
- In comparison to vertical rays, oblique rays must travel through a thicker section of the atmosphere. The oblique rays must therefore go farther than the vertical rays.

2. Length of the Day:

- The longer the duration of sunshine more will be the quality of solar radiation received. Days in the Equatorial region are longer than in the polar region so the Equatorial region receives more insolation than the polar region.
- The circle of illumination, also known as the light circle, constantly divides the equator into two equal halves, resulting in a 12-hour day at the equator. However, the length of the day lengthens poleward with the northward march of the sun in the northern hemisphere whereas it shortens around the summer solstice in the southern hemisphere (21 June). On the other hand, during the time of the winter solstice (22 December), the length of the day grows from the poleward in the southern hemisphere but decreases in the northern hemisphere (southward march of the sun).
- It's crucial to note that during the sun's northward movement from 21 March to 23 September, the length of the day at the north pole increases to 6 months, while the length of the night at the south pole increases to 6 months.) In contrast, during the southward migration of the sun, the length of the day becomes 6 months at the South Pole (23 September to 21 March), while the duration of the night becomes 6 months at the North Pole.
- Even though the days are the longest at the poles, insolation is at its lowest because:
    - o the sun's beams are more or less parallel to the ground
    - o the majority of the solar radiation is reflected by the ice cover.
- It is clear that the length of the day does not have as much of an impact on the amount of insolation received as the angle of the sun's beams.

3. **Transparency of Atmosphere:** The atmosphere is not transparent to the radiation coming from the Sun. Different gases present in the atmosphere respond in different ways to the sun's Radiation. Some absorb it and cause the greenhouse effect.

4. **Distance from Sun:**

   - The earth should receive the most insolation at the period of perihelion when it is closest to the sun, and the least at the time of aphelion, when it is farthest from the sun. In fact, because of the low amount of insolation received in January, when the earth is closest to the sun, the northern hemisphere experiences winter rather than summer. On the other hand, due to the great quantity of insolation received in July, when the earth is furthest from the sun, summer instead of winter occurs in the northern hemisphere.
   - Insolation is distributed differently depending on the angle of the sun's rays and the duration of the day than depending on how far the earth is from the sun.
   - Insolation received varies in a year due to variations in its distance from the sun. Insolation received on perihelion (3rd January) is more than the amount of insolation on Aphelion (4th July).

5. **Earth's Rotation on its Axis and Revolution of the Article Around the Sun:**

   - Perihelion – When the earth is nearest to the Sun (147 million km), usually on 3rd January.
   - Aphelion – When the earth is farthest from the Sun (152 million km), usually on 4th July.
   - On average, the Earth receives 1.94 calories per sq cm per minute.

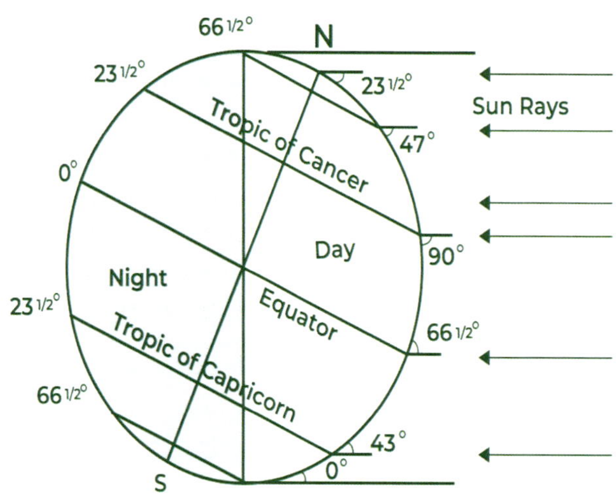

Source: https://www.geeksforgeeks.org/insolation-and-heat-budget/

3.2.1.3. Sunspot

- Periodic disruptions and explosions cause sunspots to form on the solar surface. Every year has a different number of sunspots. Studies have revealed that the variation in sunspot number is cyclical. In other words, a cycle of 11 years completes the increase and decrease in the number of sunspots. The highest number of sunspots occurs every eleventh year.

- The amount of insolation received at the earth's surface likewise increases as the energy radiated from the sun increases along with the number of sunspots. On the other hand, as the number of sunspots declines, less insolation reaches the surface of the planet.

3.2.2. Albedo

Albedo, which ranges from 0 to 1, represents a black body that absorbs all incident radiation and 1 represents a body that reflects nearly all incident radiation. Albedo is the amount of diffuse solar radiation reflected from the total solar radiation.

Surface albedo is the ratio of radiosity to irradiance (flux per unit of area) that a surface achieves. The characteristics of the Earth's surface, in addition to the spectral and angular spread of solar radiation approaching it, determine the proportion reflected. These factors vary over time, space, and in response to the makeup of the atmosphere.

3.2.2.1. Albedo Effect

The albedo effect is the directional integration of reflectance over all solar angles at a given time, while bi-hemispheric reflectance is measured for a given angle of incidence or Sun location. When determining the temporal resolution, flux measurements can range from seconds to regular, weekly, or yearly averages.

If solar radiation isn't directed toward a specific wavelength, the albedo effect affects a large spectrum of light (spectral albedo). Due to measurement limitations, it is typically given for the range where the majority of solar energy reaches the surface, which is between 0.3 and 3m.

Such spectrum includes visible light (0.4–0.7 m), which describes why surface albedo with such a low albedo (– for example, trees) look dark and surfaces with a large albedo look bright (e.g., snow reflects major radiation).

An important concept in astronomy, climatology, and sustainable development is albedo (see, for instance, the LEED program, which rates the sustainability of buildings). The Earth's planetary albedo, or average albedo from the upper atmosphere, is between 30 and 35% because of cloud cover, though it varies widely locally from across the surface due to different geological and environmental factors.

Johann Heinrich Lambert's work Photometria, published in 1760, was the first to use the word albedo in optics.

3.2.2.2. Terrestrial Albedo

The albedo of some of the darkest materials, such as charcoal, ranges from nearly 0.04 in visible light to about 0.9 for fresh snow. The black body's active albedo will be zero in deeply shadowed cavities.

From a distance, the ocean surface exhibits a low albedo, similar to that of most trees; on the other hand, desert regions have some of the highest albedos of any landform. The albedo of most land areas ranges from 0.1 to 0.4. Earth's albedo has historically averaged 0.3. The contribution of clouds is significantly higher than that of the seas. NASA regularly measures the Earth's surface albedo using the MODIS instruments on board the Terra and Aqua satellites, the CERES instrument on the Suomi NPP, and the JPSS. To predict bi-hemispheric reflectance and directional-hemispherical reflectance from a sample collection of satellite reflectance measurements, a mathematical model is being used, as satellites could only calculate the amount of reflected radiation in one direction, not all directions.

For such computations, the bidirectional reflectance distribution function (BRDF) is utilized, which describes how well the reflectance of a given surface varies depending on the solar angle and the observer's view angle. BDRF could help translate reflectance measurements into albedo.

Examples of Terrestrial Albedo Effects:

1. **Illumination:** Albedo is in fact not exclusively dependent on illumination, with the exception of circumstances where a change in illumination results in a change in the Earth's surface at that location. This is because increasing the amount of incident light directly influences the amount of reflected light (for example, through the melting of reflective ice).
2. **Insolation Effects:** Albedo concentration and local insolation (solar irradiance) both affect albedo temperature to varying degrees. For example, high albedo regions in the Arctic and Antarctic appear cold due to low insolation, while high albedo areas in the Sahara Desert, which also have significantly higher albedo, appear warmer due to increased insolation.
3. **Albedo–Temperature Feedback:** A snow-temperature input happens when snowfall modifies a region's albedo. Snowfall increases the albedo of the area, reflecting sunlight and causing a cooling effect. Theoretically, the lower temperature and higher albedo will retain the existing snowpack and encourage additional snowfall, strengthening the snow-temperature response, assuming that the outside temperature stays constant.
4. **Snow:** Snow albedo varies dramatically, varying from 0.9 for freshly fallen snow to 0.4 for snow melt and even as low as 0.2 for dirty snow. Ice albedo in Antarctica measures somewhat more than 0.8. As a marginally snow-covered region warms, the snow melts, reducing the albedo and thereby causing more snowmelt as the snowpack absorbs additional radiation.
5. **Solar Photovoltaic Effects:** Albedo may have an impact on solar photovoltaic

systems' ability to produce electricity. When compared to the standard spectral-integrated albedo predictions, differences in the spectrally weighted albedo of solar photovoltaic techniques, for example, demonstrate the effects of a spectrally sensitive albedo. These differences are dependent on hydrogenated amorphous silicon (a-S H) and crystalline silicon (c-Si). Research indicates that the effects may reach up to 10%. The study was recently extended to include the effects of albedo on the output of seven photovoltaic materials, including three common photovoltaic system topologies: commercial flat rooftops, industrial, and residential pitched-roof installations, as well as the consequences of spectral bias due to specular reflectivity of 22 frequently occurring surface materials.

3.2.2.3. Astronomical Albedo

The albedo of satellites, planets, and minor planets like asteroids can be used to infer a great deal about their characteristics. Photometry is a branch of astronomy that focuses primarily on albedos, which are dependent on lighting angle, wavelength, and time variation. The majority of our knowledge about small, far-off objects that telescopes are unable to resolve comes from studying their albedos. For example, absolute albedo will show the composition of surface ice on bodies in the outer Solar System; phase angle variation in albedo will show properties of regolith; and extremely high radar albedo will show high metal concentrations in asteroids.

Enceladus, a moon of Saturn, has one of the highest albedos ever measured—0.99—among all the planets in the Solar System. Several small objects in the asteroid belt and outer Solar System have albedos as small as 0.05. A typical comet nucleus has an albedo of 0.04. This type of dark surface is thought to originate from a simple, highly space-weathered layer that contains certain organic compounds.

The total albedo of the Moon is thought to be approximately 0.14, but it has a significant opposition impact and is highly directional and non-Lambertian. These reflectance characteristics are common, even though they differ from those of terrestrial terrains on airless Solar System bodies (Barry et al., 2010; Burroughs et al., 2007).

3.3. EARTH'S HEAT BUDGET

One type of energy that the earth receives from the sun is heat. The quantity of heat energy that the sun emits and how it is used by the earth's surface and atmosphere is known as the "heat budget." This explains why, despite the massive heat transfer that occurs, the earth neither warms nor cools (Figure 3.9).

The equilibrium between energy coming in and going out of the Earth is referred to as the "heat budget," which is like a financial budget except that it deals with energy rather than cash. It's essential to comprehend the climate of our planet and any future changes to it.

Here's a breakdown of the key points:

1. **Incoming Energy:**
   - **Shortwave Radiation:** This is the main source, coming directly from the Sun as high-energy, short-wavelength radiation.
   - **Longwave Radiation:** This originates from other sources like Earth's internal heat and the atmosphere, with lower energy and longer wavelengths.
2. **Outgoing Energy:**
   - **Reflected Shortwave Radiation:** About 30% of incoming sunlight bounces back to space, mainly from clouds and ice.
   - **Longwave Radiation Emitted by Earth:** This is the primary heat loss, primarily radiating back into space.
   - **Longwave Radiation Absorbed by the Atmosphere:** Some Earth radiation is trapped by greenhouse gases, contributing to warming.
3. **Balance and Importance:**
   - The heat budget is balanced when incoming and outgoing energy is equal, maintaining a stable average temperature.
   - An imbalance can lead to temperature changes, potentially impacting climate. Understanding this balance is crucial for studying climate change.
   - Changes in greenhouse gases or other factors can alter the balance, leading to warming or cooling trends.

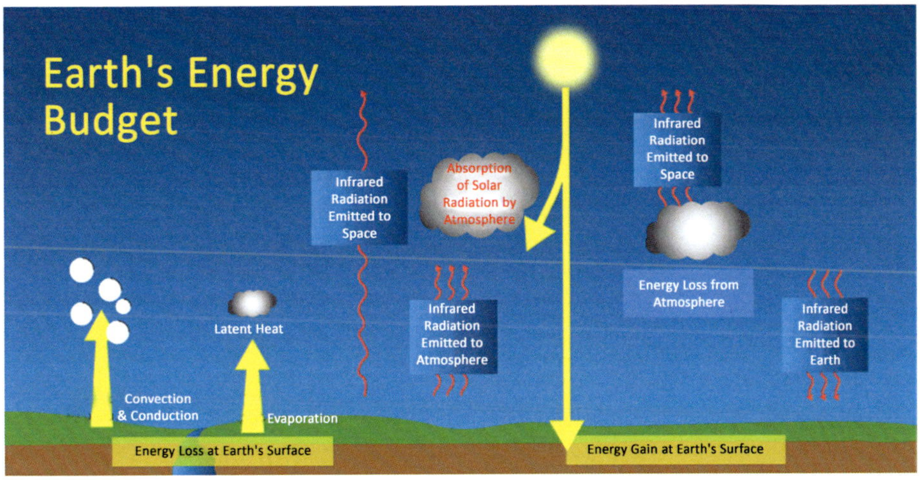

Figure 3.9. *Heat budget of earth diagram.*

Source: https://www.geeksforgeeks.org/heat-budget-of-earth/.

3.3.1. Heat Budget of Earth – Explained with an Example

The amount of solar radiation that the earth's surface and atmosphere receive, as well as the amount of heat that is lost through outgoing terrestrial long-wave radiation from the earth's surface and the atmosphere, are thought to be in balance. The earth's heat budget is the term used to describe the balance.

Even before it reaches the surface of the earth, 35 of the 100 units of heat that originate from the sun are reflected into space. Of these 35 units, the upper atmosphere reflects and scatters six of them, clouds scatter 27 of them, and the earth's snow-covered surface primarily absorbs two of them. The Earth's albedo is the portion of solar radiation that is reflected to space. The atmosphere absorbs 65 of the remaining units, the earth's surface absorbs 51 of the remaining units, and the atmosphere absorbs 14 of the remaining units.

The various types of terrestrial radiation reflect the 51 units that the earth has absorbed. The remaining 34 units are absorbed by the atmosphere as latent heat of condensation, turbulence, and convection, with 17 of them being radiated back into space. In conclusion, the atmosphere emits back 48 units of radiation. i.e., 34 from terrestrial radiation back into space and 14 from solar radiation.

As a result, 65 units of heat escape from the earth overall i.e., 48 from the atmosphere and 17 from the surface of the earth. The ratio of the incoming 65 units to the outgoing 65 units is balanced in this way. The phrase "Heat Budget of the Earth" refers to this delicate and dynamic balance of heat.

3.3.2. Components of Heat Budget

The components of the heat budget are as follows:

- **Reflection:** This happens when incoming solar radiation hits a surface in the sky, on land, or water and bounces back without producing heat.
- **Absorption:** Radiation absorption is the process of transforming electromagnetic radiation into heat energy.
- **Scattering:** When solar radiation collides with tiny things in the Earth's atmosphere, such as air molecules, water droplets, or aerosols, the radiation scatters, spreading the solar waves in all directions.
- **Terrestrial Radiation:** Long-wave radiation emitted by the Earth's surface or its atmosphere is called terrestrial radiation. The processes that Terrestrial Radiation uses to keep the temperature balanced include the following:
- **Latent Heat Transfer:** Latent heat is the heat or energy absorbed or released during a substance's phase change. It could transition from a solid to a liquid, from a liquid to a gas, or vice versa. The amount of heat transferred when one substance is prepared to change its condition is known as latent heat transfer.
- **Sensible Heat Transfer:** The energy needed to increase a substance's temperature without inducing a phase shift is known as sensible heat. Sensible heat transfer occurs when energy is supplied to an object as heat without causing the object's state to change. The earth or the air itself may absorb sunlight, causing a shift in temperature.
- **Emission by Vapor and Clouds:** The clouds and water vapor also emit significant amounts of terrestrial radiation.

3.3.3. Importance of Heat Budget of Earth

- One of the main things that keeps the Earth habitable is its thermal balance, which is achieved via the planet's heat budget.
- It maintains Earth's temperature.
- Improving the production of solar panels that absorb and transform this energy is imperative.
- It is responsible for the temperature differential between the equator and the poles.
- It facilitates photosynthesis, which in turn promotes plant development.
- It is also in charge of the differences in rainfall patterns between the equator and the poles.

3.3.4. Effects of Heat Budget of Earth

The Earth's heat budget, which is the equilibrium between thermal radiation released by the planet and solar radiation received from space, has a big influence on the climate system. Changes in this energy balance may affect the local and global climate. The following are the main ways that the Earth's heat budget affects the climate system:

- **Global Warming and Climate Change:** As a result of human actions like deforestation and the combustion of fossil fuels, the atmosphere now contains more greenhouse gases. The enhanced greenhouse effect is a result of these gases' increased ability to trap outgoing heat radiation. Global warming is the outcome, which causes variations in precipitation regimes, temperature patterns, and climate zones.

- **Melting of Ice and Sea Level Rise:** Changes in the Earth's heat budget contribute to the warming of the planet, leading to the melting of polar ice caps and glaciers. This meltwater contributes to rising sea levels, with implications for coastal areas and low-lying islands.
- **Ocean Currents and Heat Transport:** The heat budget influences ocean currents, which play a crucial role in redistributing heat around the globe. Changes in these currents can impact regional climates and have cascading effects on marine ecosystems and fisheries.
- **Impacts on Agriculture and Water Resources:** The heat budget's effects on climate patterns can have an impact on water supplies and agricultural productivity. Crop yields and the availability of water for use in agriculture and human consumption can be impacted by changes in precipitation patterns, temperature extremes, and modified growing seasons.

3.3.5. Variation in the Heat Budget of Earth

- The Earth still maintains a balance between terrestrial radiation and insolation, although this is not the case at any latitude.
- In the tropical zone, the heat budget and insolation are greater than the terrestrial radiation. It is therefore an area with an abundance of heat.
- In the arctic zone, heat gain is smaller than heat loss. It is therefore a region experiencing a heat shortage.
- This imbalance is partially mitigated by winds and ocean currents, which transfer heat from hot spots to areas with excess heat.
- The term "latitudinal heat balance" refers to this process of latitudinal heat redistribution and balancing.

3.3.6. How Is the Heat Budget of Earth Analyzed and Calculated?

Assume that all insolation reaches the top of the atmosphere by 100%. The majority of energy that enters the atmosphere is returned to the Earth's surface by being absorbed, scattered, and reflected, leaving behind the remaining bit. A little over 35 units of radiation are reflected into space before it even reaches the Earth's surface. Totaling 27, these come from cloud-top reflections and 2 from snow- and ice-covered regions, making up Earth's albedo.

The Earth emits 51 units of terrestrial radiation. Thirteen of them are released directly into space, and thirty-four are absorbed by the atmosphere. Nine of these are absorbed instantly through turbulence and convection, nineteen through the latent heat of condensation, and six through direct absorption. The remaining 65 units are absorbed by the Earth and its atmosphere. Out of this, the Earth's surface absorbs 51 units, and the atmosphere absorbs 14 units.

34 units come from terrestrial radiation and 14 units come from insolation, making up the 48 units that the atmosphere emits back into space. The total radiation returning from the atmosphere and Earth is equal to 17 and 48 units, respectively, so the 65 units of solar insolation received are offset by this total radiation.

3.4. MECHANISMS OF HEAT TRANSFER

Like all stars, the Sun is a ball of hot plasma that never stops burning and emitting energy into space. The primary energy source for Earth and every other planet in the solar system is the energy that the Sun radiates. Insolation is the term used to describe the solar radiation that the Earth's surface receives. But only a tiny portion of this energy that the Sun radiates is received by the Earth.

Surfaces of land and water, clouds, atmospheric gases, and dust all absorb energy from the incoming insolation by intercepting it. The absorbed energy subsequently materializes as vegetation, strong ocean currents, and a variety of weather patterns. These energy manifestations, which include ice-covered landscapes, plains, mountain tops, rainforests, deserts, and oceans, vary depending on the location. Aside from interferences, the amount of energy that reaches the Earth's surface can vary by up to 75% depending on whether clouds are present in the atmosphere. There are the following three mechanisms of heat transfer (Figure 3.10):

- Conduction;
- Convection;
- Radiation.

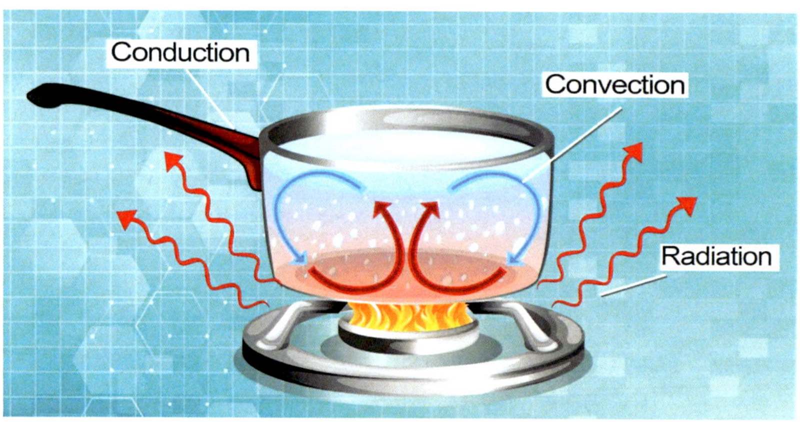

Figure 3.10. *Mechanisms of heat transfer.*

Source: https://www.studyiq.com/articles/heat-transfer/.

3.4.1. Heat Transfer in Atmosphere

The main source of heat comes from solar radiation. Its impact is not immediately felt, though. For instance, contrary to what one might anticipate, the temperature steadily decreases as one ascends a mountain or ascends through the atmosphere. This specific phenomenon is caused by a complex and uneven solar insolation heating mechanism in the atmosphere.

3.4.2. Types of Heat Transfer

There are four common types of energy transfer involved in heating the atmosphere. They are Radiation, Conduction, Convection, and Advection.

3.4.3. Heat Transfer Radiation

Heat waves or energy that are emitted through electromagnetic radiation without the use of a medium result in heat transfer through the process of radiation. When the Sun's electromagnetic waves (EMRs) fall on Earth's surface after traveling through space in a vacuum, it is a common occurrence to witness the phenomenon of heat transfer through radiation. Energy is transferred when photon particles in the radiation collide with atmospheric air molecules. Every object constantly emits radiation, regardless of temperature. A body's wavelength of radiation is influenced by its surface temperature. Shorter wavelengths increase the radiation's energy.

Due to its extremely high surface temperature, the Sun emits light at relatively short wavelengths (visible light, ultraviolet light), some of which is visible light and some of which is felt as warmth. The surface of the Earth, however, is chilly. At far longer wavelengths, the Earth tends to reradiate heat that it had previously absorbed from the Sun (thermal infrared radiations). Terrestrial radiation is the term used to describe the heat that the Earth's surface reradiates.

Short waves can escape into space because the atmosphere is transparent to them. However, long waves absorbed by atmospheric gases—especially carbon dioxide, water vapor, and other greenhouse gases—absorb no light from the atmosphere. Therefore, it is reasonable to conclude that the Earth's surface energy loss heats the atmosphere more than solar radiation entering the atmosphere.

3.4.4. Heat Transfer Conduction

Heat energy transfers from the warmer to the colder of two objects when they come into contact. Conduction is the term for this method of heat transmission. The flow keeps going until the two objects' temperatures equalize or the contact breaks.

At the point where the atmosphere and the earth's surface meet, terrestrial radiations cause conduction in the atmosphere. But since it only affects the air in contact with the earth's surface, this is only a small-scale method of heat transfer to warm the atmosphere. This is a result of the air's poor heat conductivity.

3.4.5. Convection Heat Transfer

Convection is the movement of particles within a liquid or gas to transfer heat energy from one area to another. When the lower portion of a liquid mass heats up, molecules warm and vibrate more, which leads to collisions that move the molecules next to them. It gradually rises to carry the heat as it expands and loses density. For instance, a convection current is said to rise when air moves upward. In this way, heat is transferred from warmer to colder materials.

3.4.6. Heat Transfer Advection

Advection is the term for the horizontal movement of air that transfers heat. When these atmospheric winds pass over a region, they typically carry the features of that region with them. A location that is nearer warmer winds experiences an increase in temperature. If the location is in the path of winds coming from cold climates, the temperature will drop. The summertime "Loo" wind from north India and the summertime "Sirocco" wind from the Mediterranean region carry the heat from the Sahara desert. Advection alone is responsible for the majority of the diurnal (day and night) variation in daily weather in the middle latitudes (Ciais et al., 2013; Dessler et al., 2016) (Table 3.3).

Table 3.3. *Comparison Among Radiation, Conduction, Convection, and Advection*

Radiation	Conduction	Convection	Advection
Heat transfer by the process of radiation occurs when heat waves or energy are emitted through electromagnetic radiation.	When two objects of varying temperatures come in contact, heat energy flows from the warmer object to the cooler one until equilibrium is achieved.	Convection is the transfer of heat energy from one part of a fluid (liquid or gas) to another by the movement of particles themselves.	Heat transfer results from the movement of warm molecules of fluid to the cooler molecules of a fluid in a horizontal manner.
Does Not require any medium for heat transfer. Heat can be transferred in a vacuum space also.	Requires a medium for heat transfer.	Requires a medium for heat transfer.	Requires a medium for heat transfer.
Heat transfers through electromagnetic radiation. Meaning no physical contact is required between the bodies.	Heat transfers through direct contact. Physical contact is required.	Heat transfers through the intermediary substances due to temperature differences.	Heat transfers through the intermediary substances due to pressure and temperature differences.
The rate of heat transfer is fast.	The rate of heat transfer is slow.	The rate of heat transfer is slow.	The rate of heat transfer is slow.
Heat transfer mostly occurs between two bodies with or without a medium.	Occurs in solids through molecular collisions.	Occurs in fluids by the flow of heat in the molecules in a vertical direction.	Occurs in fluids by the flow of heat in the molecules in a horizontal direction.

Source: https://www.studyiq.com/articles/heat-transfer/.

CASE STUDY

Heat Exchangers Help Turn Landfill Biogas to Energy: Emissions of biogas from landfills can be a significant source of greenhouse gas (GHG) emissions. However, they can also be captured and utilized for renewable energy generation.

The Client: LMS Energy (LMS) is Australia's largest and most experienced landfill biogas company. It operates 60 biogas sites, generating more than 550,000 MW of base load power each year and abating over 4 million tons of CO2, making the company Australia's largest emissions reducer.

The company captures potentially harmful methane gas which is generated in landfill as the contents decompose and burns it in generators to produce electricity. To enable the gas to be burned, the water content has to be reduced from fully saturated 100% relative humidity to around 80%. In the past LMS built its own generation modules which included on-board gas treatment, including dehumidification.

Source: https://www.hrs-heatexchangers.com/case-study/heat-exchangers-help-turn-landfill-biogas-to-energy/.

The Challenge: In 2018 the company switched to using containerized generator sets from Jenbacher, these lacked the space for on-board gas conditioning. This meant the company needed a new method to dehumidify biogas. "The solution was to create a gas delivery skid containing an on-board blower, a heat exchanger (to condense the water), and condensate removal vessel," explains Michael Lee, a senior mechanical engineer at LMS.

Having read an article by HRS and done some more research, LMS approached HRS Heat Exchangers to quote for the supply of identical biogas moisture removal (dehumidifier) units.

The Solution: The engineering team at HRS realized that the K Series of stainless-steel shell and tube heat exchangers was ideal for the purpose and that HRS's corrugated tube technology would increase heat transfer and efficiency over standard smooth tube designs. During the lengthy design and approval process, HRS undertook some design revisions to achieve an optimum bespoke design that met the strict dimension and footprint limitations, while maintaining velocities and heat transfer coefficients which were higher than those specified.

The first two units were installed and commissioned and quickly proved their performance and reliability. Following the success of these units, HRS has supplied a further ten K Series heat exchangers to LMS, with the most recently currently being incorporated into a gas delivery skid for on-site installation later this year.

"As well as making the biogas suitable for the generator set, reducing humidity also helps prolong engine maintenance and component life, extends oil change schedules, and generally helps with operations," continues Michael. "We've found HRS to be easy to deal with and communication has always been very professional. Their drawings have been very high quality and we have never had any issues with installing the heat exchangers into the gas delivery skids."

CLASS ACTIVITY

Working in small groups, have students observe the radiometer. Give students some time to investigate the radiometer and then have them answer the following question:

- Can you spin the vanes of the radiometer without touching it? How?
- Have each group share their ideas with the class.

Next, have students put the radiometer near a light source and observe it spin. Have them move the light source closer and farther away from the radiometer and observe its motion. Then have them answer the following question:

- Have each student to write an explanation about how they think a radiometer works.

SUMMARY

- The Earth has a blanket of atmospheric gases around it. Most of these gases, such as carbon dioxide, are formed naturally but are now also being produced by activities such as the burning of fossil fuels. Many gases such as CFCs are produced exclusively due to human activities. Some of these gases have the capacity to absorb infrared radiation that is reflected from the earth's surface so that it is not lost to space but retained within the atmosphere.
- Solar radiation is indeed in the short-wavelength range i.e., less than 4 μm. It is reported that "about 18% of the insolation is absorbed directly by ozone and water vapor."
- Solar radiation, often called the solar resource or just sunlight, is a general term for the electromagnetic radiation emitted by the sun. Solar radiation can be captured and turned into useful forms of energy, such as heat and electricity, using a variety of technologies. However, the technical feasibility and economical operation of these technologies at a specific location depends on the available solar resource.
- The amount and spectral nature of this solar radiation changes as it passes through the atmosphere. Gases such as ozone, oxygen, and water vapor absorb radiation and in so doing warm the atmosphere, while air molecules scatter radiation and particulates (aerosols) both scatter and absorb radiation.
- In simple terms, insolation can be defined as the incoming solar radiation from the sun, which is received on Earth in short waves. As the size of the earth and its distance from the sun are both relatively small, it receives significantly less quantity of insolation.
- Albedo is a calculation of diffuse solar radiation reflected from the total solar radiation, and it's determined on a range of zero to 1, with 0 corresponding to a black body that absorbs every incident radiation and 1 corresponding to a body that reflects almost every incident radiation.
- Heat is a form of energy, received by the earth from the sun. Heat Budget deals with the amount of heat energy received from the sun and its utilization in the atmosphere and the surface of the earth. This explains why the earth neither warms up nor cools down despite the huge transfer of heat that takes place.
- The Earth's heat budget refers to the balance between incoming and outgoing energy, similar to a financial budget but dealing with energy instead of money. It's a crucial aspect of understanding our planet's climate and its potential changes.
- Solar energy is the primary source of heat. However, its effect is not directly realized. For example, as one climbs a mountain or ascends in the atmosphere, the temperature steadily lowers, rather than increases, which one might expect. This particular phenomenon is experienced as the mechanism of heating by solar insolation in the atmosphere is not very simple or uniform.

REVIEW QUESTIONS

1. What is solar radiation?
2. What are the infrared absorption bands of water vapor?
3. What are the important greenhouse gases?
4. What is insolation?
5. What three factors affect insolation?
6. What is Albedo?
7. What is the heat budget of the Earth?
8. List the components of the heat budget.
9. What is heat transfer?
10. Name the four types of heat transfer.

REFERENCES

1. Barry, R. G., & Chorley, R. J. (2010). *Atmosphere, Weather, and Climate* (8th ed.). Routledge.
2. Burroughs, W. J. (2007). *Climate Change: A Multidisciplinary Approach* (2nd ed.). Cambridge University Press. ISBN: 978-0-521-69033-1.
3. Ciais, P., Sabine, C., Bala, G., et al. (2013). Carbon and other biogeochemical cycles. In T. F. Stocker, D. Qin, G. K. Plattner, M. Tignor, S. K. Allen, J. Boschung, A. Nauels, Y. Xia, V. Bex, & P. M. Midgley (Eds.), *Climate Change 2013: The Physical Science Basis*. Contribution of Working Group I to the Fifth Assessment Report of the Intergovernmental Panel on Climate Change. Cambridge University Press.
4. Dessler, A. (2016). *Introduction to Modern Climate Change* (2nd ed.). Cambridge University Press. ISBN: 978-521-17315-5.
5. Duffie, J. A., & Beckman, W. A. (1991). *Solar Engineering of Thermal Processes* (2nd ed.). John Wiley & Sons.
6. Fletcher, C. (2018). *Climate Change: What The Science Tells Us* (2nd ed.). John Wiley & Sons. ISBN: 978-1-118-79306-0.
7. Houghton, J. T. (2015). *Global Warming: The Complete Briefing* (5th ed.). Cambridge University Press. ISBN: 978-0-521-70916.
8. Hoyt, D. V., & Schatten, K. H. (1997). *The Role of The Sun in Climate Change*. Oxford University Press.
9. Intergovernmental Panel on Climate Change (IPCC). (2001a). *Climate Change 2001: The Scientific Basis. Contribution Of Working Group I to the Third Assessment Report of The Intergovernmental Panel on Climate Change* (J. T. Houghton, Y. Ding, D. J. Griggs, et al., Eds.). Cambridge University Press.
10. Intergovernmental Panel on Climate Change (IPCC). (2001b). *Climate Change 2001: Mitigation. Contribution Of Working Group III to The Third Assessment Report of*

The Intergovernmental Panel on Climate Change (B. Metz, O. Davidson, R. Swart, & J. Pan, Eds.). Cambridge University Press.

11. Intergovernmental Panel on Climate Change (IPCC). (2007). *Climate Change 2007: Synthesis Report. Contribution Of Working Groups I, II, And III to The Fourth Assessment Report of The Intergovernmental Panel on Climate Change*. Intergovernmental Panel on Climate Change.
12. Intergovernmental Panel on Climate Change (IPCC). (2013). *Climate Change 2013: The Physical Science Basis. Contribution of Working Group I to The Fifth Assessment Report of The Intergovernmental Panel on Climate Change* (T. F. Stocker, D. Qin, G. K. Plattner, M. Tignor, S. K. Allen, J. Boschung, A. Nauels, Y. Xia, V. Bex, & P. M. Midgley, Eds.). Cambridge University Press.
13. Intergovernmental Panel on Climate Change (IPCC). (2014a). *Climate Change 2014: Synthesis Report. Contribution Of Working Groups I, II, And III to The Fifth Assessment Report of The Intergovernmental Panel on Climate Change* (Core Writing Team: R. K. Pachauri & L. A. Meyer, Eds.). IPCC.
14. Intergovernmental Panel on Climate Change (IPCC). (2014b). *Climate Change 2014: Impacts, Adaptation, and Vulnerability. Part A: Global and Sectoral Aspects. Contribution Of Working Group II to The Fifth Assessment Report of The Intergovernmental Panel on Climate Change* (Eds. C. B. Field, V. R. Barros, D. J. Dokken, K. J. Mach, M. D. Mastrandrea, T. E. Bilir, M. Chatterjee, K. L. Ebi, Y. O. Estrada, R. C. Genova, B. Girma, E. S. Kissel, A. N. Levy, S. MacCracken, P. R. Mastrandrea, & L. L. White). Cambridge University Press.
15. Intergovernmental Panel on Climate Change (IPCC). (2014c). *Climate Change 2014: Mitigation of Climate Change. Contribution Of Working Group III to The Fifth Assessment Report of The Intergovernmental Panel on Climate Change* (Eds. O. Edenhofer, R. Pichs-Madruga, Y. Sokona, E. Farahani, S. Kadner, K. Seyboth, A. Adler, I. Baum, S. Brunner, P. Eickemeier, B. Kriemann, J. Savolainen, S. Schlömer, C. von Stechow, T. Zwickel, & J. C. Minx). Cambridge University Press.
16. Interlaboratory Working Group. (2000). *Scenarios for a Clean Energy Future* (ORNL/CON-476, LBNL-44029, NREL/TP-620-29379). Oak Ridge National Laboratory; Lawrence Berkeley National Laboratory; National Renewable Energy Laboratory.
17. International Energy Agency (IEA). (1998). *World Energy Outlook–1998 Update*. IEA/OECD.
18. Iqbal, M. (1983). *An Introduction to Solar Radiation*. Academic Press.
19. LeNoble, J. (1993). *Atmospheric Radiative Transfer*. A. Deepak.
20. Liou, K. N. (1980). *An Introduction to Atmospheric Radiation*. Academic Press.
21. Lipper, L., Thornton, P., Campbell, B., et al. (2014). Climate-smart agriculture for food security. *Nature Climate Change, 4*(12), 1068–1072. https://doi.org/10.1038/nclimate2437.
22. Matson, P., Parton, W., Power, A., & Swift, M. (1997). Agricultural intensification and ecosystem properties. *Science, 277*(5325), 504–509. https://doi.org/10.1126/science.277.5325.504.

23. Ohmura, A., Dutton, E. G., Forgan, B., et al. (1998). Baseline Surface Radiation Network (BSRN/WCRP): New precision radiometry for climate research. *Bulletin of the American Meteorological Society, 79*(10), 2115–2136. https://doi.org/10.1175/1520-0477(1998)079<2115>2.0.CO;2.
24. Oke, T. R. (1987). *Boundary Layer Climates*. Methuen.
25. Ranjana. (2002). *Making Sense of Climate Change*. Tata Energy Research Institute.
26. Smith, P., Bustamante, M., Ahammad, H., et al. (2014). Agriculture, forestry, and other land use (AFOLU). In *Climate Change 2014: Mitigation of Climate Change. Contribution of Working Group III to the Fifth Assessment Report of the Intergovernmental Panel on Climate Change* (O. Edenhofer, R. Pichs-Madruga, Y. Sokona, E. Farahani, S. Kadner, K. Seyboth, A. Adler, I. Baum, S. Brunner, P. Eickemeier, B. Kriemann, J. Savolainen, S. Schlömer, C. von Stechow, T. Zwickel, & J. C. Minx, Eds.). Cambridge University Press.
27. Weart, S. (2008). The carbon dioxide greenhouse effect. In *The Discovery of Global Warming*. American Institute of Physics.
28. Wells, N. (1997). *The Atmosphere and Ocean: A Physical Introduction*. John Wiley & Sons.
29. Willson, R. C., & Mordvinov, A. V. (2003). Secular total solar irradiance trend during solar cycles 21–23. *Geophysical Research Letters, 30*(5), 1199. https://doi.org/10.1029/2002GL016038.

CHAPTER 4

Temperature and Its Distribution

LEARNING OBJECTIVES

After studying this chapter, you will be able to:

- Explain the insolation (solar radiation);
- Define the heat budget;
- Identify the temperature and its distribution;
- Discuss temperature scales and measurements.

KEY TERMS FROM THIS CHAPTER

Atmospheric composition	Carbon cycle	Carbon dioxide
Climate system	Earth's rotation	Global warming
Earth's climate system	Heat balance	Heat budget
Earth's energy balance	Heat energy	Hydrological cycle
Insolation	Ocean currents	Radiation
Sea and land breezes	Temperature	Terrestrial radiation

INTRODUCTION

Temperature, a fundamental aspect of climatology, plays a pivotal role in shaping Earth's climate system. It refers to the measure of heat energy in the atmosphere, influenced by the sun's radiation and distributed across the globe through various atmospheric and oceanic processes. Temperature distribution is far from uniform, influenced by latitude, altitude, land and sea contrasts, ocean currents, and atmospheric circulation patterns.

At the core of temperature distribution is the concept of solar radiation, which varies with latitude. Equatorial regions receive direct sunlight year-round, resulting in consistently high temperatures. In contrast, Polar Regions receive oblique solar rays, leading to lower temperatures and marked seasonal variations. This latitudinal gradient creates distinct climate zones: tropical, temperate, and polar, each characterized by unique temperature ranges and seasonal patterns.

Altitude also significantly affects temperature. The troposphere, the lowest layer of the atmosphere, experiences a temperature decrease with height, typically about 6.5°C per kilometer, known as the environmental lapse rate. This phenomenon is why mountainous regions exhibit cooler temperatures than surrounding lowlands, impacting local climates and ecosystems.

Land and sea contrasts further modulate temperature distribution. Land heats and cools more rapidly than water due to differences in specific heat capacity. Consequently, coastal regions often have milder climates with smaller temperature ranges compared to continental interiors, where temperature extremes are more pronounced. This effect is evident in phenomena like maritime and continental climates, where proximity to large water bodies moderates temperature fluctuations.

Ocean currents, driven by wind patterns and the Earth's rotation, play a crucial role in redistributing heat. Warm currents, such as the Gulf Stream, transport heat from the equator towards higher latitudes, influencing coastal temperatures and contributing to milder winters in regions like Western Europe. Conversely, cold currents, like the California Current, bring cooler temperatures to adjacent coastal areas, impacting local climate conditions.

Atmospheric circulation, including the movement of air masses and the jet streams, also affects temperature distribution. The Hadley, Ferrel, and Polar cells facilitate the heat from the equator toward the poles, contributing to the global heat balance. Jet streams, fast-flowing air currents in the upper atmosphere, influence weather patterns and temperature distribution by steering storm systems and influencing the movement of air masses. In summary, temperature distribution in climatology is a complex interplay of solar radiation, latitude, altitude, land-sea contrasts, ocean currents, and atmospheric circulation. Understanding these factors is crucial for comprehending regional and global climate patterns, predicting weather events, and addressing climate change impacts. This intricate balance shapes the diverse climates experienced across the Earth, from the scorching deserts to the frigid polar ice caps.

4.1. INSOLATION (SOLAR RADIATION)

The earth's main source of energy is the sun. Short waves are used to radiate this energy into space in all directions. We call this solar radiation. Out of the 1,00,00,00,000 units of energy that the sun radiates, only two billionths, or two units of energy, make it to the surface of the earth. Since solar radiation is the only significant source of energy for the majority of physical and biological phenomena on Earth, even this tiny fraction of solar radiation has enormous significance. The term "insolation" refers to the short waves of incoming solar radiation. Due to the earth's small size and distance from the sun, the amount of insolation received on its surface is far less than that which is radiated from the sun. Moreover, a tiny quantity of insolation is absorbed by ozone, dust, water vapor, and other gases in the atmosphere.

4.1.1. Factors Influencing Insolation

There are regional variations in the amount of insolation that the earth receives. It varies periodically and from one location to another. The maximum annual insolation is received in the tropical zone. In the direction of the poles, it steadily drops. There is more insolation in the summer and less in the winter.

The following factors influence the amount of insolation received.

- The angle of incidence.
- Duration of the day (daily sunlight period)
- Transparency of the atmosphere

4.1.1.1. The Angle of Incidence

The sun's rays hit the surface of the round earth at different angles and in different places. The angle of incidence is the angle created at a given location by the sun's ray and the earth's circle's tangent. It has two effects on insolation. First, the sun's rays are vertical when it is almost overhead. Because of the large angle of incidence, they are concentrated in a smaller area, which increases the amount of insolation there. When the sun's rays are oblique, their angle of incidence is small, requiring them to heat a larger region and resulting in less insolation being received in that area. Second, compared to rays striking at a large angle, sunrays traveling at a small angle cover a greater area of the atmosphere. The amount of heat that the atmosphere reflects and absorbs increases with the length of the sun's rays. Consequently, a location experiences reduced insolation intensity (refer to Figure 4.1).

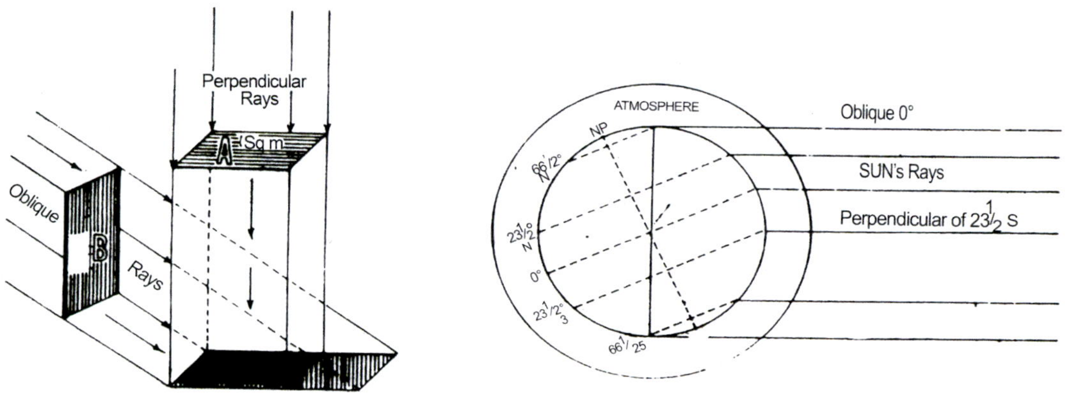

Figure 4.1. Effect of angle of incidence on insolation.

Source: https://www.nios.ac.in/media/documents/316courseE/ch10.pdf.

4.1.1.2. Duration of the Day

The length of the day varies depending on the location and season. It determines how much insolation the surface of the earth receives. The amount of insolation received increases with the length of the day. On the other hand, less insolation is received during shorter days.

4.1.1.3. Transparency of the Atmosphere

The amount of insolation that reaches the earth's surface is also determined by the transparency of the atmosphere. Because they reflect, absorb, or transmit insolation, dust particles, water vapor, and cloud cover all affect transparency. While a clear sky allows insolation to reach the surface, thick clouds prevent it from doing so. Less insolation reaches the surface as a result of absorption by water vapor.

4.1.1.3.1. Heating and Cooling of the Atmosphere

The sun is the primary source of energy and heat in the atmosphere, but its impact is indirect. For instance, contrary to what one might anticipate, the temperature steadily drops as we ascend a mountain or ascend through the atmosphere. This is due to the complicated nature of the atmosphere's heating mechanism. The atmosphere is heated directly by four different heating processes. They are:

1. Radiation: The process by which solar energy enters the earth and is lost to space is called radiation. The radiation process is the term used to describe the direct transfer of heat from a heat source to an object through heat waves. Heat moves through the space during this process. Radiation is the enormous amount of heat energy that enters and exits the earth. The following facts about radiation are worth noting.

- All objects whether hot or cold emit radiant energy continuously.
- Hotter objects radiate more energy per unit area than colder objects.
- The temperature of an object determines the wavelength of radiation. Temperature and wavelength are inversely related. The hotter the object shorter the length of the wave.
- Insolation reaches the earth's surface in short waves and heat is radiated from the earth in long waves.

It's interesting to note that the atmosphere is permeable to short waves but not to long waves. Consequently, the energy emitted from the Earth's surface, known as terrestrial radiation, warms the atmosphere more than the incoming solar radiation, or insolation.

2. Conduction: It is the term for the process of heat transfer that occurs when two objects with different temperatures come into contact with one another. Heat energy moves from the warmer object to the cooler object. The flow keeps going until the two objects' temperatures are equal or the contact breaks. At the point where the atmosphere and the earth's surface meet, conduction takes place in the atmosphere. However, because it only affects the air near the earth's surface, this method of heat transfer is insignificant in terms of warming the atmosphere.
3. Convection: It is the term used to describe the movement of a mass or substance, usually vertically, to transfer heat. Either conduction or earth's radiation heats the air in the lower atmosphere's layers. The air expands as a result of heating. It rises and experiences a decrease in density. A vacuum is created in the lower atmosphere by the heated air's continuous ascent. Convection results from cooler air entering the vacuum and rising to the surface. The atmosphere warms due to the cyclical movement brought on by the convectional process, which transfers heat from the lower to the upper layers of the atmosphere.
4. Advection: The temperature of a place is transported by winds. If a location is in the path of winds from warmer regions, its temperature will rise. If the location is in the path of cold-weather winds, the temperature will drop. Advection is the term for this method of heat transfer by winds that occurs horizontally.

4.1.2. Source of Energy on the Earth

The sun is the main energy source for the surface of the earth. It manifests as electromagnetic radiation in various electromagnetic spectrum bands or wavelengths. A few of these musical groups. The earth's surface and atmosphere are being heated by some of these bands, both visible and thermal. Although geothermal energy from the earth's interior also reaches the surface, it has very little effect on how temperatures are distributed. As a result, we will talk about the temperature distribution that results from the sun alone.

4.2. HEAT BUDGET

The energy that is dispersed and that is directly transmitted through the atmosphere makes up the insolation. The quantity of solar radiation known as insolation travels through shortwaves to reach the earth's surface. The earth, like all other hot objects, emits heat energy. We refer to this as "terrestrial radiation." The earth's surface always has a constant annual mean temperature. Because of the equilibrium between insolation and terrestrial radiation, it has been feasible. This equilibrium is known as the earth's heat budget.

Let us suppose that the total heat (incoming solar radiation) received at the top of the atmosphere is 100 units (see Figure 4.2) Roughly 35 units of it are reflected into space even before reaching the surface of the earth. Out of these 35 units, 6 units are reflected to space from the top of the atmosphere, 27 units are reflected by clouds, and 2 units from snow and ice-covered surfaces. Out of the remaining 65 units (100–35), only 51 units reach the earth's surface and 14 units are absorbed by the various gases, dust particles and water vapors of the atmosphere.

Terrestrial radiation, or 51 units, is radiated back by the earth. Thirteen units of these 51 units of terrestrial radiation are directly sent into space, while the remaining 34 units are absorbed by the atmosphere. Along with absorbing 34 units of outgoing radiation and 14 units of incoming radiation, the atmosphere also radiates 48 units of radiation back into space. 65 units of solar radiation that enter the atmosphere are reflected back into space as a result. The earth's surface heat balance is constantly maintained by this accounting of incoming and outgoing radiation.

Nonetheless, the balance between the sun's incoming radiation and the earth's outgoing radiation is maintained throughout. However, the observations we make at various latitudes refute this. As was previously mentioned, latitude has a direct correlation with the amount of insolation received. The amount of insolation exceeds the amount of terrestrial radiation in the tropical region. It is therefore an area of excess heat. Heat gain is less than heat loss in the Polar regions. It is therefore a region of heat deficit. As a result, the insolation causes an imbalance in heat at various latitudes (see Figure 4.3). However, winds and ocean currents, which move heat from surplus heat regions to deficit heat regions, partially offset this. This is referred to as latitudinal heat balance in general.

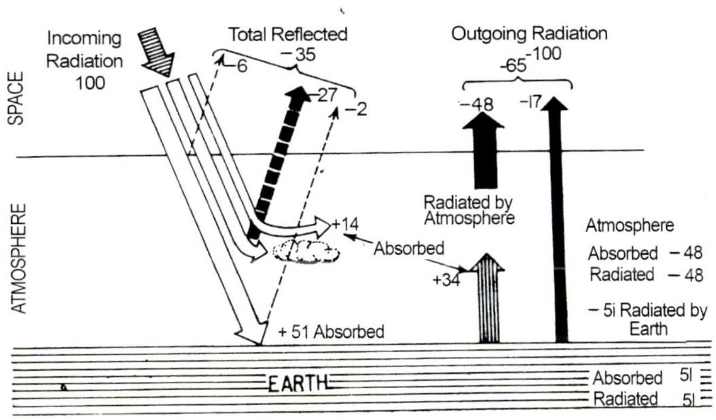

Figure 4.2. Heat budget (balance between insolation and terrestrial radiation).

Source: https://www.nios.ac.in/media/documents/316courseE/ch10.pdf.

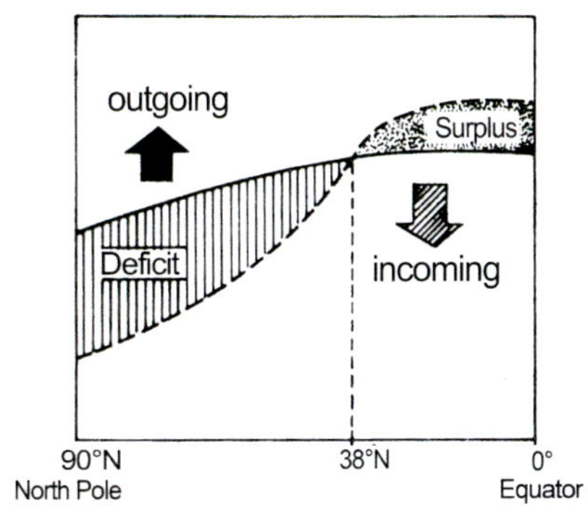

Figure 4.3. Latitudinal heat balance.

Source: https://www.nios.ac.in/media/documents/316courseE/ch10.pdf.

4.2.1. Variation in the Net Heat Budget at the Earth's Surface

The quantity of radiation that the earth's surface receives varies. There is a surplus of radiation in one area of the Earth and a deficit in another. The latitudinal variation in the Earth-atmosphere system's net radiation balance is shown in Figure 4.4. The figure indicates that the regions close to the poles have a deficit in net radiation balance, while there is an excess of net radiation balance between 40° north and south. The tropics do not gradually warm up as a result of the accumulation of excess heat or the high latitudes freezing permanently as a result of excess deficit because the excess heat energy from the tropics is redistributed poleward.

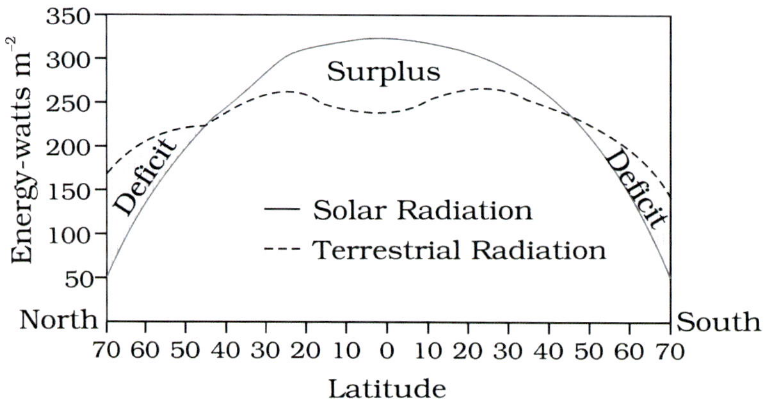

Figure 4.4. Latitudinal variation in net radiation balance.

Source: https://ncert.nic.in/ncerts/l/kegy209.pdf.

4.2.2. Global Warming

One of the most significant environmental issues facing our planet right now is global warming. Scientists observe a strong correlation between it and the rise in atmospheric carbon dioxide and the thinning of the stratosphere's ozone layer. As you are aware, there is an ozone gas layer in the upper stratosphere. The sun's UV radiation can be greatly absorbed by ozone, preventing it from penetrating the earth's surface. Scientists have discovered that the ozone layer is getting thinner. This is raising the quantity of UV radiation that reaches Earth and upsetting the atmosphere's gas balance. In addition to seriously burning human skin, increasing the risk of skin cancer, destroying some microscopic life forms, and harming plants, ultraviolet radiation is the cause of the planet's rising global temperature. The atmosphere's concentration of carbon dioxide is steadily rising. The amount of carbon dioxide in the atmosphere is thought to have risen by 25% over the past 100 years. While carbon dioxide absorbs terrestrial radiation, it permits insolation to pass through. The temperature of the atmosphere rises when the amount of carbon dioxide in the atmosphere increases. The atmosphere's temperature is thought to have risen by roughly 0.50°C over the past millennium. The combustion of waste, large-scale deforestation, burning of fossil fuels, factory combustion processes, and volcanic eruptions are a few of the causes of the increase in atmospheric carbon dioxide.

The time will come when the temperature of the atmosphere rises to the point where the polar ice caps melt, raising the sea level and submerging coastal areas and islands, provided the ozone layer and carbon dioxide content continue to diminish. Global warming is the phenomenon of an increase in atmospheric temperature that occurs all over the world as a result of ozone layer depletion and rising carbon dioxide concentration (Chrystie et al., 2017; Winston et al., 2018).

4.3. TEMPERATURE AND ITS DISTRIBUTION

A substance's relative degree of heat is indicated by its temperature. Temperature is a measurement of the intensity of heat, whereas heat is the energy that causes things or objects to become hot. Heat and temperature are closely related even though they are very different from one another because heat gain or loss is required to raise or lower the temperature. The Swedish astronomer is honored by the name of the Celsius scale. Scientists worldwide agree that the Celsius scale is the most accurate way to measure air temperature. The historical temperature records of several English-speaking countries include values on the Fahrenheit scale, Fahrenheit temperatures may be converted to their Celsius equivalents by the formula $C = \frac{5}{9}(F - 32)$ Moreover, temperature difference determines the direction of the flow of heat. This we can understand by studying temperature distribution. The distribution of temperature varies both horizontally and vertically. Let us study it under:

- The horizontal distribution of temperature;
- The vertical distribution of temperature.

4.3.1. Horizontal Distribution of Temperature

The horizontal distribution of temperature on Earth's surface refers to how it is distributed among the latitudes. "Isotherms," or lines connecting points with equal temperatures, are a common way for cartographers to depict the horizontal temperature distribution. The terms "therm" and "iso," which stand for "temperature" and "equal," respectively, make up an isotherm. An isotherm map will show you that the temperature distribution is not uniform.

The factors responsible for the uneven distribution of temperature are as follows:

- Latitude;
- Land and sea contrast;
- Relief and altitude;
- Ocean currents;
- Winds;
- Vegetation cover;
- Nature of the soil;
- Slope and aspect.

4.3.1.1. Latitude

Latitude, the angular distance of a place north or south of the Earth's equator, is a crucial factor in climatology, significantly influencing temperature and climate patterns. It is measured in degrees, with the equator at 0°, the poles at 90° north and south, and varying climatic zones falling in between. The latitude of a region determines the angle and intensity of solar radiation it receives, directly impacting its temperature and weather.

At the equator, the sun's rays strike the Earth perpendicularly throughout the year, leading to consistently high temperatures and minimal seasonal variation. This results in the tropical climate zone, characterized by warm temperatures year-round, high humidity, and significant precipitation, often in the form of daily thunderstorms. The tropics, extending roughly from 23.5°N to 23.5°S latitude, experience little temperature fluctuation but can have wet and dry seasons due to the shifting Intertropical Convergence Zone (ITCZ).

Moving away from the equator towards the mid-latitudes, solar radiation becomes less direct, causing a decrease in average temperatures and more pronounced seasonal variations. The temperate zones, located between approximately 23.5° and 66.5° in both hemispheres, experience four distinct seasons: winter, spring, summer, and autumn. These regions have a greater range of temperatures over the year, with hotter summers and colder winters compared to the tropics. The temperate zones include diverse climates, from the Mediterranean, characterized by hot, dry summers and mild, wet winters, to the humid continental climate, with significant temperature variations and precipitation throughout the year.

At higher latitudes, closer to the poles, solar radiation strikes the Earth at a much lower angle, spreading the energy over a larger surface area and resulting in lower temperatures. The Polar regions, extending from about 66.5° to 90° in both hemispheres, experience extreme variations in daylight, with long, dark winters and short, cool summers. The Arctic and Antarctic climates are characterized by frigid temperatures, ice, and snow, with the polar ice caps playing a crucial role in Earth's albedo effect, reflecting a significant portion of solar radiation into space.

Latitude also influences atmospheric circulation patterns, contributing to the distribution of heat and moisture around the globe. The Hadley, Ferrel, and Polar cells are large-scale atmospheric circulation patterns that redistribute warm air from the equator the poles and cold air from the poles toward the Equator. This circulation helps moderate temperatures and influences weather patterns, such as trade winds, westerlies, and polar easterlies, which further affect regional climates.

In summary, latitude is a fundamental factor in climatology, determining the amount and angle of solar radiation a region receives, which in turn affects its temperature and climate. From the consistent warmth of the tropics to the severe cold of the Polar regions, latitude shapes the diverse climatic zones on Earth, influencing weather patterns, ecosystems, and human activities. Understanding latitude's impact on climate is essential for predicting weather, planning agricultural activities, and addressing the challenges posed by climate change.

4.3.1.2. Land and Sea Contrast

The contrast between land and sea influences temperature significantly. On a sunny day, land warms more quickly and thoroughly than water. Additionally, during the night, it cools down faster than water. As a result, during the day the temperature is generally higher on land, and during the night it is higher in the water. Similar seasonal variations exist in temperature. The air above land is warmer in the summer than the ocean. However, during the winter, the air above the oceans is warmer than that of landmasses. Despite the stark differences between the surfaces of land and water, different land surfaces heat at different rates. Because of the significant amount of solar energy reflection, a land covered in snow, like in Polar regions, warms very slowly. Due to the significant amount of insolation required to evaporate the water from the plants, land covered in vegetation does not become overly heated.

Land heats and cools more rapidly than water due to its lower specific heat capacity. Specific heat capacity is the amount of heat required to raise the temperature of a substance by one degree Celsius. Water has a high specific heat capacity, meaning it can absorb and store more heat without undergoing a significant temperature change. In contrast, land, with its lower specific heat capacity, heats up quickly during the day and cools down rapidly at night. This disparity causes significant differences in temperature between coastal and inland areas, known as the maritime and continental effects.

Coastal regions, influenced by the presence of large water bodies, typically experience maritime climates. These areas have milder temperatures with smaller daily and seasonal temperature ranges compared to inland regions. The ocean acts as a thermal reservoir, absorbing heat during the summer and releasing it during the winter, moderating temperature fluctuations. For instance, cities like San Francisco exhibit relatively stable temperatures year-round, with cool summers and mild winters, a direct result of the Pacific Ocean's moderating influence.

In contrast, inland areas, far from the moderating effects of large water bodies, experience continental climates characterized by more significant temperature extremes. Summers can be intensely hot, and winters can be extremely cold, as seen in cities like Moscow or Winnipeg. The lack of a nearby ocean to buffer temperature changes results in a more considerable seasonal variation in temperature, leading to distinct hot and cold periods throughout the year.

The land-sea contrast also influences atmospheric circulation patterns and weather systems. During the day, land heats up faster than water, creating a temperature gradient that causes air to rise over the land, drawing in cooler air from the sea. This process generates sea breezes, which bring cooler air and often increase humidity in coastal areas. Conversely, at night, the land cools down more quickly than the sea, resulting in land breezes as cooler air flows from the land towards the warmer sea. These local wind patterns are essential for regulating coastal climates and can significantly impact weather conditions, including the development of fog and localized precipitation.

On a larger scale, the land-sea contrast drives monsoon systems, especially in regions like South Asia. During the summer, intense heating of the Indian subcontinent creates a low-pressure area that draws in moist air from the Indian Ocean, leading to heavy rainfall and the characteristic monsoon climate. In winter, the reverse occurs, with the land cooling rapidly and creating high-pressure systems that push dry air toward the ocean, resulting in dry conditions.

Ocean currents further enhance the land-sea contrast by redistributing heat across the globe. Warm currents, such as the Gulf Stream in the North Atlantic, transport warm water from the tropics towards higher latitudes, influencing coastal climates by warming adjacent land areas. Cold currents, like the California Current, bring cooler water from Polar regions towards the equator, cooling coastal regions and often leading to arid conditions.

The land-sea contrast is a fundamental aspect of climatology, driven by the differing heat capacities of land and water. This contrast leads to significant temperature differences between coastal and inland areas, influences local wind patterns, drives monsoon systems, and is modulated by ocean currents. Understanding this contrast is crucial for predicting regional climate variations, weather patterns, and their broader impacts on ecosystems and human activities.

4.3.1.3. Relief and Altitude

Relief features like plains, mountains, and plateaus alter the distribution of temperature to control it. Mountains operate as wind-blocking structures. In the winter, the Himalayan ranges keep India protected from the chilly winds that originate in Central Asia.

Relief influences climate through its effects on wind patterns and precipitation. Mountains act as barriers to air movement, forcing air to rise along their slopes. As air ascends, it cools and condenses, leading to increased precipitation on the windward side of the mountain. This process is known as orographic lifting. The western slopes of the Sierra Nevada in the United States and the windward side of the Andes in South America experience heavy rainfall due to this effect. Conversely, the leeward side of mountains, often referred to as the rain shadow region, receives significantly less precipitation. This is because the descending air warms and dries, creating arid conditions. The Atacama Desert in Chile, one of the driest places on Earth, lies in the rain shadow of the Andes.

Altitude has a direct impact on temperature. As altitude increases, the atmosphere becomes less dense, and there is a corresponding decrease in temperature. This phenomenon, known as the environmental lapse rate, typically results in a temperature drop of about 6.5°C for every 1,000 meters gained in elevation. Consequently, mountainous regions are generally cooler than adjacent lowlands. For example, the peaks of the Andes and the Himalayas are perpetually capped with snow and ice, even though they are located near the equator. This cooling effect of altitude also influences local ecosystems and human activities, as seen in the unique flora and fauna adapted to high-altitude environments and the specific agricultural practices suited to mountainous areas.

Valleys and basins, shaped by relief, also exhibit unique climatic characteristics. These areas can trap cold air, leading to temperature inversions where the temperature increases with height rather than decreases. This inversion layer can result in severe frost conditions and limit the dispersion of pollutants, contributing to poor air quality in urban basins like Los Angeles. Similarly, valleys may funnel winds, creating local wind patterns that influence temperature and precipitation distribution.

The influence of altitude and relief extends to broader climatic patterns, impacting regional weather systems. Mountain ranges can alter the paths of prevailing winds and storm tracks, affecting the climate over large areas. For example, the Rocky Mountains in North America play a significant role in directing the movement of air masses, contributing to the distinct climatic differences between the western and eastern United States. The Himalayas influence the Indian monsoon by acting as a barrier to the northward movement of moist tropical air, intensifying rainfall over the Indian subcontinent during the summer monsoon season.

As we move upwards from sea level, we experience a gradual decrease in temperature. Temperature decreases at an average rate of 60C per 1000 m. altitude. It is known as the normal lapse rate. The air at lower elevations is warmer than that of higher elevations because it is closest to the heated surface of the earth. As a result, mountains are cooler than the plains even during summer. It is worth remembering that the rate of decrease of temperature with altitude varies with time of day, season, and location.

Quito and Guayaquil are two cities of Ecuador (South America) situated near the equator and relatively close to each other. Quito is at 2800 meters high from mean sea level while Guayaquil is just at 12 meters altitude. However, because of the difference in altitude. Quito experiences an annual mean temperature of 13.3°C while in Guayaquil it is 25.5°C.

4.3.1.4. Ocean Currents

Ocean currents, the continuous, directed movement of seawater, play a pivotal role in climatology by redistributing heat and moisture around the globe, thereby influencing regional climates and weather patterns. These currents are driven by factors such as wind patterns, Earth's rotation (Coriolis Effect), temperature differences, salinity gradients, and the configuration of coastlines and ocean basins.

One of the primary ways ocean currents impact climate is through the transfer of heat from equatorial regions to higher latitudes. Warm currents, like the Gulf Stream in the North Atlantic, carry warm water from the tropics toward the poles. The Gulf Stream, for instance, significantly moderates the climate of Western Europe, making it much milder in the winter than other regions at similar latitudes. This current not only warms the coastal regions it flows past but also influences atmospheric circulation patterns, contributing to the development of weather systems.

Conversely, cold currents, such as the California Current in the Pacific Ocean and the Humboldt Current along the western coast of South America, bring cooler water from Polar regions toward the equator. These currents can lead to cooler coastal temperatures and, in some cases, arid conditions. The Humboldt Current, for example, contributes to the aridity of the Atacama Desert by cooling the air and reducing the ability of the atmosphere to hold moisture, resulting in very low precipitation levels.

Ocean currents also influence the distribution of marine life and ecosystems by transporting nutrients and affecting water temperature. Areas where warm and cold currents converge, such as the Grand Banks off the coast of Newfoundland, are often rich in marine biodiversity and highly productive fisheries. These nutrient-rich waters support a variety of marine species and contribute to the global food supply.

In addition to their role in heat distribution, ocean currents play a crucial part in the global water cycle. They help regulate the distribution of heat and moisture, which affects precipitation patterns. For example, the North Atlantic Drift, an extension of the Gulf Stream, brings warm, moist air to Northwestern Europe, resulting in higher precipitation levels and milder winters.

The El Niño-Southern Oscillation (ENSO) is a prime example of how ocean currents can influence global climate patterns. During an El Niño event, the normal patterns of ocean currents and winds in the Pacific Ocean are disrupted, leading to significant changes in weather patterns worldwide. Warm water accumulates in the central and eastern Pacific, which can lead to increased rainfall and flooding in some regions, such as the western coast of South America, and droughts in others, such as Australia and Southeast Asia. La Niña, the opposite phase of ENSO, is characterized by cooler-than-average sea surface temperatures in the central and eastern Pacific, leading to different but equally significant impacts on global weather patterns.

The thermohaline circulation, also known as the global conveyor belt, is another critical ocean current system that influences climate. This deep-ocean current is driven by differences in water density, which are affected by temperature (thermo) and salinity (haline). The thermohaline circulation moves warm surface water from the equator towards the poles, where it cools, becomes denser, sinks, and then flows back towards the equator at deeper levels. This process plays a crucial role in regulating Earth's climate by redistributing heat and maintaining the balance of the global climate system.

4.3.1.5. Winds

Winds, the movement of air from high-pressure to low-pressure areas, are integral to climatology, shaping weather patterns, distributing heat and moisture, and influencing regional climates. They arise due to the unequal heating of the Earth's surface by the sun, which creates pressure differences. These winds can be categorized into several types based on their scale, origin, and effects.

4.3.1.5.1. Global Wind Patterns

1. Trade Winds: These steady winds blow from the subtropical high-pressure belts (around 30° latitude) towards the equatorial low-pressure zone. In the Northern Hemisphere, they blow from the northeast, and in the Southern Hemisphere, from the southeast. Trade winds are crucial for maritime navigation and significantly influence the climate of tropical regions by promoting the formation of cumulus clouds and thunderstorms, which contribute to the heavy rainfall typical of these areas.
2. Westerlies: Prevailing in the mid-latitudes (between 30° and 60°), westerlies blow from the west towards the east. They are responsible for much of the weather movement across continents in these latitudes, bringing moist air from oceans inland and influencing the climate of regions such as Western Europe and the eastern United States. Westerlies can lead to frequent cyclones and frontal systems, contributing to the variable and often stormy weather experienced in these zones.
3. Polar Easterlies: These cold, dry winds blow from the polar high-pressure areas (around 90° latitude) towards the subpolar low-pressure zones (around 60° latitude). They

flow from the east and are crucial in shaping the polar climates, contributing to the extreme cold conditions of the Arctic and Antarctic regions. Polar easterlies interact with the westerlies at the polar front, often creating significant weather phenomena such as polar vortices and cold outbreaks.

4. Regional and Local Winds:
   - Monsoons: Seasonal wind patterns that result from the differential heating of land and sea. In summer, the land heats up more quickly than the ocean, creating low pressure over the continents and drawing in moist air from the oceans, leading to heavy rainfall. This is evident in the South Asian monsoon, which brings copious amounts of rain to India and surrounding regions. In winter, the process reverses, with high pressure over the cooler land pushing dry air towards the ocean, resulting in dry conditions over the continent.
   - Sea and Land Breezes: These are local winds that occur due to temperature differences between land and water bodies. During the day, land heats up faster than water, causing air to rise over the land and draw cooler air from the sea, creating a sea breeze. At night, the land cools down more quickly than the sea, causing the air over the sea to rise and be replaced by cooler air from the land, resulting in a land breeze. These breezes moderate coastal temperatures and can influence local weather, often leading to the development of clouds and precipitation.
   - Mountain and Valley Breezes: Similar to sea and land breezes, these winds occur due to temperature differences between mountain slopes and valleys. During the day, the sun heats the mountain slopes faster than the valleys, causing air to rise along the slopes and draw cooler air from the valleys, creating an upslope (valley) breeze. At night, the slopes cool down rapidly, causing the cooler air to descend into the valleys, creating a downslope (mountain) breeze. These breezes can lead to unique weather patterns, such as afternoon thunderstorms in mountainous regions.
   - Katabatic and Anabatic Winds: Katabatic winds are strong, cold winds that flow downhill from high-altitude areas, often intensified by the topography. They can be particularly strong in regions with significant elevation differences, such as Antarctica and Greenland. Anabatic winds, on the other hand, are warm, upward-moving winds that occur during the day as the sun heats the slopes.

4.3.1.5.2. Impact on Climate and Weather

Winds play a crucial role in the redistribution of heat and moisture around the globe. They transport warm air from the tropics towards higher latitudes and bring cooler air towards the equator, helping to maintain the Earth's energy balance. Winds also influence ocean currents through the process of wind-driven circulation, which includes phenomena like upweling and downweling, affecting marine ecosystems and climate.

4.3.1.6. Vegetation Cover

Soil devoid of vegetation cover receives heat more rapidly than the soil under vegetation cover. Because vegetation cover absorbs much of the sun's heat and then prevents quick radiation from the earth whereas the former radiates it more rapidly. Hence the temperature variations in dense forested areas are lower than those in desert areas. For example, annual range of temperature in equatorial regions is about 5°C while in hot deserts, it is as high as 38°C.

The presence and type of vegetation affect the Earth's energy balance, carbon cycle, hydrological cycle, and atmospheric composition, all of which are essential components of the climate system.

1. Energy Balance and Albedo: Vegetation cover significantly impacts the Earth's energy balance through its effect on albedo, the measure of reflectivity of the Earth's surface. Vegetated areas, such as forests, grasslands, and crops, typically have lower albedo compared to barren land or urban areas. This means they absorb more solar radiation and convert it into heat, which influences local temperatures. Dense forests, for example, can absorb a substantial amount of sunlight, leading to warmer temperatures in those regions. Conversely, areas with sparse vegetation or snow cover reflect more sunlight, contributing to cooler temperatures.
2. Carbon Cycle: Vegetation plays a crucial role in the carbon cycle by acting as a major carbon sink. Through the process of photosynthesis, plants absorb carbon dioxide (CO_2) from the atmosphere and convert it into organic matter, storing carbon in their biomass and the soil. Forests, particularly tropical rainforests, are significant carbon reservoirs. Deforestation and land-use changes can release stored carbon back into the atmosphere, increasing greenhouse gas concentrations and contributing to global warming. Conversely, reforestation and afforestation can enhance carbon sequestration, mitigating climate change.
3. Hydrological Cycle: Vegetation cover influences the hydrological cycle by affecting processes such as evapotranspiration, infiltration, and runoff. Plants absorb water from the soil and release it into the atmosphere through transpiration, contributing to cloud formation and precipitation. Forested areas typically have higher rates of evapotranspiration compared to non-vegetated areas, which can increase local humidity and rainfall. Vegetation also enhances soil infiltration and reduces surface runoff, which helps in maintaining groundwater levels and reducing the risk of floods. The presence of vegetation, especially forests, thus plays a vital role in maintaining regional water cycles and hydrological balance.
4. Climate and Weather Patterns: Vegetation cover can modify local and regional weather patterns.

For example, the extensive forests in the Amazon basin influence the South American climate by generating moisture and promoting rainfall through evapotranspiration. Similarly, large-scale agricultural activities can create microclimates that affect local weather. Urban areas, with less vegetation cover, often experience the urban heat island effect, where temperatures are higher than surrounding rural areas due to reduced evapotranspiration and increased heat absorption by buildings and pavement.

5. Soil Health and Land Degradation: Vegetation helps maintain soil health by preventing erosion, improving soil structure, and enhancing nutrient cycling. Root systems of plants bind the soil, reducing erosion caused by wind and water. Vegetation also adds organic matter to the soil through leaf litter and root decay, enhancing soil fertility and water retention capacity. Loss of vegetation cover, through deforestation or overgrazing, can lead to land degradation, desertification, and a decline in agricultural productivity, which can have severe climate implications by altering land-atmosphere interactions.
6. Biodiversity and Ecosystem Services: Vegetation supports biodiversity and provides essential ecosystem services that contribute to climate regulation. Diverse plant communities create habitats for various species, promoting ecological balance and resilience. Healthy ecosystems offer services such as carbon sequestration, water purification, and temperature regulation, which are vital for sustaining human life and mitigating climate change impacts.
7. Feedback Mechanisms: Vegetation interacts with climate through complex feedback mechanisms. For instance, increased CO_2 levels can enhance plant growth (CO_2 fertilization effect), potentially increasing carbon sequestration. However, climate change-induced stress, such as higher temperatures and altered precipitation patterns, can negatively impact vegetation health and reduce its carbon sequestration capacity. Additionally, changes in vegetation cover, such as deforestation, can alter regional climate patterns by modifying surface albedo, moisture availability, and atmospheric circulation.

4.3.1.7. Nature of the Soil

The nature of soil, encompassing its composition, structure, and properties, plays a crucial role in climatology and environmental science. Soil influences the Earth's energy balance, hydrological cycle, carbon cycle, and vegetation cover, all of which are integral to climate regulation. Composed of minerals, organic matter, water, and air, soil's physical properties, such as texture and structure, affect water retention, drainage, and aeration. Sandy soils drain quickly but hold less water and nutrients, while clayey soils retain water well but have poor drainage. Loamy soils, with balanced proportions of sand, silt, and clay, are ideal for plant growth. Soil's chemical properties, including pH and nutrient content, affect nutrient availability and microbial activity, with organic matter enhancing fertility and water-holding capacity. Microorganisms decompose organic matter, recycle nutrients, and improve soil structure, maintaining soil health. Soil's albedo influences the Earth's energy balance, with dark soils absorbing more solar radiation and light soils reflecting it. Soil texture and structure affect water infiltration, retention, and drainage, influencing local and regional hydrology. Soils store more carbon than the atmosphere and vegetation combined, playing a significant role in the carbon cycle. Land-use changes can disrupt soil carbon storage, contributing to greenhouse gas emissions. Soil fertility and health directly influence vegetation cover, which affects climate through processes like evapotranspiration, albedo modification, and carbon sequestration. Degraded soils, caused by erosion, compaction, salinization, and nutrient depletion, pose significant threats to climate regulation and agricultural productivity. Sustainable soil management practices, such as conservation tillage, cover cropping, and organic amendments, are essential for maintaining soil health and mitigating climate change. The nature of soil is fundamental to climatology, affecting energy balance, hydrological and carbon cycles, and vegetation dynamics. Understanding and managing soil sustainably is vital for addressing climate change, ensuring food security, and preserving environmental quality.

4.3.1.8. Slope and Aspect

The slope's angle and direction determine how much insolation is received. Along a gentler slope, the angle of incidence of the sun's rays is greater; along a steeper slope, it is smaller. In both situations, the ray contains the same quantity of solar energy. The temperature rises with greater solar energy concentration per unit area along gentler slopes and decreases with lesser concentration along steeper slopes. The Himalaya's southern slopes are warmer than its northern ones for these reasons. In addition, slopes that face the sun receive more insolation and experience warmer temperatures than those that are shielded from the sun's direct rays. For instance, the Himalaya's northern slopes, which face away from the sun and are thus exposed to chilly northern winds, are unquestionably colder. However, because they face the sun and provide protection from the chilly winds from the north, the southern slopes of the Himalayas are warmer. As a result, we see that the southern Himalayan slopes are primarily inhabited and agricultural, while the northern slopes are primarily covered in forests.

4.3.2. Vertical Distribution of Temperature

Temperatures are seen to steadily drop as one travels from the equator to the poles. In the same manner, as atmospheric elevation increases, the temperature steadily drops. The term "vertical temperature gradient" refers to this drop in temperature with height. A complex interplay exists between the various factors influencing the vertical temperature gradient. Heat transfer from the earth, radiation-induced air cooling, and latent heat from condensation are examples of energy transfers. The vertical temperature gradient decreases as a result of air rising through high-pressure systems and warming large air masses. Conversely, systems with low temperatures cause ascending air currents, which heighten the vertical temperature gradient. It is said that moisture is the extra element that complicates the vertical temperature distribution the most.

4.3.2.1. Lapse Rate

"Lapse rate" refers to the temperature

decrease that occurs vertically. The lapse rate is not constant; rather, it changes according to the season, height, and location. Only up to the tropopause does the temperature continue to drop vertically. It ends there. The lapse rate is typically high in tropical regions with high insolation—it can reach 160 meters most afternoons. The lapse rate can vary depending on the atmospheric conditions, from low to steep. The lower air is warmer than the upper layer because it is denser and contains more dust and water vapor than the upper layer. This is not because the lower air is closer to the earth's surface. Conversely, there is less CO_2 and less water vapor in the upper strata due to their dryness. The temperature is relatively low and the upper air is more transparent. As a result, the temperature drops as we move away from the surface of the earth. The difference between the actual and normal lapse rates is constant. Oceans and continents affect the vertical distribution in addition to the horizontal distribution. Normal circumstances cause the temperature to drop with altitude. Nonetheless, there are instances when altitude and temperature rise. We call this phenomenon "temperature inversion." In Polar Regions, layers of near-surface air get so incredibly cold during the winter that elevation causes the temperature to rise to a certain point. The inversion of temperature across continents is a frequent occurrence outside of the polar region. However, summertime temperature inversions occur in oceans. The fact that the temperature drops with elevation in tropical regions up to a height of 16 to 18 km in the troposphere is another significant aspect of the lapse rate. The temperature at the troposphere's outer boundary drops to –80° in this zone, but the lapping rate in the Polar Regions persists at a distance of up to 6 km. The troposphere rises to a height of 10 km in summer and 9 km in winter above latitudes 60°N and S. In higher latitudes, the troposphere is comparatively lower in height.

4.3.3. Temperature Inversion

The temperature in the lowest portion of the atmosphere typically drops with altitude up to 8–10 km above the earth's surface (6°C per km). However, on rare occasions, the opposite occurs, and it is discovered that, rather than falling with elevation, the temperature rises. The gradient of temperature in the other ward is reversed. It's referred to as temperature inversion.

Temperature Inversion near the surface may be produced under the following conditions:

- When there is a long clear winter nights-clear sky with high clouds, dry and calm air, the inversion of temperature occurs.
- When there is little wind movement near the ground or wind movement is too little mixing of air, inversion takes place.
- In higher latitudes- snow-covered areas- where solar radiation is reflected loss of heat-air near the surface goes rapid cooling-temperature inversion develops.
- In Polar Regions temperature inversion is common- all year round Snow covered land surfaces in temperate regions- witness temperature inversion at night in winter.
- The inversion layer in Polar Regions is thicker than in middle latitudes.
- It is greater on the continent in winter- and on oceans in summer except in the Arctic Ocean. During winter there is the absence of

temperature inversion.

Type of inversion of temperature:

- Radiation inversion;
- Subsidence;
- Inversions trade;
- Wind inversion;
- Frontal inversion;
- Ground inversion;
- Valley inversion.

1. Radiation Inversion: It is the term used to describe the surface inversion created when lower air cools through radiation. At around ninety meters above sea level, the inversion layer forms. The coldest air lies on the ground and is covered by warm air during nighttime because the land surface radiates more heat than the air. As a result, the ground cools more quickly than the air at high altitudes. Radiation and condensation cool the air layer closest to the earth's surface more quickly than they do the upper air layer. Thus, the temperature rises with altitude above a certain point (90 meters). It goes on for 300 meters.
2. Subsidence Inversions: Inversions of subsidence occur in both valleys and mountains. The air is heated as it blows over the hills and is compressed into their sides by the warm air that rises to the top; this warm air is warmer than the valley's cooler air. Additionally, a greater inversion causes the air in the valley to be compressed as it passes over the hill's summit, heating the valley's cooler air from above. We call these Subsidence Inversions.
3. Trade-Wind Inversion: The temperature inversion that is typically found over the eastern regions of the tropical oceans in the trade-wind streams. Although it can be seen in some weather patterns, the inversion does not exist as a mean condition over the western parts of the trade-wind belt and in the equatorial zone. The inversion's strength varies greatly; it can occasionally be more than 10°C over a distance of one kilometer, but it can also occasionally be completely absent, particularly in the Northern Hemisphere. When the base's height is at its lowest, the inversion is typically strongest, and vice versa. The inversion layer has an average thickness of roughly 400 meters and ranges in thickness from a few meters to over 1000 meters.
4. Frontal Inversions: A shallow cold front bringing in warmer air is what causes frontal inversions. Put another way, an inversion occurs when a mass of cold air, known as an air mass, is occasionally carried by the wind from one location to a warmer one and ends up blown underneath the warm air. We call this frontal inversion.
   - Ø Effects of Inversions: Smog is created when inversions trap pollution near the ground, which is one of their most dangerous effects. This is a serious issue in cities like Salt Lake City and Los Angeles due to the emissions that are released into the atmosphere by power plants, automobiles, and other factories and heat sources. Additionally, sound

waves are trapped by inversions. As a result, when sound waves refract off the inversion layer and return to the ground, the loud noises made by airplanes taking off will appear louder. Yet weather inversions can also bend light due to temperature fluctuations, so it's not just sound waves that are affected.

5. Ground Inversion: When the earth's surface cools quickly due to solar radiation under a clear sky, ground inversion happens. In this sense, the air above the ground continues to be warmer than the air below. The likelihood of fog forming increases when the temperature close to the surface approaches the dew point. In higher latitude regions and even in tropical areas, ground inversion is a common occurrence during winter months on the plain (Figure 4.5).

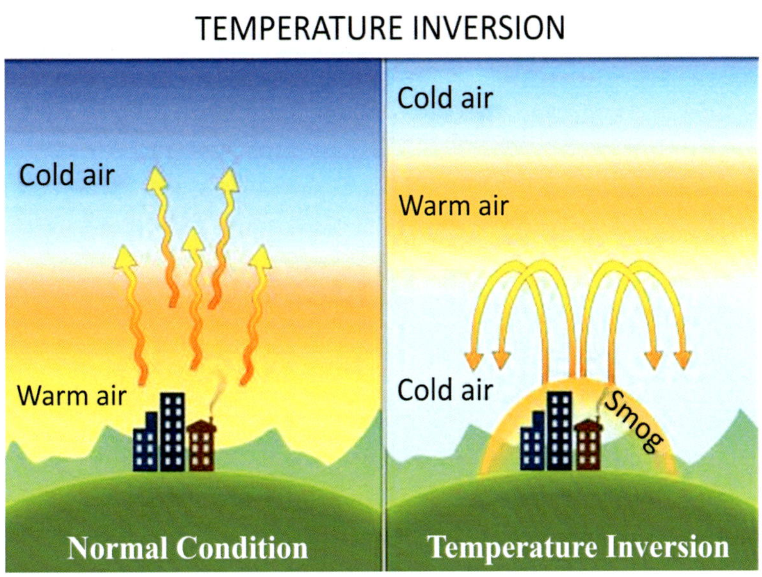

Figure 4.5. Inversion of temperature on the ground.

Source: https://www.mlsu.ac.in/econtents/491_Oceanography%20and%20Climatology%20Unit-4%20TEMPERATURE%20AND%20ITS%20DISTRIBUTION%20uploaded%20by%20DR.%20ANKUSH%20SHRIVASTAVA.pdf.

6. Valley Inversion: In areas with undulating terrain, especially those with hills, valley inversion occurs. In this case, the air near the mountainside cools at night and the slope itself becomes cooler. The valley is filled with cool air as it moves down the hillside. The temperature inversion is noticeable because the valley's warm air is forced upward.

4.4. TEMPERATURE SCALES AND MEASUREMENTS

Temperature measurement is a relatively recent idea. Although early scientists could distinguish between hot and cold, it wasn't until the seventeenth century that they were able to measure different heat levels. The basic water thermoscope was created in 1597 by the Italian astronomer Galileo Galilei. It was a long glass tube that was inverted and sealed inside a jar that held water and air. The liquid was forced up the tube by the expanding air in the heated jar. To demonstrate the relative changes that occur when heat is added or removed, the water level in the tube could be compared at various temperatures. The thermoscope did not, however, have a simple means of measuring temperature directly. A few years later, Santorio, an inventor and physician from Italy, enhanced Galileo's design by giving the thermoscope a numerical scale. The development of the fluid-filled thermometers that are widely used today was influenced by these early thermoscopes. The way that some fluids expand when heated is how modern thermometers work. The fluid in a thermometer expands when it absorbs heat, taking up more space and raising the fluid level in the tube. The fluid level drops as a result of the fluid contracting as it cools and takes up less volume. An object's temperature is a measurement of its heat energy content (for additional information on this idea, refer to our Energy module). Scales based on reference points are required to measure temperature precisely because it is a relative quantity. The Fahrenheit (°F), Celsius (°C), and Kelvin (K) scales are the three primary scales that are widely used to measure temperature in the modern world.

1. Fahrenheit: Daniel Gabriel Fahrenheit (1686–1736) was a German physicist who is credited with the invention of the alcohol thermometer in 1709 and the mercury thermometer in 1714. The Fahrenheit temperature scale was developed in 1724. Fahrenheit originally established a scale in which the temperature of an ice-water-salt mixture was set at 0°. The temperature of an ice-water (no salt) mixture was set at 30° and the temperature of the human body was set at 96°. Using this scale, Fahrenheit measured the temperature of boiling water as 212°F on his scale. He later adjusted the freezing point of water from 30°F to 32°F, thus making the interval between the freezing and boiling points of water even 180° (and making body temperature the familiar 98.6°F). The Fahrenheit scale is still commonly used in the United States.

   Converting between Fahrenheit and Celsius is a frequent necessity due to the global use of the Celsius scale. The conversion formulas are:

- From Fahrenheit to Celsius: C=5/9(F−32);
- From Celsius to Fahrenheit: F=5/9C+32.

These conversions are essential for scientists, travelers, and those involved in international communication to ensure accurate and understandable temperature readings.

The Fahrenheit scale is less convenient than Celsius or Kelvin due to its non-metric nature. Celsius and Kelvin scales are part of the International System of Units (SI), which facilitates scientific calculations and international data sharing. For example, water's freezing and boiling points are 0°C and 100°C, respectively, making Celsius more straightforward for scientific use. Kelvin, used primarily in scientific disciplines like physics, starts at absolute zero (0 K), the point at which molecular motion ceases, with 1 Kelvin being equivalent to 1°C.

2. Celsius: Anders Celsius (1701–1744) was a Swedish astronomer credited with the invention of the centigrade scale in 1742. Celsius chose the melting point of ice and the boiling point of water as his two reference temperatures to provide a simple and consistent method of thermometer calibration. Celsius divided the difference in temperature between the freezing and boiling points of water into 100° (thus the name centi, meaning one hundred, and grade, meaning degrees). After Celsius's death, the centigrade scale was renamed the Celsius scale and the freezing point of water was set at 0°C and the boiling point of water at 100°C. The Celsius scale takes precedence over the Fahrenheit scale in scientific research because it is more compatible with the base ten format of the International System (SI) of metric measurement. In addition, the Celsius temperature scale is commonly used in most countries of the world other than the United States.

3. Kelvin: The Kelvin temperature scale, denoted by the symbol K, is the primary unit of temperature measurement in the International System of Units (SI) and is widely used in scientific research, particularly in physics and chemistry. Named after the Scottish physicist William Thomson, 1st Baron Kelvin, the Kelvin scale is an absolute temperature scale based on the concept of absolute zero—the point at which all molecular motion ceases, defined as 0 K. This absolute zero corresponds to −273.15° Celsius (°C), making the Kelvin scale directly related to the Celsius scale but starting from a fundamentally different point.

   One of the key advantages of the Kelvin scale is its simplicity in scientific calculations, as it avoids negative numbers when dealing with thermodynamic temperatures. Each increment on the Kelvin scale is equivalent to one degree Celsius, simplifying the conversion between the two scales: K=°C+273.15K = °C + 273.15K=°C+273.15. This direct relationship allows for the seamless integration of Kelvin in various scientific formulas and equations involving temperature.

   The Kelvin scale is essential in fields such as thermodynamics, where it is used to describe temperatures about absolute zero, enabling precise measurements and comparisons. For instance, in gas laws and equations of state, temperatures must be expressed in Kelvins to ensure accuracy and consistency. Kelvin is also crucial in the study of blackbody radiation, where the temperature of objects emitting thermal radiation is measured in Kelvins.

In addition to its scientific applications, the Kelvin scale is used in many engineering and industrial processes, where precise temperature control and measurement are critical. The scale's absolute nature provides a universal standard that facilitates international collaboration and data comparison across various scientific disciplines.

Overall, the Kelvin temperature scale's foundation on absolute zero, its direct correlation with the Celsius scale, and its critical role in scientific and industrial applications underscore its importance as a fundamental unit of temperature measurement.

4. Absolute Zero: Temperature is a measure of how much thermal energy a system has. This measurement means that all of the atoms and molecules which are moving around have a certain amount of kinetic energy (and less obviously potential energy). When all of the molecules (or atoms) in a system stop moving completely, that's as cold as they can get. This temperature, where there's no thermal energy at all, is called absolute zero (Winston Smith (2018)).

   Numerically, this is written as 0 K, –273.15°C, or –459.67°F.

   The idea of absolute zero plays into understanding how much energy is available from gas molecules in the ideal gas law since the temperature has to be measured on an absolute scale (like Kelvin), for the ideal gas law to make sense. Additionally, the idea of absolute zero plays into the physics of blackbody radiation (how much energy radiates out from an object at a particular temperature) and the maximum possible efficiency of a heat engine (called the Carnot efficiency).

   The concept of absolute zero is also part of the physics of climate change. The average temperature of the Earth, which is about 15°C, would be 288 K. If greenhouse gases increase the temperature of the planet by 1% then it wouldn't go up 0.15°, it would go up 2.88°. Kelvin and Celsius both have the same degree increment but Kelvin is an absolute scale (meaning that it's 0. really is 0) and Celsius is a relative scale (it's 0. is arbitrary – it was chosen by a scientist). This is why the temperature would increase by 2.88° instead of 0.15°. Understanding how these small percent changes in the Earth's temperature can lead to drastic consequences for the planet is an important part of climate science (Janet Larsen (2018)).

4.4.1. Technologies

Numerous techniques have been devised to measure temperature. The majority of these rely on measuring a temperature-dependent physical property of a working material. The glass thermometer is among the most widely used instruments for taking temperature. This is made up of a glass tube that serves as the working fluid and is filled with mercury or another liquid. The fluid expands when its temperature rises, so the volume of the fluid can be used to calculate its temperature. These thermometers are typically calibrated so that the fluid level in the thermometer can be used to determine the temperature. The gas thermometer is a different kind of thermometer that is less common in use but has theoretical significance.

Other important devices for measuring temperature include:

- Thermocouples;
- Thermistors;
- Resistance temperature detector (RTD);
- Pyrometer;
- Langmuir probes (for electron temperature of a plasma);
- Infrared thermometer;
- Other thermometers.

When measuring temperature, one must exercise caution to make sure that the thermometer, thermocouple, or other measuring device is accurate and is the same temperature as the substance being measured. The measured temperature may differ from the system's actual temperature in certain situations due to heat from the measuring device creating a temperature gradient. In this scenario, the system's temperature and its heat transfer characteristics will both affect the measured temperature.

There is more to the thermal comfort that people, animals, and plants feel than just the temperature recorded on a glass thermometer. A room's relative humidity can cause evaporative cooling to occur to varying degrees. This humidity effect is normalized through wet-bulb temperature measurement. Thermal comfort can also be influenced by mean radiant temperature. Even though a glass thermometer reads the same temperature in windy conditions as in calm ones, the wind chill factor causes the weather to feel colder. The rate at which heat is transferred from or to the body is accelerated by airflow, which causes a greater variation in body temperature for a given ambient temperature.

The zeroth law of thermodynamics, which states that if you have three bodies, A, B, and C, and if A and B, and B and C are at the same temperature, then A and C are at the same temperature, is the theoretical foundation for thermometers. Naturally, B is the thermometer. Triple point cells are the practical foundation of thermometry. Triple points are pressure, volume, and temperature settings where three phases—solid, vapor, and liquid—are present at the same time. There are no degrees of freedom for a single component at a triple point, and any modification to any one of the three variables causes one or more of the phases to disappear from the cell. Triple point cells can therefore be used as general temperature and pressure references (see Gibbs phase rule). Temperature can be directly measured under certain circumstances using Planck's law of black-body radiation. The measurement of the cosmic microwave background temperature, for instance, has been made using the photon spectrum obtained from satellite observations like the WMAP. Single particle spectra can occasionally be used as a thermometer in the heavy-ion collision study of the quark-gluon plasma.

- Ø Non-Invasive Thermometry: Numerous thermometric techniques have been developed in the last few decades. In the context of biotechnology, the most popular and promising non-invasive thermometric methods are those that rely on the analysis of echotomography, CT, and magnetic resonance imaging data. These methods enable tissue temperature monitoring without the need for a sensing component. Laser-induced fluorescence (LIF), CARS, and laser absorption spectroscopy have been used to measure the temperature inside engines, gas turbines, shock tubes, synthesis reactors, and other reactive flows (such as combustion and plasmas) (Chrystie, Robin S. M.; Feroughi, Omid M.; Dreier, Thomas; Schulz, Christof (2017–03–21)). The capability of such optical-based techniques includes rapid measurement (down to nanosecond timescales), notwithstanding the ability to not perturb the subject of measurement (e.g., the flame, shock-heated gases) (Bradley et al., 1993; Diaz et al., 1994).

CASE STUDY

UNDERSTANDING OUR OCEANS:

Oceans are the largest active carbon sinks on Earth, absorbing more than a quarter of anthropogenic carbon emissions. The ocean's interaction with the atmosphere, and its ability to absorb carbon dioxide, is strongly influenced by properties of seawater, such as salinity and acidity. Reliable and comparable measurements of these properties are of crucial importance to climate researchers, enabling them to detect small changes in ocean dynamics over decades and even centuries.

Oceans play a key role in regulating the global climate system. The interaction of oceans with the Earth's atmosphere is strongly linked to seawater properties such as salinity and pH, which must be accurately monitored to identify long-term climate trends. However, measurements of these properties are challenging without a traceability chain to link them to units defined in the SI, which would ensure they are comparable regardless of where and when they are made.

Challenge:

The oceanography community most commonly measures ocean salinity using the 'Practical Salinity Scale,' which is based on comparing seawater electrical conductivity with the conductivity of a commercially available standard solution made from North Atlantic seawater. This salinity is then used to calculate other properties, like density and sound speed. However, small chemical changes in the standard solution, and in the ocean itself due to climate change, make relationships between Practical Salinity and these other properties uncertain. One such change arises from increases in dissolved carbon dioxide. This means small changes in ocean properties cannot be reliably identified. The equation of state for seawater, from which the thermodynamic properties of seawater can be derived, is another key component of global climate models. The inclusion of density and speed of sound data in the equation of state will assist in the determination of other properties, including salinity and temperature, especially for deep-sea regions where precise measurements are difficult to make, and further improve the reliability of the climate models they feed into. Ocean-going vessels equipped with acoustic sensors could provide a cost-effective, extensive seawater temperature measurement network, via speed of sound, to supplement satellite data for climate models. However, improved SI traceability for the probes used to measure the speed of sound in seawater is needed before this type of monitoring system can be implemented.

Solution:

The EMRP project Metrology for ocean salinity and acidity has provided a reference method for ocean salinity, which makes Practical Salinity measurements traceable to the SI units through density measurements. This means that possible long-term drifts in the chemical composition of the standard seawater solution can now be accounted for and eliminated. In addition, probes used to measure the speed of sound in seawater can now be traceably calibrated under typical operating conditions at newly-developed facilities at project partners PTB and INRIM – the National Measurement Institutes of Germany and Italy. Measurements at sea have confirmed the performance of a prototype transfer standard developed at INRIM, bringing easily achievable traceability to ship-based probes, and have assisted in the generation of in-situ measurement best practices.

Impact:

As a result of the EMRP project, NMI members have for the first time been invited to sit on the prestigious Joint Committee for the Properties of Seawater, the organization responsible for maintaining and improving seawater standards (including the Thermodynamic Equation of Seawater – 2010: TEOS-10). The committee is keen to promote measurement traceability and ensure data compatibility in the years to come. Ocean Scientific International Ltd (OSIL) is going to incorporate density measurements into the preparation of its standard seawater, which is the only internationally-recognized calibration standard for the measurement of practical salinity. This will provide SI traceability to ocean salinity measurements across the globe and allow the oceanography community to reliably identify even small changes.

CLASS ACTIVITY

Insolation and Temperature

1. Give a geographical term for each of the following statements:

a. The process of horizontal transport of heat by winds.

....

b. Imaginary lines on a map joining the places of equal temperature, reduced to sea level.

....

c. Difference between the mean temperatures of the hottest and that of the coldest month.

....

2. Select a meteorological observatory located in your city or near your town. Tabulate the temperature data as given in the climatological table of observatories:

i. Note the altitude, latitude of the observatory and the period for which the mean is calculated.

ii. Define the terms related to temperature as given in the table.

iii. Calculate the daily mean monthly temperature.

iv. Draw a graph to show the daily mean maximum, the daily mean minimum and the mean temperature.

v. Calculate the annual range of temperature.

vi. Find out in which months the daily range of temperature is the highest and the lowest.

vii. List out the factors that determine the temperature of the place and explain the possible causes for temperature variation in January, May, July, and October

Example

Observatory: New Delhi

Latitude: 28°35°' N

Based on observations: 1951 – 1980

Altitude above mean sea level: 216 m

Month	Mean of Daily Max. (°C)	Mean of Daily Min. (°C)	Highest Recorded (°C)	Lowest Recorded (°C)
January	21.1	7.3	29.3	0.6
May	39.6	25.9	47.2	17.5

SUMMARY

- Isotherms while passing from the continents to oceans get distorted. The distortion is larger in the northern hemisphere due to the larger percentage of the land surface.
- Absorption of solar radiation takes place if the obstructing particles happen to be water vapors, ozone molecules, carbon dioxide molecules, or clouds.
- The highest temperatures are found over the continents in subtropical latitudes where descending air maintains clear skies, not in equatorial regions where frequent cloudiness prevents the highest temperatures.
- The annual range of temperature at a particular place is highly controlled by height. At high elevations the rarity of the air, large amount of precipitation and cloudiness combine together to lower down the average temperature. Thus, places situated at higher elevations have a lower annual range of temperature.
- The warm ocean currents, prevailing winds help to raise the temperature of the adjoining regions. The effect of warm ocean currents is more pronounced in winter.
- Latitude is one of the most important and primary factors affecting the unequal heating of the Earth's surface and its atmosphere. The intensity of heating determine the degree of temperature on the Earth and so is its distribution. Overall annual maximum energy is received on or near the equator.
- In low latitudes particularly near the equator, the solar beam of light travels for lesser distances in the atmosphere. Hence, less reflection, less obstruction, and less absorption are observed in the atmosphere. This results in a greater amount of solar energy received on the earth and records higher temperatures.
- The direction of the slope and its angle control the amount of solar radiation received locally. Slopes more exposed to the sun receive more solar radiation than those away from the sun's direct rays.
- Slope is one of the factors which affect the temperature of a place. Slopes facing the sun have higher temperature during summer months, and the slopes protected by the sun have lower temperature during winter. Thus, the local factors also affect an annual range of temperature.
- Water is heated or cooled in a longer period than land. Because of this characteristic of water, the coastal areas enjoy a moderate climate and the variations in temperature between the warmest and coldest months are not very large.
- The decrease of temperatures is known as vertical temperature gradient or normal lapse rate which is 1000 times more than the horizontal lapse rate.
- Winds and ocean currents affect the temperature of any region. Winds are the horizontal movement of air from high air pressure region to low pressure. Ocean currents also transport the heat energy from one region to another. The prevailing winds blowing towards land from the oceans modify the temperature.

REVIEW QUESTIONS

1. What are the factors controlling the temperature distribution?
2. What is the current situation of temperature and climate?
3. How does the angle of incidence vary with latitude and time of year?
4. Describe the impact of atmospheric transparency on the amount of insolation reaching the Earth's surface.
5. Explain the greenhouse effect and its role in heating the atmosphere.
6. Describe the process of solar radiation absorption and its distribution on Earth.
7. Explain the concept of the Earth's radiation budget.
8. Define global warming and distinguish it from climate change.
9. How do latitude, altitude, and proximity to water bodies affect horizontal temperature patterns?
10. Describe the conditions under which temperature inversions typically occur.

REFERENCES

1. Bradley, R. S. (1996). In P. D. Jones, R. S. Bradley, & J. Jouzel (Eds.), Climatic Variations and Forcing Mechanisms of The Last 2000 Years (pp. 603–624). Springer.
2. Bradley, R. S., & Jones, P. D. (1993). 'Little Ice Age' summer temperature variations: Their nature and relevance to recent global warming trends. Holocene, 3(4), 367–376. https://doi.org/10.1177/095968369300300409.
3. Chrystie, R. S. M., Feroughi, O. M., Dreier, T., & Schulz, C. (2017). SiO multi-line laser-induced fluorescence for quantitative temperature imaging in flame-synthesis of nanoparticles. Applied Physics B, 123(4), 104. https://doi.org/10.1007/s00340-017-6685-0.
4. D'Arrigo, R. D., Cook, E. R., Jacoby, G. C., & Briffa, K. R. (1993). NAO and sea surface temperature signatures in tree-ring records from the North Atlantic sector. Quaternary Science Reviews, 12(6), 431–440. https://doi.org/10.1016/S0277-3791(05)80006-5.
5. Diaz, H. F., & Pulwarty, R. S. (1994). An analysis of the time scales of variability in centuries-long ENSO-sensitive records in the last 1000 years. Climatic Change, 26(3), 317–342. https://doi.org/10.1007/BF01092441.
6. Jones, P. D., & Briffa, K. R. (1992). Global surface air temperature variations during the 20th century: Part 1—Spatial, temporal and seasonal details. Holocene, 1(2), 165–179. https://doi.org/10.1177/095968369200100208.
7. Kushnir, Y. (1994). Interdecadal variations in North Atlantic Sea surface temperature and associated atmospheric conditions. Journal of Climate, 7(1), 141–157. https://doi.org/10.1175/1520-0442(1994)007<0141>2.0.CO;2.
8. Smith, T. M., Reynolds, R. W., Livezey, R. E., & Stokes, D. C. (1996). Reconstruction of historical sea surface temperatures using empirical orthogonal functions. Journal of Climate, 9(6), 1403–1420. https://doi.org/10.1175/1520-0442(1996)009<1403>2.0.CO;2.

CHAPTER

5

Atmospheric Moisture and Precipitation

LEARNING OBJECTIVES

After studying this chapter, you will be able to:

- Describe water vapor in the atmosphere;
- Explain the concept of precipitation;
- Discuss the sources of atmospheric moisture;
- Know the variations in absolute humidity;
- Define the diurnal and seasonal changes in relative humidity.

KEY TERMS FROM THIS CHAPTER

Absolute humidity	Atmospheric water vapor	Bergeron process
Chemical reactions	Cloud droplets	Convective rain
Extratropical cyclones	Global warming	Ice pellets
Moisture content	Precipitation	Relative humidity
Solar atmosphere	Vapor molecules	Vapor pressure

INTRODUCTION

The amount of moisture in the atmosphere plays a crucial role in fire weather. It directly impacts how easily forest fuels can catch fire, and also influences other aspects of fire behavior through its connection to other weather conditions. Water vapor is continuously exchanged between the atmosphere and dead wildland fuels. Dry fuels will absorb moisture when the air is humid and release moisture when the air is dry. In dry periods, low humidity can also affect the moisture levels in green fuels. When atmospheric moisture condenses and falls as precipitation, it increases the moisture content of dead fuels and helps green vegetation grow by replenishing soil moisture.

Moisture plays a crucial role in determining surface temperatures, including those of surface fuels, by affecting radiation in its vapor form and reflecting and radiating heat when condensed into clouds. The energy released during condensation fuels thunderstorms and the strong winds they bring. Moisture is essential for lightning development, which in many mountainous regions poses a significant wildfire risk.

Precipitation is the result of atmospheric water vapor condensing and falling from clouds due to gravity. The main types of precipitation are drizzle, rain, sleet, snow, ice pellets, graupel, and hail. Precipitation occurs when a portion of the atmosphere reaches full saturation with water vapor, causing the water to condense and fall. Fog and mist are not considered precipitation but colloids because the water vapor does not condense enough to fall. Air can become saturated with water vapor through cooling or adding more water vapor. Precipitation forms as smaller droplets merge within a cloud through collision with other raindrops or ice crystals. Sporadic, intense periods of rain are known as showers.

When moisture is lifted or forced to rise over a layer of cold air at the surface, it can condense into clouds and rain, particularly during freezing rain events. In these situations, a stationary front is usually located near the area experiencing freezing rain, causing air to rise. If there is enough moisture in the atmosphere, the rising air will form clouds like nimbostratus and cumulonimbus, which can lead to significant precipitation. Eventually, the cloud droplets will grow large enough to become raindrops and fall to the Earth, often freezing upon contact with objects. In areas near warm water bodies, such as lakes, lake-effect snowfall can be a concern as warm water evaporates into the cold cyclonic flow around extratropical cyclones.

Lake-effect snowfall can be quite heavy in certain areas. Thundersnow may occur within a cyclone's comma head and in bands of lake-effect precipitation. In mountainous regions, heavy precipitation can occur where wind is forced to rise on the windward side of the terrain at higher elevations. On the opposite side of mountains, desert conditions can be found due to the warm, dry air produced by compression. The majority of rainfall happens in tropical regions and is a result of convection. The monsoon trough, or intertropical convergence zone, brings rainy seasons to savannah areas.

5.1. WATER VAPOR IN THE ATMOSPHERE

Vaporized moisture behaves similarly to any other gas, mixing with other gases in the air while retaining its distinct characteristics. Moisture acts as the primary agent in condensation, storing significant amounts of energy acquired through evaporation, which is then released during condensation. The heat produced during the condensation of water vapor provides the energy for severe weather events such as thunderstorms, tornadoes, and hurricanes. Water vapor availability is crucial for precipitation, which in turn determines a region's ability to support vegetation, ultimately serving as fuel for wildfires.

The moisture in the atmosphere is constantly undergoing physical changes such as condensation, freezing, melting, evaporation, and re-condensation. These changes are all influenced by temperature, which dictates the level of molecular activity in a substance. At an extremely low temperature of about -460°F (absolute zero), all substance molecules are completely still. However, as the temperature increases, the molecules become more energetic and move at higher speeds. Water molecules, for example, move slowly at below-freezing temperatures, faster at melting temperatures, and even more rapidly during boiling. Despite the temperature, individual molecules – whether solid, liquid, or gas – do not have the same speeds or directions of movement. Continuous collisions occur that alter their speeds and directions (Figure 5.1).

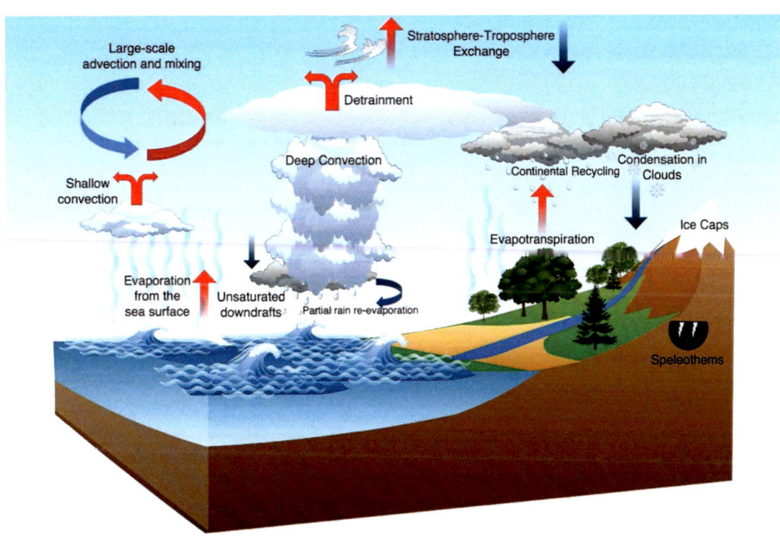

Figure 5.1. Key processes governing the isotopic composition of atmospheric water vapor.

Source: https://eos.org/editors-vox/uncovering-the-hidden-secrets-of-water-vapor.

Water vapor is a common component of the atmosphere, found in the solar atmosphere, all planets in the Solar System, and various astronomical objects like natural satellites, comets, and large asteroids. If water vapor is detected in other planetary systems, it suggests a similar distribution. Additionally, the presence of water vapor can provide indirect evidence of liquid water on some planetary mass objects of extraterrestrial origin.

Water vapor is classified as 'feedback' because it reacts to temperature changes and intensifies the impact of factors that contribute to warming, making it a greenhouse gas.

5.1.1. Evaporation

Some molecules can gain high speeds from collisions with other molecules, causing them to escape into the air if near the surface of liquid water. This process is known as evaporation, where a liquid water molecule turns into a water vapor molecule. As molecules with higher energy content evaporate, the liquid left behind has a lower average energy level, leading to a decrease in temperature. Therefore, evaporation acts as a cooling process. Each molecule that changes state and escapes into the air carries approximately 1,000 times the energy needed to raise the temperature of a water molecule by 1°F.

The vapor pressure of water is the pressure at the water-air boundary caused by molecular movement toward escaping from the liquid. This pressure is dependent solely on the temperature of the water and impacts the speed at which water molecules transform into vapor molecules in the air. As water vapor molecules escape into the air, they displace air molecules and add to the overall atmospheric pressure. This contribution is known as the partial pressure due to water vapor, or commonly referred to as vapor pressure for simplicity. (Li, Choy, Zaminpardaz et al., 2023) (Figure 5.2).

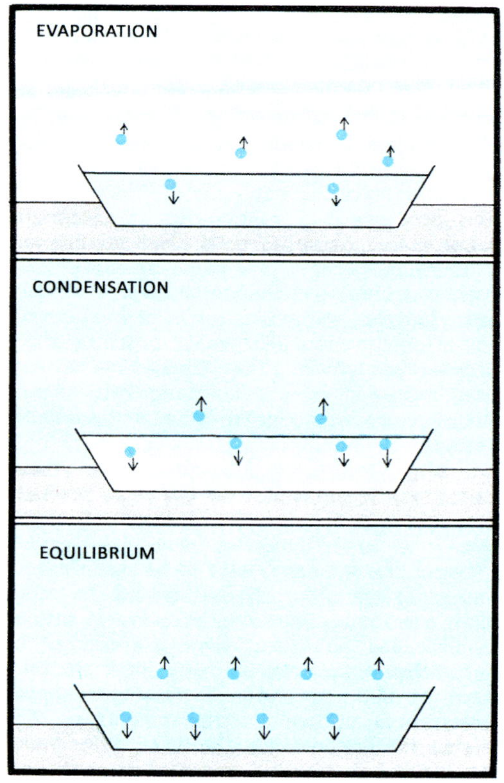

Figure 5.2. Evaporation occurs when an excess of water molecules leaves a water surface, and condensation occurs when an excess of molecular arms the liquid water. In an equilibrium condition, there is no net exchange in either direction, or the atmosphere is saturated.

Source: https://www.frames.gov/documents/behaveplus/publications/Schroeder_and_Buck_1970_PMS-425-1_NFES-1174_Fire_Wx_ch_03.pdf.

The vapor pressure is determined by the amount of water vapor present in the air, ranging from minimal levels in cold, dry air to approximately 2 inches of mercury in warm, humid air. High levels are typically found in the warm, lower parts of the troposphere. As the vapor pressure increases, some water vapor molecules condense back into liquid form on surfaces. This releases the same amount of heat energy that was required for evaporation, heating up the surface of condensation.

Molecules are constantly exchanged between water and air at the boundary, with a greater exchange typically occurring in one direction. Evaporation happens when more molecules leave the water than enter it, while condensation occurs when the opposite occurs. Both condensation and evaporation happen simultaneously. Sublimation, where molecules are exchanged between water vapor and ice, also occurs. Ice has a lower vapor pressure compared to water at the same temperature, making sublimation on ice easier than condensation on a water surface at low temperatures.

When the vapor pressure of the atmosphere matches the vapor pressure of a water or ice surface, there is no overall exchange of water molecules in any direction, and the atmosphere is considered saturated. A saturated volume of air contains the maximum amount of vapor it can hold. The saturation vapor pressure is the vapor pressure at saturation, which depends on the temperature of the air and is the same as the vapor pressure of water at that temperature. As temperature increases, the amount of water vapor that air can hold also increases, leading to a higher saturation vapor pressure. On the contrary, lower temperatures result in a lower saturation vapor pressure.

Evaporative cooling is influenced by atmospheric conditions, particularly humidity which is the measure of water vapor in the air. Hygrometers are used to measure the vapor content of air, with readings given either as specific humidity or percent relative humidity. The equilibrium vapor pressure is determined by the temperatures of the atmosphere and water surface, with 100% relative humidity occurring when the partial pressure of water vapor equals the equilibrium vapor pressure, also known as complete saturation. Humidity can range from 0 grams per cubic meter in dry air to 30 grams per cubic meter when the vapor is saturated at 30°C. (Bauer M, Del Genio AD, Lanzante JR. 2002)

5.1.2. Sublimation

Sublimation is the process in which water molecules transition directly from ice to water vapor without passing through the liquid phase. This phenomenon is responsible for the gradual disappearance of ice and snow in the middle of winter at temperatures too cold for melting to occur. Antarctica showcases sublimation on a remarkable scale due to its minimal precipitation levels, making it the continent with the lowest rate of precipitation on our planet. As a result, vast areas of Antarctica feature layers of snow that have sublimed over thousands of years, leaving behind any non-volatile substances they may have contained. This is particularly advantageous for various scientific fields, with a noteworthy example being the abundant and well-preserved meteorite specimens left exposed as a result of sublimation. Sublimation is a crucial step in the processing of specific types of biological specimens for scanning electron microscopy. The specimens are typically cryofixed and freeze-fractured, followed by freeze-etching to reveal intricate details. Through eroding the broken surface under vacuum, this technique can accurately display protein molecules, organelle structures, and lipid bilayers with minimal distortion.

5.1.3. Condensation

Condensation occurs when water vapor changes into liquid form by coming into contact with a cooler surface or when the air's capacity to hold water vapor is exceeded. This process results in a warming effect on the surface it condenses on, as the water molecule carries heat energy. Simultaneously, the temperature of the surrounding atmosphere slightly decreases. In the atmosphere, condensation leads to the formation of clouds, fog, and precipitation, typically facilitated by cloud condensation nuclei. The dew point temperature is the point at which air must cool before water vapor starts to condense. Cloud droplets are formed through condensation in the atmosphere.

Moreover, when the surface temperature drops to or below the dew point temperature of the surrounding atmosphere, water vapor condenses on surfaces in the form of a thin layer. Deposition, a distinct phase transition, occurs when water vapor directly transforms into ice without first condensing. Frost and snow are common examples of deposition.

There are multiple ways in which cooling can occur through condensation:

- Heat loss through conduction or radiation.
- Adiabatic cooling, which happens when air is lifted, causing a drop in air pressure. This can be caused by mountains, convection, and cold or warm fronts.
- Advective cooling, which is cooling due to horizontal air movement.

5.1.4. Dew Point

Saturation occurs when the air is cooled to the point where its saturation vapor pressure matches the actual vapor pressure. This temperature is known as the dew-point temperature, or simply the dew point. Additional cooling can lead to condensation of vapor into liquid droplets, which form clouds, fog, or dew. Cooling near the surface is typically caused by contact with colder ground or water. Alternatively, moist air can be lifted to higher altitudes, where it cools adiabatically to reach the dew point.

When the temperature of the air drops below its dew point, condensation happens because the air holds more water vapor than it can at a lower temperature. Typically, the actual vapor pressure cannot surpass the saturation vapor pressure by a significant margin.

5.1.5. Chemical Reactions

There are chemical reactions that produce water as a product. When these reactions occur at temperatures above the dew point of the surrounding air, the water will be generated as vapor, increasing the local humidity. If the temperature is below the dew point, local condensation will take place. Common reactions that lead to the formation of water include the burning of hydrogen or hydrocarbons in air or other oxygen-containing gas mixtures, as well as reactions with oxidizers.

Just like with rust forming on iron or steel when exposed to water vapor, other chemical or physical reactions can occur in the presence of water vapor, leading to the creation of new chemicals. For example, polymerization can take place in certain polyurethane foams and cyanoacrylate glues when they come into contact with atmospheric humidity. Additionally, anhydrous chemicals may absorb enough water vapor to form a crystalline structure or change an existing one, sometimes causing characteristic color changes that can be used for measurement.

5.1.6. Absolute Humidity

Absolute humidity is the weight of water vapor in a certain volume of air, typically measured in pounds per 1,000 cubic feet. There is a direct link between the dew point, vapor pressure, and absolute humidity, as they all depend on the actual amount of water vapor in the air. When the air is saturated, the dew point matches the temperature, the vapor pressure equals the saturation vapor pressure, and the absolute humidity is at its saturation level.

5.1.7. Relative Humidity

Surface air saturation is ideal for fire weather conditions, resulting in low fire danger. On the other hand, unsaturated conditions allow for evaporation from forest fuels, making them more flammable and increasing the risk of fire. Relative humidity is a valuable indicator of atmospheric moisture, representing the percentage of moisture in the air compared to its total capacity at a certain temperature and pressure. It is also the ratio of actual vapor pressure to saturation vapor pressure multiplied by 100. Relative humidity ranges from 100% when saturated to close to zero for very dry air and is influenced by the air's moisture content, temperature, and pressure. It is important to consider the relationship between relative humidity and temperature. For example, air at 80°F with 24% relative humidity (Figure 5.3).

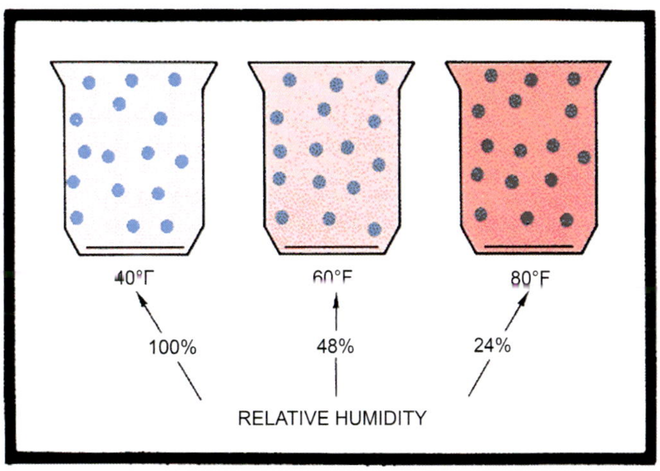

Figure 5.3. Relative humidity decreases as temperature increases even though the amount of water vapor in the air remains the same.

Source: https://www.frames.gov/documents/behaveplus/publications/Schroeder_and_Buck_1970_PMS-425-1_NFES-1174_Fire_Wx_ch_03.pdf.

5.2. PRECIPITATION

Precipitation is the process by which liquid or frozen water forms in the atmosphere and descends to the earth's surface. It can take the form of rain, sleet, or snow. Together with evaporation and condensation, precipitation is a key component of the worldwide water cycle.

When water vapor in the clouds condenses into larger droplets, precipitation forms. As the droplets become heavier, they eventually fall to the ground. In colder clouds, such as those at higher altitudes, water droplets may freeze and form ice crystals. These ice crystals can then fall to the ground as snow, hail, or rain, depending on the temperature within the cloud and at the Earth's surface. Most rain starts as snow at high altitudes in the clouds. As the snowflakes descend through warmer air, they melt and transform into raindrops.

Tiny particles of dust or smoke in the air are necessary for rain to form. Known as "condensation nuclei," these particles serve as a foundation for water vapor to condense. As a result, water droplets can merge and grow in size, eventually leading to precipitation.

One common misconception is that raindrops have a teardrop shape when falling. In reality, smaller raindrops are nearly perfectly spherical at around one millimeter in diameter. Larger raindrops, between two to three millimeters in diameter, also have a rounded shape with a small indent on their bottom side, resembling kidney beans as they fall. Very large raindrops, larger than 4.5 millimeters, have a significant indent and appear more like parachutes. These extra-large drops often split into two smaller droplets. The indents on raindrops are a result of air resistance.

All precipitation is fresh water, even if it comes from the ocean, as sea salt does not evaporate with water. However, pollutants in the atmosphere can contaminate water droplets, leading to acid rain. While acid rain may not directly harm humans, it can make bodies of water more acidic, which in turn can harm aquatic ecosystems by affecting the ability of plants and animals to adapt to the increased acidity. (Alexander LV 2016)

Precipitation plays a crucial role in the water cycle by supplying fresh water to the planet. Around 505,000 cubic kilometers (121,000 cu mi) of water falls as precipitation annually, with 398,000 cubic kilometers (95,000 cu mi) over oceans and 107,000 cubic kilometers (26,000 cu mi) over land. This results in a global average annual precipitation of 990 millimeters (39 in), but over land, it is only 715 millimeters (28.1 in). Climate classification systems like the Köppen system utilize mean annual rainfall to distinguish between different climate types. The effects of global warming are already altering weather patterns, leading to increased precipitation in some regions and decreased precipitation in others, resulting in more frequent extreme weather events (Table 5.1).

Table 5.1. Precipitation Size and Speed

	Intensity Inches/Hour (Cm/Hour)	Median Diameter (Millimeters)	Velocity of Fall Feet/ Second (Meters/ Second)	Drops Per Second Per Square Foot (Square Meter)
Fog	0.005 (0.013)	0.01	0.01 (0.003)	6,264,000 (67,425,000)
Mist	0.002 (0.005)	0.1	0.7 (.21)	2,510 (27,000)
Drizzle	0.01 (0.025)	0.96	13.5 (4.1)	14 (151)
Light rain	0.04 (0.10)	1.24	15.7 (4.8)	26 (280)
Moderate rain	0.15 (0.38)	1.60	18.7 (5.7)	46 (495)
Heavy rain	0.60 (1.52)	2.05	22.0 (6.7)	46 (495)
Excessive rain	1.60 (4.06)	2.40	24.0 (7.3)	76 (818)
Cloudburst	4.00 (10.2)	2.85	25.9 (7.9)	113 (1,220)

Source: https://www.usgs.gov/special-topics/water-science-school/science/precipitation-and-water-cycle#:~:text=Precipitation%20is%20water%20released%20from,Most%20precipitation%20falls%20as%20rain.

5.2.1. Types of Precipitation

Precipitation plays a crucial role in the water cycle by contributing the majority of the planet's fresh water. Annually, around 505,000 km^3 (121,000 cu mi) of water falls as precipitation, with approximately 398,000 km^3 (95,000 cu mi) of it occurring over the oceans. With the Earth's surface area taken into account, this results in an average global annual precipitation of 990 millimeters (39 in).

Different mechanisms for precipitation include convective, stratiform, and orographic rainfall. Convective processes involve strong vertical motions that can result in rapid overturning of the atmosphere and heavy precipitation within a short period. On the other hand, stratiform processes involve weaker upward motions and less intense precipitation. Precipitation can be classified into three categories based on whether it falls as liquid water that freezes upon contact with the surface, or ice. It is also possible for different types of precipitation to fall simultaneously. Liquid forms of precipitation include rain and drizzle, while frozen forms include snow, ice needles, ice pellets, hail, and graupel. Freezing rain or freezing drizzle occurs when rain or drizzle freezes upon contact with a subfreezing air mass.

1. Liquid Precipitation: Rainfall is typically measured using a rain gauge and can be expressed in millimeters (mm) of height or depth. Alternatively, it can be described as a volume of water per collection area in liters per square meter (L/m^2). This can be simplified to just "mm" because 1 liter is equivalent to 1 cubic decimeter, or 1 millimeter Times Square meter. The measurement can also be expressed as an area density in kg/m^2, assuming that 1 liter of water weighs 1 kg. In the English system, rainfall is usually measured in inches. In Australia pre-metrication, rainfall was also measured in "points," where each point represented one-hundredth of an inch.
2. Solid Precipitation: A snow gauge is commonly used to measure the amount of solid precipitation, typically snowfall. This is done by allowing snow to collect in a container and then measuring the height in centimeters. The snow can then be melted if desired to determine the water equivalent in millimeters, similar to measuring liquid precipitation. The relationship between snow height and water equivalent may vary based on the snow's water content, so the water equivalent measurement may only give a rough estimate of snow depth. Other types of solid precipitation, like snow pellets, hail, and sleet, can also be melted and measured for their respective water equivalents in millimeters, similar to liquid precipitation.

5.2.2. Air Becomes Saturated

5.2.2.1. Cooling Air to Its Dew Point

The dew point is the temperature at which air becomes saturated and condenses into water. Water vapor typically condenses on particles like dust, ice, and salt to create clouds. The concentration of cloud condensation nuclei affects cloud microphysics. An area of high pressure in the frontal zone causes widespread lifting, resulting in the formation of cloud layers like altostratus or cirrostratus. Stratus clouds, on the other hand, develop when a cool and stable air mass is trapped beneath a warmer air mass, or when advection fog is lifted during windy conditions.

The four primary methods of bringing the air down to its dew point are radiational cooling, evaporative cooling, conductive cooling, and adiabatic cooling. As air rises and expands, adiabatic cooling takes place. Convection, broad atmospheric motions, or a physical barrier like a mountain can cause the air to rise (orographic lift). When air is blown from one surface to another, such as from a liquid water surface to colder land, it comes into contact with a colder surface and conducts cooling. Either the air or the surface beneath it emits infrared radiation, which causes radiational cooling. When moisture is introduced to the air through evaporation, the air temperature is forced to drop to its wet-bulb temperature or until saturation, resulting in evaporative cooling.

5.2.2.2. Adding Moisture to the Air

The main processes that contribute to the addition of water vapor to the atmosphere are transpiration from plants, precipitation or virga falling from above, wind convergence into areas of upward motion, daytime heating that evaporates water from the surface of oceans, water bodies, or wetland, and lifting air over mountains.

5.2.3. Forms of Precipitation

5.2.3.1. Raindrops

Water droplets can coalesce to form larger droplets by fusing or by freezing onto an ice crystal, a process called the Bergeron process. Very small droplets do not fall from the sky because their fall rate is so small; precipitation only happens when these droplets combine to form larger drops. Different-sized droplets will collide and produce larger droplets due to their differing terminal velocities. Turbulence will intensify the collision process. Coalescence continues as these bigger water droplets descend, making the drops heavy enough to overcome air resistance and fall as rain.

Raindrops have a mean diameter that ranges from 5.1 to 20 millimeters (0.20 to 0.79 in), beyond which they tend to fragment. Cloud droplets are smaller droplets with a spherical shape. A raindrop's shape becomes more oblate as it gets bigger, with its largest cross-section facing the direction of the airflow. Their shape is not like a teardrop, in contrast to cartoon depictions of raindrops. Rainfall duration and intensity are typically inversely correlated, i.e., low-intensity storms can last for a long time, while high-intensity storms are usually brief. Compared to other raindrops, hail-related raindrops are often larger. Rainfall is coded with METAR as RA, and rain showers are coded with SHRA.

5.2.3.2. Ice Pellets

A type of precipitation known as sleet or ice pellets is made up of tiny, transparent ice balls. Hailstones are larger than ice pellets, though this isn't always the case. They usually do not freeze into a solid mass unless combined with freezing rain, and they frequently bounce when they hit the ground. Ice pellets have the METAR code PL.

When there is a layer of air above freezing and air below freezing, ice pellets are created. This results in any snowflakes that fall through the warm layer melting completely or partially. They refreeze into ice pellets as they descend back into the sub-freezing layer nearer the surface. On the other hand, freezing rain will occur at the surface if there is insufficient time for the precipitation to refreeze if the sub-freezing layer beneath the warm layer is too thin. During the cold season, a temperature profile indicating a warm layer above the ground is usually found ahead of a warm front, though it is also sometimes found behind a passing cold front.

5.2.3.3. Hail

Hail is produced in storm clouds, just like other precipitation, when supercooled water droplets freeze when they come into contact with condensation nuclei like dirt or dust. Hailstones are propelled to the upper clouds by the storm's updraft. As the updraft disappears, the hailstones descend, re-enter the updraft, and then rise once more. A hailstone is at least five millimeters (0.20 in) in diameter. When referring to larger hail, METAR code GR designates hail with a minimum diameter of 6.4 millimeters (0.25 in). Grêle is the French word from which GR is derived. The code for smaller hail and snow pellets is GS, a contraction of the

French word grésil. One of the most common hail sizes that are reported is stones that are slightly bigger than a golf ball. Hailstones can weigh more than 500 grams (1 lb) and reach a maximum size of 15 centimeters (6 in). Large hailstones may have their outer shell melted by latent heat released by additional freezing. The hailstone may then experience "wet growth," in which smaller hailstones are gathered by the liquid outer shell. With each ascent, the hailstone accumulates an ice layer and gets bigger. A hailstone drops from the cloud when it gets too heavy for the storm's updraft to support.

5.2.3.4. Snowflakes

Tiny droplets of supercooled clouds, approximately 10 μm in diameter, freeze and eventually form snow crystals. A droplet grows in the supersaturated environment after it has frozen. The ice crystals can enlarge to hundreds of micrometers at the expense of the water droplets because the water droplets are more numerous than the ice crystals. The Wegener-Bergeron-Findeisen process is the name given to this procedure. The droplets evaporate as a result of the corresponding decrease in water vapor, which means that the ice crystals expand at the droplets' expense. Because of their mass, these massive crystals fall through the atmosphere and have the potential to collide and form aggregates or clusters when they do. This makes them an effective source of precipitation. These aggregates, which fall to the ground as ice particles, are called snowflakes. According to Guinness World Records, the largest snowflakes ever recorded occurred in January 1887 at Fort Keogh, Montana; one of them is said to have measured 38 cm (15 in) in width. Research is still ongoing to determine the precise workings of the sticking mechanism.

Even though the ice is clear, the small ice particles' diffuse reflection of the entire spectrum of light causes the crystals to frequently appear white due to light scattering by the crystal facets and hollows/imperfections. The temperature and humidity at which a snowflake forms largely determine its shape. Rarely, angular snowflakes with threefold symmetry can form at a temperature of about –2°C (28°F). Although near-perfect snowflakes may appear more frequently in photographs due to their greater visual appeal, the most common snow particles are asymmetrical. Since snowflakes grow at different rates and in different patterns based on the temperature and humidity changes in the atmosphere they fall through on their way to the ground, no two snowflakes are alike. Snowfall is coded SN on the METAR, and snow showers are coded SHSN.

5.2.3.5. Diamond Dust

At temperatures approaching –40°C (–40°F), diamond dust, also referred to as ice needles or ice crystals, forms when slightly more humid air from above mixes with colder air from the surface. They are composed of basic hexagon-shaped ice crystals. In international hourly weather reports, diamond dust is identified by the METAR code IC.

5.2.4. Causes

5.2.4.1. Frontal Activity

Stratiform or dynamic precipitation develops when air rises slowly in synoptic systems (about cm/s), over cold fronts that are on the surface and over warm fronts that are ahead of them. Comparable ascent is observed in comma-head precipitation patterns surrounding mid-latitude cyclones and around tropical cyclones outside of the eyewall. Along an occluded front, a wide range of weather conditions can be found. Thunderstorms are possible, but their passage is typically accompanied by a drying of the air mass. Mature low-pressure areas are typically the sites of occluded front formation. On celestial bodies other than Earth, precipitation can happen. Instead of rain or snow, Mars likely experiences ice needles as precipitation during cold weather.

5.2.4.2. Convection

Convective clouds, such as cumulonimbus or cumulus congestus, produce convective rain, also known as showery precipitation. Rainfall occurs in sudden bursts of intensity. Convective clouds have a limited horizontal extent, so convective precipitation only lasts a short while over a given area.

Figure 5.4. Convective precipitation.

Source: https://upload.wikimedia.org/wikipedia/commons/5/57/Konvektionsregen.jpg.

Though it has been proposed that stratiform precipitation also happens, convective precipitation seems to account for the majority of precipitation in the tropics. Hail and Graupel are signs of convection. Convective precipitation occurs sporadically in mid-latitudes and is frequently linked to baroclinic boundaries like warm fronts, cold fronts, and squall lines. Mesoscale convective systems make up the majority of convective precipitation, which results in torrential downpours, thunderstorms, wind damage, and other severe weather conditions (Figure 5.4).

5.2.4.3. Orographic Effects

With an average annual rainfall of 12,000 millimeters (460 in), Mount Wai⊠ale⊠ale on the Hawaiian island of Kauai is renowned for its intense precipitation. Storm systems bring heavy rains to the state from October through March. Each island's topography, which divides it into leeward (Kona) and windward (Ko⊠olau) regions according to location about the higher mountains, causes significant variations in local climates. Leeward sides are sunnier and drier with less rain and cloud cover than windward sides, which face the east to northeast trade winds and receive significantly more rainfall.

5.2.4.4. Snow

Extratropical cyclones, also known as windstorms in Europe, can produce extremely cold and hazardous weather along with significant snowfall and rain. Their winds can reach speeds of over 119 km/h (74 mph). As a result of the weak upward vertical motion of air over the frontal boundary, which condenses as it cools and produces precipitation within an elongated band that is wide and stratiform—that is, falling out of nimbostratus clouds—the band of precipitation associated with their warm front is frequently extensive. Within the poleward side of the elongated precipitation band, overrunning snow may occur when moist air attempts to push aside an arctic air mass. Poleward in the Northern Hemisphere refers to the direction of the North Pole, or north. Poleward in the Southern Hemisphere refers to a direction that is south of the South Pole.

Narrow lake-effect snow bands can occur southwest of extratropical cyclones due to curved cyclonic flow that carries cold air across comparatively warm water bodies. Strong localized snowfall is brought about by those bands for the following reason: large bodies of water, like lakes, effectively store heat, resulting in notable temperature differences (greater than 13°C or 23°F) between the water's surface and the air above. This temperature differential causes warmth and moisture to be carried upward, where they condense into clouds that are oriented vertically (see satellite image), creating snow showers. Both the water's temperature and the surrounding environment have a direct impact on the temperature drop with height and cloud depth. Deeper clouds and a higher rate of precipitation result from a stronger temperature drop with height.

When air is forced to ascend mountains and squeeze out precipitation along their windward slopes, which falls as snow in cold weather, heavy snowfall accumulates in mountainous areas. The terrain is so rough that predicting where heavy snowfall will fall is still very difficult.

5.2.4.5. Within the Tropics

The time of year that includes one or more months when the majority of a region's average yearly rainfall occurs is known as the wet, or rainy, season. Tourist bureaus occasionally use the term "green season" as a euphemism. Parts of the tropics and subtropics are home to a variety of locations with wet seasons. Rainy summers and dry winters are common in savanna climates and regions subject to monsoon patterns. Since their rainfall is evenly distributed throughout the year, tropical rainforests technically do not have dry or wet seasons. While the intertropical convergence zone or monsoon trough moves poleward of their location in the middle of the warm season, some places with prominent rainy seasons will experience a break in rainfall in the middle of the season. Rainfall mostly occurs in the late afternoon and early evening during the wet season, which falls during the warm season, or summer. During the wet season, freshwater quality, vegetation growth, and air quality all significantly improve. Erosion rises and soil nutrients decrease.

Animals have survival and adaptation mechanisms for the wetter regime. Due to the immature crops from the previous dry season, there are food shortages during the wet season. Due to food shortages experienced before the first harvest, which takes place late in the wet season, developing nations have observed that their populations exhibit seasonal weight fluctuations.

Large air masses several hundred miles across, with low pressure at its center and winds blowing inward towards it either clockwise (southern hemisphere) or counterclockwise (northern hemisphere) are the components of tropical cyclones, a source of extremely heavy rainfall. Even though cyclones can cause a great deal of damage to property and human lives, they may also play a significant role in the precipitation regimes of the areas they affect because they can provide much-needed precipitation to otherwise dry areas. A tropical cyclone can bring with it a year's worth of rain to areas it passes over.

5.2.5. Measurement

The standard rain gauge, available in 100

mm (3.9 in) plastic and 200 mm (7.9 in) metal varieties, is used to measure rainfall or snowfall. Rainfall totaling 25 mm (0.98 in) fills the inner cylinder, with overflow into the outer cylinder. Metal gauges require the use of a stick designed with the appropriate 0.25 mm (0.0098 in) markings, whereas plastic gauges have markings on the inner cylinder down to 0.25 mm (0.0098 in) resolution. The amount inside the inner cylinder is discarded after it is filled, and the outer cylinder is then filled with the remaining rainfall until all of the fluid inside is gone, adding to the total until the outer cylinder is empty. When using these gauges in the winter, the inner cylinder and funnel are removed, allowing snow and freezing rain to gather inside the outer cylinder. To avoid having to melt snow or ice that falls into their gauge, some people add anti-freeze to it. The snowfall/ice can be brought inside to melt once it has finished building up or as 300 mm (12 in) approaches. Alternatively, lukewarm water can be used to fill the inner cylinder and melt the frozen precipitation in the outer cylinder, recording the warm fluid added, which is then deducted from the total amount once all of the ice/snow has melted.

Additional gauge varieties include the widely used wedge gauge, which is also the most affordable and brittle, the tipping bucket rain gauge, and the weighing rain gauge. In the case of snow, the wedge and tipping bucket gauges malfunction. There has been little success in trying to warm the tipping bucket to compensate for snow or ice because snow can sublimate if the gauge is kept significantly above freezing. Snow should not cause any problems for antifreeze-filled weighing gauges, but once more, the funnel needs to be taken out before the event starts. If left outside, a can with straight sides and a cylindrical shape will function as a rain gauge for the least amount of money; however, the accuracy of the measurement will depend on the type of ruler used. With enough knowledge, any of the aforementioned rain gauges can be constructed at home.

There are a number of networks, including CoCoRAHS and GLOBE, that allow rainfall measurements to be submitted online when they are taken, both domestically and internationally. The closest local weather office will probably be interested in the measurement if there isn't a network in the area where the person lives.

5.2.5.1. Hydrometeor Definition

The hydrometeor is a concept used in the measurement of precipitation. Hydrometeors are any particles of liquid or solid water in the atmosphere. Hydrometeors are the building blocks of condensation-related formations like clouds, haze, fog, and mist. By definition, all forms of precipitation—including virga, or precipitation that evaporates before hitting the ground—are composed of hydrometeors. Hydrometeors are particles that are carried off the surface of the Earth by wind. Examples of these particles include snow and hail.

5.2.5.2. Satellite Estimates

While surface precipitation gauges are widely accepted as the gold standard for measuring precipitation, their use is not practical in many areas. This applies to both isolated land areas and wide stretches of ocean. In other situations, the distribution of gauge observations is hampered by administrative, technical, or social problems. As a result, satellite observations are crucial to the current global record of precipitation.

By recording different electromagnetic spectrum frequencies that are demonstrated to be correlated with both the frequency and intensity of precipitation in theory and practice, satellite sensors can remotely sense precipitation. Unlike active sensors (radar, lidar), which send out a signal and detect its impact on the area being observed, the sensors are almost entirely passive and record what they see, much like a camera.

There are two types of satellite sensors for precipitation that are currently in use. The primary information provided by thermal infrared (IR) sensors is about cloud tops. They record a channel with a wavelength of about 11 microns. Colder clouds nearly always occur at higher altitudes because of the general inverse relationship between cloud-top temperatures and cloud-top heights caused by the typical structure of the atmosphere. Furthermore, compared to clouds with smooth tops, those with a lot of small-scale variation are probably more vigorous. These and other characteristics are used by a variety of mathematical models, or algorithms, to calculate precipitation from infrared data.

The microwave region of the electromagnetic spectrum contains the second class of sensor channels. The frequencies that are utilized span from several hundred GHz to roughly 10 gigahertz. Information on liquid hydrometeors (rain and drizzle) in the lower regions of clouds is mainly available on channels up to roughly 37 GHz; larger volumes of liquid release more microwave radiant energy. While emission signals can be seen on channels above 37 GHz, solid hydrometeor activity (snow, glaciers, etc.) predominates.) to disperse radiant microwave energy. Microwave sensors are used by satellites like the Global Precipitation Measurement (GPM) mission and the Tropical Rainfall Measuring Mission (TRMM) to estimate precipitation.

It has been shown that additional sensor channels and products—such as visible, infrared, water vapor, and atmospheric sounding retrievals—provide additional useful information. The majority of precipitation data sets in use today, however, do not make use of these data sources.

5.2.5.3. Satellite Data Sets

Although the infrared estimates are not very accurate at short time and space scales, they are available from satellites in geosynchronous Earth orbit at least once every 15 minutes. In situations where stratiform (layered) precipitation predominates, such as in the tropics, infrared radiation performs best. It is less effective in areas where stratiform precipitation is less common, particularly in mid- and high-latitude regions. The microwave estimates are more accurate on short time and space scales than infrared (IR) due to the more direct physical connection between hydrometeors and microwave channels. Nevertheless, only low-Earth orbit satellites are equipped with microwave sensors, and there are so few of these that the average interval between observations is longer than three hours. Due to the transient nature of most precipitation systems and the incapacity of a single satellite to accurately capture the typical daily cycle of precipitation at a given location, this several-hour interval is insufficient to adequately document precipitation.

Since the late 1990s, numerous algorithms have been created to merge precipitation data collected by multiple satellites, focusing on highlighting the strengths and reducing the weaknesses of each dataset. The objective is to produce the most accurate estimates of precipitation on a consistent time and space grid, covering as much of the world as feasible. Some efforts also prioritize the long-term consistency of the dataset, meeting the Climate Data Record criteria.

In some cases, the main objective is to generate the most accurate immediate satellite estimate, known as the High-Resolution Precipitation Product method. In any scenario, the secondary goal is also seen as desirable. One important finding from studies involving multiple satellites is that incorporating even a small amount of surface gauge data can help mitigate the biases that often come with satellite estimates. However, utilizing gauge data presents challenges as it is limited in availability and typically takes at least two months for transmission, assembly, processing, and quality control. This results in precipitation estimates with gauge data being delayed compared to estimates without gauge data. While estimates with gauge data may offer a more precise representation of actual precipitation, they are generally not ideal for real-time or near-real-time applications.

5.2.6. Changes Due to Global Warming

Rising temperatures can cause higher evaporation rates, resulting in increased precipitation. From 1900 to 2005, precipitation has generally risen over land areas located north of 30°N but has decreased in tropical regions since the 1970s. There has been no significant global trend in precipitation over the past century, with varying trends observed by region and over time. A study conducted in 2018 using a detailed global precipitation dataset spanning over 33 years found that although there are regional fluctuations, there is no indication of a widespread increase in global precipitation in response to climate change.

Because of their particular circumstances, every part of the world will experience variations in precipitation. Wetter regions include eastern North and South America, northern Europe, and northern and central Asia. It is now drier in the Sahel, the Mediterranean, southern Africa, and some parts of southern Asia. Over the past century, many regions have seen an increase in the frequency of heavy precipitation events. Additionally, since the 1970s, droughts have become more common, particularly in the tropics and subtropics. Reduced salinity in mid- and high-latitude waters (which implies more precipitation) and increased salinity in lower latitudes (which implies less precipitation, more evaporation, or both) are indicators of changes in precipitation and evaporation over the oceans. Since 1900, total annual precipitation in the contiguous United States has increased at an average rate of 6 1% per century. The East North Central climate region has experienced the largest increases in precipitation, at 11 1% and 11 6% per century, respectively. The only area to exhibit a decline (–9.25%) was Hawaii.

5.2.7. Forecasting

The anticipated total amount of liquid precipitation accumulated over a given period over a specified area is known as the Quantitative Precipitation Forecast, or QPF for short. When any hour during a QPF valid period is predicted to have measurable precipitation reaching a minimum threshold, a QPF will be specified. Forecasts of precipitation are typically constrained by synoptic hours, which are GMT 0000, 0600, 1200, and 1800. In QPFs, the terrain is taken into account using topography or climatological precipitation patterns derived from meticulously detailed observations. QPFs were incorporated into hydrologic forecast models to simulate the effects on rivers across the United States beginning in the mid-to late-1990s. The sensitivity of forecast models to humidity levels in the lowest layers of the atmosphere, known as the planetary boundary layer, is significant and decreases with height. Quantitative probability forecasting (QPF) is based on predicting amounts, while qualitative QPF forecasts the probability of a given amount. Within six to seven hours of the radar image, forecasting techniques based on radar imagery demonstrate a higher level of skill than model forecasts. Rain gauge measurements, weather radar estimates, or a combination of the two can be used to validate the forecasts. The value of the rainfall forecast can be measured using different skill scores (Tapiador et al., 2019; Bauer et al., 2002).

5.3. SOURCES OF ATMOSPHERIC MOISTURE

Three processes account for nearly all of the water vapor in the atmosphere: transpiration from plants, evaporation from soil, and evaporation from any moist surface or body of water. Combustion produces some water vapor. Although land sources can also play a significant role locally, the oceans are the primary source of moisture because they comprise over three-fourths of the planet's surface.

Large surfaces allow for transpiration in plants; sometimes, a plant's surface area can reach 40 square yards for every square yard of ground area. Up to eight times as much moisture can be added to the atmosphere through transpiration from dense vegetation as from an equivalent area of bare ground. The growth activity has a significant impact on the amount of moisture that transpired. In turn, this growth activity typically changes according to the time of year and the availability of groundwater. At the end of the dry season, transpiration and evaporation may be almost nonexistent in regions with little vegetation and insufficient rainfall, like many parts of the arid West. At timberline and at latitudes in the Far North, this might also be typical.

The rate at which moisture is released into the air during evaporation from soil, dead plant matter, and bodies of water varies depending on the difference in vapor pressure between these surfaces and the atmospheric vapor pressure. As long as the vapor pressure at the evaporating surface is higher than the vapor pressure in the atmosphere, evaporation will continue. As the pressure differential rises, so does the rate of evaporation. The temperature of the evaporating surface affects the vapor pressure there. Therefore, assuming that the atmospheric vapor pressure remains constant, evaporation from the surfaces of warm water bodies, warm soil, and dead plant material will be greater than from cold surfaces (Figure 5.5).

The process of transpiration from living plants is identical to that of evaporation from dead plant matter. Living things typically transpire at their fastest rates in warm weather, but on very hot and dry days, an internal regulating mechanism tends to restrict the rate of water loss to the specific requirements of the plant at that moment. In the fuel moisture chapter, we will go into greater detail about transpiration from living plants and evaporation from dead plant material. Water vapor concentrates close to the evaporating surface when it evaporates in still air. Even though the surrounding air is relatively dry, further evaporation will virtually stop if this concentration approaches saturation. By removing these stagnant layers and substituting them with drier air, wind promotes evaporation. The impact of wind on evaporation lessens when a surface dries to the point where free water is no longer exposed to the air. Wind may help reverse the process for surfaces like relatively dry soil or wood by lowering the vapor pressure of moisture present on these surfaces as they cool.

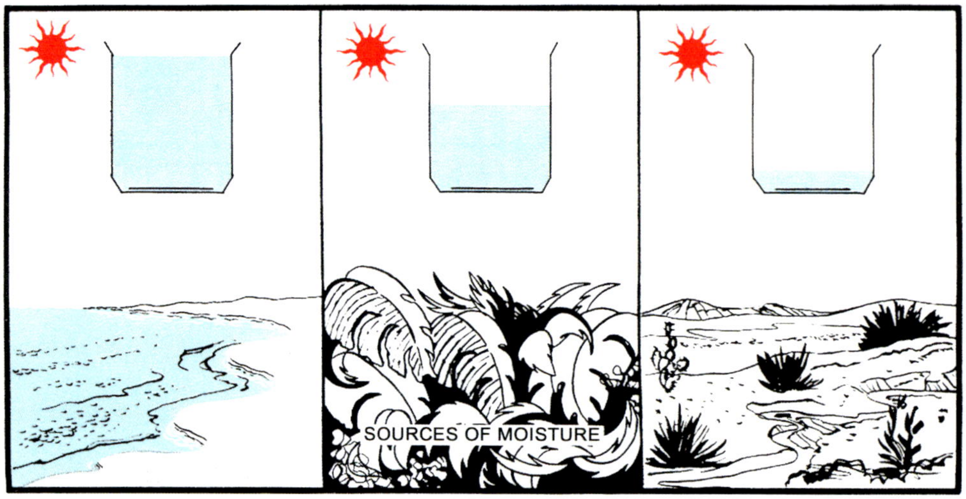

Figure 5.5. Sources of moisture.

Source: https://www.frames.gov/documents/behaveplus/publications/Schroeder_and_Buck_1970_PMS-425–1_NFES-1174_Fire_Wx_ch_03.pdf.

5.4. VARIATIONS IN ABSOLUTE HUMIDITY

There will be continuous variations in time and space within an air mass, and the actual amount of moisture in the air will differ from one air rental to the next.

The regions of origin of air masses are essentially related to the amount of moisture they contain. Air masses with continental origins tend to be dry. Those originating from the Pacific are moist or moderately moist, while those originating from the Atlantic or the Gulf of Mexico are moist. The land stations will witness sudden increases in absolute humidity as these marine air masses invade the continent. Any air mass that travels through regions other than its source experiences slow changes due to the addition or removal of moisture through evaporation, transpiration, condensation, and precipitation.

Similar to temperature, absolute humidity typically drops with height through a deep layer within an air mass. This dispersion is due to a few factors. Initially, moisture rises to the surface of the atmosphere due to convection as well as upslope and up valley winds. Second, as air is raised, both the air and the water vapor expand proportionately, reducing the amount of moisture in each volume. As the air is raised, the absolute humidity correspondingly drops. Third, the air's ability to hold moisture decreases upwards as temperature typically rises. Ultimately, condensed moisture from higher atmospheric levels is removed by the precipitation process and deposited at the surface.

Now and then, horizontal flow at intermediate altitudes above can bring in moist air, changing the typical pattern of moisture decrease with altitude. Much of the summer thunderstorm activity over much of the West is caused by this kind of flow. The sinking air above has extremely low absolute humidity. This arid air begins close to the troposphere's summit and gradually descends to lower altitudes. It may cause an abrupt rise in the risk of fire and extremely low humidity close to the surface if it falls to the ground or is mixed downward.

If we only take into account a very thin layer of air close to the surface, we discover that the absolute humidity's vertical variation with height will vary every 24 hours as conditions that favor condensation and evaporation alternate. On clear days, evaporation from warm surfaces typically adds moisture to the air, which causes the absolute humidity to rise. The absolute humidity can rise through a very thin layer at night because moisture is typically removed from the air near the surface by condensation on cold surfaces and absorption by cold soil and other materials (Bretherton et al., 2004; Ahmed et al., 2015).

5.5. DIURNAL AND SEASONAL CHANGES IN RELATIVE HUMIDITY

Compared to absolute humidity, relative humidity is far more variable. It frequently varies significantly and quickly between locations and from hour to hour. Because it is dependent on both air temperature and absolute humidity, relative humidity is far more erratic. It is inversely correlated with temperature and directly varies with moisture content. These relationships make it difficult to draw broad conclusions about variations in relative humidity, especially when those variations are vertical and occur at relatively short heights above the ground.

Both the temperature and the absolute humidity typically drop with height during the day near the surface, especially when there are clear skies. The relative humidity is affected by these two variables in opposing ways. The dominating effect is contingent upon the level of dryness on the surface. Over normal surfaces, the relative humidity typically rises with height because the effect of temperature decrease outweighs the effect of absolute humidity decrease. Relative humidity in the surface layer will drop with height on a moist surface, though the effect of the temperature drop may be greater than that of the absolute humidity drop.

At night, the temperature changes with height are typically more pronounced, and through the lowest layers, the relative humidity will decrease with height.

Over most of the lower troposphere above the lowest layers, the relative humidity typically rises with daytime height. Only convection could explain this increase. The temperature drops by 5.5°F for every 1,000 feet of airlift, while the dew point drops by roughly 1°F. As a result, the temperature and dew point get closer by 4.5°F for every 1,000 feet, and the relative humidity rises until saturation is attained.

A subsidence inversion is created when a layer of air in the troposphere that is sinking warms due to adiabatic processes. The temperature inversion at the base of the subsiding layer will cause the relative humidity to rise. For instance, the marine inversion along the west coast is an inversion caused by subsidence. The adiabatically heated subsiding air mass above has higher temperatures and lower humidities than the marine air below, which is colder and more humid. When the marine inversion is present, the slopes of coastal mountains exhibit a noticeable change in temperature and humidity. The layer immediately below the surface, where it affects both fuels and fire behavior, is where relative humidity plays a major role in the fire weather. Significant variations in relative humidity near the ground are caused by air moisture content, season, time of day, slope, aspect, elevation, clouds, and vegetation.

We will take into account variations at that level and deduce the conditions near the surface around forest fuels based on our knowledge of surface temperatures since hourly and daily changes in relative humidity are typically measured in a standard instrument shelter.

On a hygrothermograph exposed in a shelter at a valley station or one in flat terrain, a typical fair-weather pattern of relative humidity is almost exactly the opposite of the temperature pattern. When the temperature is at its lowest, around daybreak, maximum humidity usually happens. Humidity rapidly decreases after sunrise, reaching a minimum around the moment of maximum temperature. Through the night, it rises more gradually in the late afternoon. When the daily temperature range is largest, the daily range of humidity is also typically greatest. Humidity traces within an air mass typically vary little from day to day, primarily due to temperature variations. However, as the air mass progressively absorbs or releases moisture, there might be discernible cumulative variations in humidity over a few days.

The shift in humidity patterns throughout the seasons is visible. In regions experiencing fire-prone weather, the transition from a wet spring to a hot summer and early autumn is striking. The largest temperature fluctuations occur at the beginning of the fire season, with clear night skies and intense sunlight overhead. During this time, the abundant moisture in the soil and vegetation results in high nighttime humidity levels, sometimes reaching close to 100%. However, as the day progresses, intense surface heating and the upward movement of moisture lead to a decrease in relative humidity levels, often reaching very low levels in the afternoon. As the season goes on, the sun moves further south, drying out the soil and vegetation and reducing solar heating. Late in the season, daytime humidity drops even more, but the daily range narrows and fire weather gets even more intense due to a larger decrease in nighttime humidity. Summertime showers may halt this progression occasionally, but they don't significantly alter the seasonal pattern as a whole.

The daily temperature swings are usually less noticeable in places with distinct spring and fall fire seasons. Additionally, aside from unusual droughts, the cumulative drying of the soil and vegetation is not very consistent. The seasons typically bring with them sporadic rains, which makes humidity variations fairly erratic. Seasonal increases in relative humidity reduce summertime fire danger in certain places. Daytime relative humidities do not fall to as low values in the same air mass types as they do in the spring and fall, especially in the Great Lakes region where the numerous small lakes get quite warm during the summer and transpiration from vegetation peaks.

Particularly in areas that are not shaded, where soil and surface fuels exposed to the sun heat up intensely and warm the air around them, the relative humidity that impacts fuels on the forest floor is frequently very different from that in the instrument shelter. Although the air in the instrument shelter has a slightly higher dew point than this one, the air here is much warmer, resulting in a significantly lower relative humidity.

Field instruments are not practical for measuring humidity near the ground; however, tables can be used to estimate the humidity from psychrometric readings at the standard height and a surface dry-bulb temperature reading.

5.6. FORMATION AND TYPES OF CLOUDS

Clouds play a crucial role in determining how the environment's climate changes. Furthermore, clouds play a crucial role in preserving and regulating the earth's radiation. Both short- and long-wavelength radiation varies according to cloud cover and other climate-related factors. Many scientific disciplines have focused heavily on cloud formation, demonstrating that clouds are much more complex phenomena than just structures that determine the kind of rainfall that can occur in a given area.

When the atmospheric conditions are just right, condensation happens when water vapor and aerosol particles mix. As a result, clouds form because the warmer air rises with the vapor to higher altitudes where the temperature is lower.

5.6.1. The Physics Behind Cloud Formation

The basic analysis of weather based on cloud structure is far less complex than the physics of cloud formation. The concept of Cloud Physics is derived from two main perspectives.

- Cloud-Scale: The cloud evolution is determined based on the size, the environment surrounding the cloud, wind shear, turbulence, and temperature change.
- Micro Scale: The changes occurring within a few cm of the cloud surface help in determining the cloud scale. This small fractional scale of measurement and observation is known as the Micro Scale

5.6.2. Mechanism of Formation of Clouds

Tiny water droplets and ice particles make up clouds. They can float in the atmosphere because they are sufficiently light. Many different types of atmospheric phenomena can result in clouds.

- Evaporation: Water from the surface of the earth evaporates due to rising temperature and starts to travel higher with the wind. With rising altitude, the temperature falls along with the air pressure. The water is unable to retain its vaporized state and changes into ice crystals and droplets. The combination of these tiny icicles and water droplets from the clouds.
- Condensation Nuclei: Sometimes water vapor carries particulates like dust and pollens. When they reach high altitude, the vapor condenses on the particulate

which we know as condensation nuclei. Clouds then form around these nuclei.

- Sometimes the wind in mountainous areas is so strong that it forces the moist air to ascend to higher altitudes, which leads to the formation of clouds. This is known as the Leeward Side Phenomenon and the Windward Side Phenomenon.
- When a wind blows into an area of low pressure and is unable to leave, low-pressure clouds are created. This results in a drop in temperature and pressure, which produces clouds at mid-altitudes.
- Massive Clouds and Thunderstorms: These are created when cold air forces a significant amount of warmer air to ascend to higher altitudes, resulting in an abrupt shift in atmospheric pressure.

5.6.3. Classification of Cloud

The shape, size, and color of the clouds all contribute to the prediction of the next weather pattern. Generally speaking, clouds can be divided into ten categories according to the three altitudes at which they typically form (Figure 5.6) (Table 5.2).

Table 5.2. Classification of Cloud

High-Level Altitude	Mid-Level Altitude	Low-Level Altitude
Cirrostratus	Nimbostratus	Stratus
Cirrus	Altostratus	Stratocumulus
Cirrocumulus	Altocumulus	Cumulus
		Cumulonimbus

Source: https://www.noaa.gov/jetstream/clouds/four-core-types-of-clouds#:~:text=From%20his%20Essay%20of%20the,a%20fourth%20special%20type%2C%20nimbus.

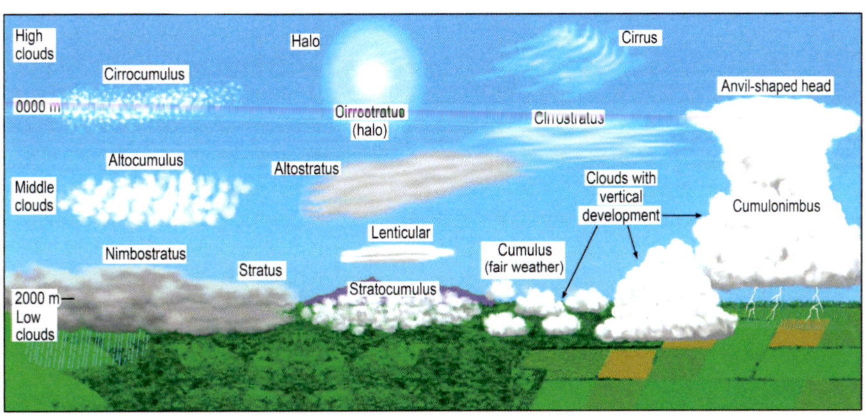

Figure 5.6. Different types of clouds.

Source: http://buzzar-brandi.blogspot.com/2016/03/cloud-types.html.

5.6.3.1. Cirrostratus

Cirrostratus is a type of stratiform cloud that is high aloft, extremely thin, and usually uniform. Ice crystals, which are fragments of frozen water, are used to make it. It can produce halos and is hard to detect. These are created when cirrostratus nebulosus, a thin form of cloud, forms. If the cloud is thicker cirrostratus fibratus, it has a fibrous texture without any halos. The cirrostratus frequently starts nebulous and transforms into fibratus as a frontal system approach. A weak front is typically indicated if the cirrostratus starts as scattered clouds in the sky. Typically, Cirrostratus can be found above 5,525 km (18,000 ft). Its existence suggests that the upper troposphere contains a significant amount of moisture. Sometimes, in the polar regions of the lower stratosphere, clouds that resemble cirrostratus form. Tiny supercooled water or nitric acid droplets can give the appearance of polar stratospheric clouds.

5.6.3.2. Cirrus

A genus of high clouds composed of ice crystals is called Cirrus. Typically, cirrus clouds have delicate, wispy white strands. Cirrus are typically created when dry, warm air rises and deposits water vapor on top of high-altitude rocky or metallic dust particles. Their global formation occurs at elevations ranging from 4,000 to 20,000 meters (13,000 to 66,000 feet) above sea level; the higher elevations are typically found in tropical regions, while the lower elevations are found in more polar areas.

The tops of thunderstorms and tropical cyclones can produce cirrus clouds, which can occasionally be used to forecast the arrival of storms or rain. Cirrus clouds themselves are nothing more than falling streaks of ice crystals, but they are a sign of impending rain and possibly storms. As they descend through drier, warmer air, these crystals disperse, melt, and evaporate; they never touch the ground. Earth is warmed by cirrus clouds, which may be a factor in climate change. Earth's warming temperature will probably lead to an increase in cirrus clouds, which could create a vicious cycle.

As light interacts with ice crystals in cirrus clouds, optical phenomena like sun dogs and halos can result. The terms "cirrostratus" and "cirrocumulus" refer to two additional high-level clouds that resemble cirrus. Whereas cirrocumulus resembles a pattern of tiny cloud tufts, cirrostratus resembles a sheet of clouds. Clouds with cirrocumulus contain droplets of supercooled (below freezing point) water, in contrast to clouds with cirrus and cirrostratus. In addition to Titan, one of Saturn's larger moons, cirrus clouds can be found in the atmospheres of Mars, Jupiter, Saturn, Uranus, and Neptune. Like water ice in cirrus on Earth, some of these alien cirrus clouds are composed of ammonia or methane ice. A subset of interstellar clouds known as cirrus is composed of dust grains that are smaller than one-thousandth of a millimeter.

5.6.3.3. Cirrocumulus

Three major genus types of high-altitude tropospheric clouds exist cirrus, cirrostratus, and cirrocumulus. Typically, they happen between 5 and 12 km (16,000 and 39,000 feet) above sea level. Cirrocumulus clouds, like stratocumulus and cumuliform clouds at lower altitudes, are indicative of convection. A tiny amount of liquid water droplets are present in cirrocumulus, in contrast to other high-altitude tropospheric clouds such as cirrus and cirrostratus, but they are supercooled. The main constituent is ice crystals, which usually cause the supercooled water droplets in the cloud to quickly freeze, turning the cirrocumulus into cirrostratus. Precipitation in the form of an ice or snow virga can also be produced by this process. Cirrocumulus clouds are therefore typically transient. They typically only form during a brief transitional phase inside a region of cirrus clouds, though they can also momentarily form when a cumulonimbus anvil splits apart.

5.6.3.4. Nimbostratus

Multilevel, amorphous, almost uniform, and frequently dark grey, nimbostratus clouds typically produce nonstop rain, snow, or sleet but not lightning or thunder.

Despite the fact that it is typically a low-based cloud, it actually forms in the middle troposphere and spreads vertically into both the low and high levels. Precipitation is typically produced over a large area by nimbostratus.

5.6.3.5. Altostratus

The cloud genus Altostratus consists of ice crystals, water droplets, or a combination of the two. It is found at intermediate altitudes. Large amounts of warm, humid air rise and cause water vapor to condense, forming altostratus clouds. Although some varieties have wavy or banded bases, altostratus clouds are typically featureless sheets of gray or blueish-gray. Thinner altostratus clouds allow for visibility of the sun, while thicker layers can be completely opaque.

Warm fronts are typically heralded by altostratus clouds. A warm front's altostratus clouds typically bring with them continuous rain or snowfall over the course of the next 12 to 24 hours. Altostratus clouds do not directly cause significant precipitation, but they do signal the arrival of warmer, wetter weather. Showers can be brought by thunderstorms that are embedded in altostratus clouds.

5.6.3.6. Altocumulus

The middle-altitude cloud genus Altocumulus is primarily classified as stratocumuli form, with globular masses or rolls in layers or patches; its constituent elements are smaller and darker than those of stratocumulus and larger than those of cirrocumulus. However, the altocumulus clouds take on a more purely cumuliform structure if the layers start to appear tufted as a result of increased airmass instability. Altocumulus clouds, like other stratocumuli form and cumuliform clouds, are indicative of convection. Sometimes, just before a weakening warm front, a sheet of partially conjoined altocumulus perlucidus is observed. This is where the altostratus is beginning to fragment, leaving patches of altocumulus perlucidus in between the altostratus areas. In a depression, altocumulus is also frequently observed between the warm and cold fronts, though it is frequently obscured by lower clouds.

Since towering cumulus clouds can transform into cumulonimbus clouds in the middle layers of the troposphere, where instability and convection occur, towering cumulus, also known as altocumulus castellanus, often heralds the arrival of thunderstorms later in the day. Thus, it is one of the three warning clouds that the aviation industry frequently records; the other two are cumulonimbus and towering cumulus. Similar to altostratus formations, altocumulus typically forms at elevations of 2,000 to 6,100 m (6,600 to 20,000 ft) above the earth. Satellite imagery has shown that these two types of clouds can produce formations that span thousands of square miles. Large-scale altocumulus formations are sometimes called altocumulus mackerel sky, especially if they resemble undulates.

5.6.3.7. Stratus

Stratus clouds are low-lying clouds that have a flat, uniform base and are not formed by rising thermals like other cloud types. These clouds are described as hazy and featureless, ranging in color from dark gray to almost white. The term "stratus" comes from the Latin word "strato-" which means "layer." Stratus clouds can bring light rain or snow and are essentially fog that has lifted off the ground or formed due to cold air at low altitudes. They are sometimes referred to as "high fog" because of their fog-like appearance.

5.6.3.8. Stratocumulus

Occasionally referred to as a cumulostratus, a stratocumulus cloud is a type of cloud genus that is characterized by large, rounded, dark masses that are typically in groups, lines, or waves. Its components are larger than those of altocumulus clouds, and its overall height is typically below 2,000 meters (6,600 feet). Because of the drier, more stable air above, weak convective currents produce shallow cloud layers (also known as a sea of clouds), which stop vertical development from continuing. Because it combines two different types of clouds, this kind of cloud has historically been referred to as a "twain cloud" in English.

5.6.3.9. Cumulus

Cumulus clouds are characterized by their flat bases and are often compared to puffy, cotton-like, or fluffy in appearance. The name "cumulus" comes from the Latin word meaning "heap" or "pile." Typically found at low altitudes, cumulus clouds are generally less than 2,000 meters (6,600 feet) high unless they take on a more vertical cumulus congestus form. These clouds can be seen individually, in lines, or clustered formations.

When weather conditions like temperature gradient, humidity, and instability are present, cumulus clouds frequently precede other kinds of clouds, like cumulonimbus clouds. Although cumulus clouds typically produce little to no precipitation, they have the ability to develop into cumulus congestus or cumulonimbus clouds, which do produce precipitation. Depending on the surrounding temperature, ice crystals, supercooled water droplets, or water vapor can all combine to form cumulus clouds. They exist in a variety of unique forms and serve to cool the planet by reflecting solar radiation back into space.

The broader class of free-convective cumuliform clouds, which also includes cumulonimbus clouds, includes cumulus clouds. The latter genus type, which frequently has a cirriform or anvil top, is more complexly structured and is therefore sometimes classified as cumulonimbi form. Additionally, stratocumulus (low-étage), altocumulus (middle-étage), and cirrocumulus (high-étage) are cumuliform clouds with limited convection. Occasionally, these final three genus types are categorized as stratocumuli form apart.

5.6.3.10. Cumulonimbus

A dense, towering vertical cloud known as a cumulonimbus usually forms when water vapor condenses in the lower troposphere and rises to the top on strong, buoyant air currents. Water vapor above the lower parts of the cumulonimbus condenses into ice crystals, such as snow and graupel, which combine to form hail and lightning, respectively. These clouds can also be referred to as thunderheads when they occur as a thunderstorm. Squall lines, clusters, or solitary clouds can all produce cumulonimbus. These clouds can produce severe weather, including tornadoes, strong winds, and big hailstones, in addition to lightning. From overdeveloped cumulus congestus clouds, cumulonimbus clouds can develop into supercells. Cumulonimbus, or Cb, is its acronym (Priego et al., 2017; Neelin et al., 2022).

5.7. EFFECTS OF TERRAIN, WIND, CLOUDS, VEGETATION, AND AIR MASS CHANGES

Humidity can differ significantly between locations, primarily due to topographical factors. The humidity recorded at a well-exposed station may be quite representative of a sizable area in terrain that is primarily level to gently sloping. Local exceptions will exist in places like fields with irrigation, wooded areas with shade, and arid regions. Particularly during the day, circulation and mixing are typically adequate to mitigate local effects over comparatively short distances.

The effects of aspect and elevation become more significant in mountainous terrain, and humidity varies more than in more level areas. Compared to high elevations, low elevations warm up and dry out earlier in the spring. Seasonally speaking, south slopes are also more advanced than north slopes. Since the surface's stored moisture is reduced as the season goes on, cumulative drying tends to even out these variations, though they do not go away.

We are aware that in the open air, daytime temperatures typically drop with altitude. The temperature and dew point decreases with height on slopes resulting in higher relative humidity at higher elevations. However, generalizations are challenging due to the complicated pattern caused by the air next to the slopes heating, moisture being transported by upslope winds, and the frequent stratification of moisture into layers.

The temperature change with height is typically the opposite when nighttime cooling sets in. At the bottom, cold air that is streaming down the slopes builds up. More cooling happens during the night, and if the air gets too saturated in the morning, fog or dew can form. At the base of the slope, relative humidity may reach 100%. At the top of the temperature inversion, or the thermal belt, it may reach a minimum value. From there, it may rise marginally higher up the slope above the inversion.

South slopes have somewhat lower day relative humidity levels than north slopes during the summer, which coincides with the fact that they dry out more quickly due to their higher day temperatures. However, because of the better air mixing at these more exposed locations, the difference between the north and south slopes becomes insignificant at higher elevations. The north and south slopes' humidity differences become marginal at night.

The daily range of relative humidity is lowest at higher elevations and greatest in valley bottoms in most mountainous regions. Because of this, even though fires on lower slopes may burn more efficiently during the day, they frequently quiet down significantly at night as humidity rises. But because of the higher temperatures, higher humidity, and faster winds at higher elevations—especially in and above the thermal belt—fires may rage through the night.

Once more, generalizations should be used with caution. For instance, in the Pacific Coast Ranges during the summer, higher humidity levels are typically observed during the day on ridge tops than at night. This anomaly is caused by moisture being carried upward by slope winds during the day from the moist marine air layer. At night, moist air that hasn't been carried away into space returns to Earth.

Wind moves air away from hot and cold surfaces, evaporating water vapor and mixing it with the surrounding air to balance out temperature extremes. As a result, during windy periods, relative humidity diurnal ranges are lower than during calm periods. By combining air with varying temperatures and moisture contents, winds also lessen regional variations. On a calm night, patches of fog signify inadequate ventilation.

Relative humidity is influenced by clouds since they have a significant impact on heating and cooling. On cloudy days, the humidity will be higher; on cloudy nights, it will be lower. Clouds therefore significantly shorten the daily range. Any kind of precipitation increases relative humidity because it cools the air and provides moisture for the air to evaporate.

By transpiring and evaporating, vegetation contributes to air moisture and lowers surface temperatures—both of which have an impact on relative humidity locally. The additional benefit of a continuous forest canopy is that it reduces surface wind speeds and air movement-induced mixing.

Humidity variations between forest stands and open spaces typically correspond with the crown canopy's density. Humidity is typically lower at night and higher during the day under a closed canopy. There is an even stronger correlation between the higher daytime humidities and a green understory. When deciduous forests are leafless, their effects on humidity are negligible.

The difference in humidity between forest stands and forest openings is lessened by two factors. Drastic conditions restrict the amount of moisture available for transpiration and evaporation, and overcast skies restrict heat and cold.

Except the heated ground surface, openings up to 20 yards in diameter do not have daytime relative humidities that are significantly different from those under the canopy. Surface air from the nearby forest is drawn into these apertures, which act as chimneys for convective airflow, as was described in the previous chapter. Strong radiation combined with stagnation at night can result in locally high humidities in small openings.

In larger clearings, the daytime humidity is comparable to that of open country. However, if the airflow is restricted, temperatures may rise marginally higher than those at stations that are exposed, and humidity levels will drop accordingly. These might be five to twenty percent lower in the afternoon in the clearing than in a forest with plenty of shade. Nighttime humidity levels are typically higher than in the woods and comparable to those at exposed locations.

Depending on crown density, the humidity characteristics of open forest stands fall somewhere between those of exposed sites and closed stands. In arid weather, particularly following extended dry spells, the relative humidity variations between open and wooded areas gradually diminish.

Up until now, the relative humidity variations have only been discussed in terms of changes within an air mass. One of the characteristics of an air mass is its moisture content, as we shall see later in the chapter on air masses and fronts. Compared to air masses originating over continents, air masses originating over bodies of water will have higher moisture content.

One can anticipate a shift in absolute humidity when a front moves through and a new air mass arrives. However, the temperature of the air mass will have a significant impact on the change in relative humidity. Relative humidity can vary significantly when a cool, moist air mass is replaced by a warm, dry one, or the other way around. However, if the temperature of the new air mass is noticeably lower, the cool, dry air mass that replaces the warm, moist one might have a higher relative humidity.

The inversion layer marks the boundary between two distinct air masses along the west coast when a warm, dry, subsiding air mass sits on top of a lower marine layer. Extremely unusual relative humidity patterns are observed inland, where the inversion crosses the Coast Ranges. The inversion is typically higher during the day and lower at night in these inland regions; nevertheless, along the coastal lowlands, the opposite is typically true. Certain places on the slopes of the nearby mountains will experience marine air during the day and dry, subsiding air at night. As dry air from above descends the slopes, the relative humidity may start to rise in the late afternoon and early evening before abruptly falling to low levels. It has been reported that in the early evening, there can be abrupt drops in humidity of up to 70% (Zhang et al., 2022; Li et al., 2023).

CASE STUDY

MEASURING HUMIDITY

Most people are all too aware of the effects humidity can have on human comfort and health, especially when combined with high temperatures. Perhaps less well-known is how humidity can impact electrical systems. For example, high humidity can cause droplets of liquid water to condense on circuits, producing problems such as shorting and corrosion.

Many geographic regions are prone to humid conditions, particularly the tropics. But even in relatively temperate climates, high humidity can occur depending on altitude, proximity to bodies of water, and seasonal effects. In addition, the "microclimate" inside and immediately around an electrical cabinet can result in condensation-related issues. As a result, standards bodies such as the International Electrical Testing Association (NETA) often require that humidity data be included in test reports.

- Absolute vs Relative:

Simply stated, humidity is the amount of water vapor in the air. This is commonly measured in either of two ways:

- o Absolute Humidity (AH): It is the mass of water vapor present in a given volume of air. This is usually expressed in grams per cubic meter (g/m^3) and changes as the volume of air changes.
- o Relative Humidity (RH): It is the ratio of water vapor density (mass per unit volume) to the water vapor density at the saturation vapor pressure (the point at which the air can hold no more water vapor and liquid droplets begin to precipitate out; this is also known as the dew point). This is typically expressed as a percentage, and changes with air pressure and temperature.

Ø Humidity and People:

Maintaining an appropriate level of humidity is important for ensuring a comfortable and healthy indoor environment. Excessive humidity can make working difficult, especially if physical activity is involved. Less directly (but equally important), high humidity promotes the growth of mold which can cause respiratory issues. It can also cause peeling paint, rust on iron objects, and slippery surfaces slick with condensation. These and other factors can significantly affect your health and safety. In general, a humidity level between 30 and 40% is considered ideal for maximum comfort. To ensure the humidity remains within this range, facilities employ a variety of HVAC systems, including air conditioning, controlled mechanical ventilation, and dehumidifiers. To test their efficiency, it is important to periodically check the humidity level with hygrometers and other moisture-measuring instruments. In many cases, it also can be useful to monitor humidity continuously over an extended interval, to identify potential trends and spikes.

- Electronic: Humidity can have a detrimental effect on electrical systems and components. The most obvious issue is condensation, which results when RH reaches 100%. For example, droplets of water forming on a heatsink can wick into the housings of power modules. In live conductors, liquid can compromise insulation. The damage may remain even after the RH falls and the droplets evaporate, making it very difficult to troubleshoot and identify the cause of the problem.

Several factors can cause RH to reach the saturation point. One of the more common is temperature disparity. As the air cools, the amount of water vapor it can hold decreases. Thus a sudden cooling causes RH to rise rapidly.

For instance, an electrical system can be warmed by the waste heat of its components. A sudden change in operating state (for instance a power failure) can cause the system to cool. Components such as the heatsink may cool far more rapidly than the air temperature in the enclosure, creating a condition where the heatsink temperature falls below the dew point. Therefore, it is critical to be aware of any changes from operation from full to reduced power, such as equipment entering standby mode or an unexpected downtime.

Changes in air temperature can also cause internal condensation in electrical systems. Many heatsinks are cooled by outside ambient air. As the inlet air temperature drops in the evening the heatsink may be cooled to below the dew point. We, therefore, recommend monitoring both the interior and exterior RH over a period of days or weeks to understand how weather and system operation interact.

In addition to electrical conductance, humidity can also have a corrosive effect on many materials. Corrosion progresses rapidly when humidity exceeds 60%. And since corrosion also progresses as temperature increases, facilities in high humidity locations must carefully monitor (and prepare countermeasures for) the amount of water vapor in the air.

- Organic Components: A recent trend is to fabricate electronic components from organic materials, due to their reduced costs and ease of manufacture. Unfortunately, these materials present special humidity-related issues. This is because they tend to be water-permeable, slowly absorbing moisture until their internal water concentration matches the ambient air.

Excessive levels of moisture in organic materials can severely impact performance and reliability. One obvious example is so-called "popcorn" cracking, caused when saturated organic materials are suddenly heated to a high temperature. Other longer-term effects include swelling of components, which can compromise electrical connections.

Ø AEMC Thermo-Hygrometer Data Logger Model 1246:

To help you monitor humidity in your facility, AEMC provides the Thermo-Hygrometer Data Logger Model 1246. This is a portable, compact digital meter designed for simplicity and one hand operation. The instrument enables you to perform a variety of recording tasks with an easy and intuitive setup from a computer using supplied Data View® software. The meter uses an NTC as the temperature sensor, providing excellent response time to changes in temperature, good repeatability, and accurate readings. The Model 1246 also utilizes a thin film polymer capacitive type relative humidity sensor, offering excellent recovery from 100% moisture, and fast response time and durability.

Features include:

- Monitor and record temperature, humidity, and dew point;
- Dual line display toggles between any two measurements;
- User selectable temperature units (°F, °C);
- Built-in sensor with removable protective cap;
- Min and Max's measurements stored;
- Hold function freezes the display;
- Spot or continuous recording up to 1 million measurements;
- Blue luminescent backlit display;
- Programmable alarms for temperature and humidity through software;
- USB and Bluetooth communication;
- Magnetic mount;
- Battery or USB powered;
- Data View® graphing and analysis software included.

The Model 1246 can store up to 1,000,000 measurements on each channel (4MB). Recorded data is stored in non-volatile memory and will be retained even if the battery is removed. The instrument can measure from 3 to 98% RH, with an accuracy of ± 2% RH + 1 count when measuring between 10 and 90% RH.

CLASS ACTIVITY

PRECIPITATION

- Ø Materials: Hot pot, glass bowl with ice, pie tin,
- Ø Focus Questions: What do you think a cloud is made of?
- Ø Learning Outcomes: Explain shapes of clouds are different depending on different weather conditions.
- Ø Exploration
- Concept: Water can be a liquid or a solid and can go back and forth from one form to the other.
- Procedure:
  1. Ask. Explain the different forms of water and give examples.
  2. Boil water in a hot pot. Only the teacher will handle the water and the pot. Make sure students are a safe distance away from the boiling water.
  3. Gather students at a safe distance to view the hot pot and ice bowl.
  4. Ask. What will happen when you hold the pot of ice water over the boiled water?
  5. Predict what will happen to the bowl of ice, to the steam, and the bottom of the bowl.
  6. The teacher holds a bowl of ice over the steam.
  7. Place a pie tin so that the water dripping from the bottom of the bowl will collect in the tin.
  8. The class observes and shares what they observe happening.
- Invention:
  1. Discussion on the relationship between activity and rain and clouds
  2. Ask. What they learned about the different forms of water and give examples and why this knowledge is important to them.

SUMMARY

- Moisture influences all surface temperatures, including surface fuel temperatures, by controlling radiation in its vapor state and by reflecting and radiating when it is condensed into clouds.
- Atmospheric moisture is a key element in fire weather. It has direct effects on the flammability of forest fuels, and, by its relationship to other weather factors, it has indirect effects on other aspects of fire behavior.
- Water vapor is a relatively common atmospheric constituent, present even in the solar atmosphere as well as every planet in the Solar System and many astronomical objects including natural satellites, comets, and even large asteroids.
- The pressure at the water-air boundary resulting from molecular motion in the direction of escape from the liquid is called the vapor pressure of water.
- Vapor pressure depends on the actual water vapor in the air, and it may vary from near zero in cold, dry air to about 2 inches of mercury in warm, moist air.
- Sublimation is the process by which water molecules directly leave the surface of ice without first becoming liquid water.
- The actual amount of water vapor in a given volume of air, that is, the weight per volume, such as pounds per 1,000 cubic feet, is called the absolute humidity.
- Precipitation is any liquid or frozen water that forms in the atmosphere and falls back to the earth. It comes in many forms, like rain, sleet, and snow.
- Stratiform or dynamic precipitation occurs as a consequence of the slow ascent of air in synoptic systems (on the order of cm/s), such as over surface cold fronts, and over and ahead of warm fronts.
- Water vapor in the air comes almost entirely from three sources: Evaporation from any moist surface or body of water, evaporation from soil, and transpiration from plants.
- Relative humidity is much more variable than absolute humidity. It often changes rapidly and in significant amounts from one hour to the next and from place to place.
- Clouds are integral decision-makers in the climatic changes of the environment. In addition to that, clouds are also important in maintaining and balancing the earth's radiation.

REVIEW QUESTIONS

1. What are the sources of atmospheric moisture?
2. What is evaporation? Discuss the factors that affect the rate of evaporation.
3. Describe how precipitation forms.
4. Explain the global pattern of precipitation.

5. Determine the stability of air and its likelihood of cloud development.
6. Identify common cloud types.

REFERENCES

1. Adams, D. K., Barbosa, H. M. J., & Gaitan De Los Ríos, K. P. (2017). A spatiotemporal water vapor–deep convection correlation metric derived from the Amazon dense GNSS meteorological network. Monthly Weather Review, 145(2), 279–288. https://doi.org/10.1175/MWR-D-16-0210.1.
2. Ahmed, F., & Schumacher, C. (2015). Convective and stratiform components of the precipitation-moisture relationship. Geophysical Research Letters, 42(24), 10453–10462. https://doi.org/10.1002/2015GL066957.
3. Alexander, L. V. (2016). Global observed long-term changes in temperature and precipitation extremes: A review of progress and limitations in IPCC assessments and beyond. Weather and Climate Extremes, 11, 4–16. https://doi.org/10.1016/j.wace.2015.10.007.
4. Ambaum, M. H. P. (2020). Accurate, simple equation for saturated vapor pressure over water and ice. Quarterly Journal of the Royal Meteorological Society, 146(733), 4252–4258. https://doi.org/10.1002/qj.3897.
5. Bauer, M., Del Genio, A. D., & Lanzante, J. R. (2002). Observed and simulated temperature-humidity relationships: Sensitivity to sampling and analysis. Journal of Climate, 15(2), 203–215. https://doi.org/10.1175/1520-0442(2002)015<0203>2.0.CO;2.
6. Bretherton, C. S., Peters, M. E., & Back, L. E. (2004). Relationships between water vapor path and precipitation over the tropical oceans. Journal of Climate, 17(7), 1517–1528. https://doi.org/10.1175/1520-0442(2004)017<1517>2.0.CO;2.
7. Hagos, S. M., Leung, L. R., Garuba, O. A., Demont, C., Harrop, B., Lu, J., & Ahn, M. (2021). The relation between precipitation and precipitable water in CMIP6 simulations and implication for tropical climatology and change. Journal of Climate, 34(5), 1587–1600. https://doi.org/10.1175/JCLI-D-20-0273.1.
8. Li, H., Choy, S., Zaminpardaz, S., Carter, B., Sun, C., Purwar, S., Liang, H., Li, L., & Wang, X. (2023). Investigating the inter-relationships among multiple atmospheric variables and their responses to precipitation. Atmosphere, 14(3), 571. https://doi.org/10.3390/atmos14030571.
9. Liu, C., & Illingworth, A. (2000). Toward more accurate retrievals of ice water content from radar measurements of clouds. Journal of Applied Meteorology, 39(7), 1130–1146. https://doi.org/10.1175/1520-0450(2000)039<1130>2.0.CO;2.
10. Neelin, J. D., Martinez-Villalobos, C., Stechmann, S. N., Ahmed, F., Chen, G., Norris, J. M., Kuo, Y., & Lenderink, G. (2022). Precipitation extremes and water vapor. Current Climate Change Reports, 8(1), 17–33. https://doi.org/10.1007/s40641-021-00187-3.
11. Priego, E., Jones, J., Porres, M. J., & Seco, A. (2017). Monitoring water vapor with GNSS during a heavy rainfall event in the Spanish Mediterranean area. Geomatics,

Natural Hazards and Risk, 8(2), 282–294. https://doi.org/10.1080/19475705.2016.1238850.

12. Tapiador, F. J., Roca, R., Del Genio, A., Dewitte, B., Petersen, W., & Zhang, F. (2019). Is precipitation a good metric for model performance? Bulletin of the American Meteorological Society, 100(2), 223–233. https://doi.org/10.1175/BAMS-D-17-0218.1.
13. Zhang, Z., Lou, Y., Zhang, W., Liang, H., Bai, J., & Song, W. (2022). Correlation analysis between precipitation and precipitable water vapor over China based on 199–2015 ground-based GPS observations. Journal of Applied Meteorology and Climatology, 61(12), 1663–1677. https://doi.org/10.1175/JAMC-D-21-0068.1.

CHAPTER 6

Atmospheric Pressure and Wind Systems

LEARNING OBJECTIVES

After studying this chapter, you will be able to:

- Understand the measurement and distribution of atmospheric pressure;
- Know wind formation and factors affecting wind;
- Explain global wind systems;
- Focus on cyclones, anticyclones, and local wind systems.

KEY TERMS FROM THIS CHAPTER

Altitudes	Atmospheric pressure	Barometer
Climate patterns	Coriolis effect	Density
Earth	Gravity	Humidity
Ideal gas law	Isobars	Kinetic energy
Molecules	Monsoon	Polar easterlies
Pressure gradients	Prevailing westerlies	Trade winds
Water vapor	Weather	Wind

INTRODUCTION

Even though it is invisible to us, air surrounds us. Gravity keeps the layer of gases that cover the Earth close to the surface. Air pressure is produced by the gravitational attraction of gases to the surface. The air's ability to rise or sink is also influenced by a wide range of additional atmospheric and surface factors. As a result, the Earth is covered in a variety of pressure systems. Wind is the horizontal movement of air caused by variations in air pressure from one place to another. In addition to distributing energy and humidity throughout the Earth's surface, this air movement greatly affects the planet's weather and climate patterns.

6.1. MEASUREMENT AND DISTRIBUTION OF ATMOSPHERIC PRESSURE

The force per unit area that an atmospheric column—that is, the entire body of air above the designated area—exerts is known as atmospheric pressure. A mercury barometer, which measures the height of a mercury column that precisely balances the weight of the atmosphere column over the barometer, can be used to measure atmospheric pressure (thus the frequently used synonym barometric pressure). An aneroid barometer is another tool used to measure atmospheric pressure. It consists of one or more hollow, partially evacuated, corrugated metal disks that are supported against collapsing by an internal or external spring. A pen arm and a clock-driven rotating drum are used to record how the disk's shape changes as the pressure changes.

Millimeters (or inches) of mercury, pounds per square inch (psi), dynes per square centimeter, millibars (mb), standard atmospheres, or kilopascals are some of the units used to express atmospheric pressure.

By definition, standard sea level pressure is equal to each of the following:

- 760 mm (29.92 inches) of mercury;
- 14.70 pounds per square inch;
- 1,013.25 × 10^3 dynes per square centimeter;
- 1,013.25 millibars;
- one standard atmosphere;
- 101.325 kilopascals.

There are very little variations in these values. For instance, sea-level pressure records show that the highest point was 32.01 inches in the middle of Siberia, and the lowest point was 25.90 inches during a typhoon in the South Pacific. The Earth's wind and storm patterns are mostly determined by the minute differences in pressure that do occur.

Atmospheric pressure near the Earth's surface drops with height at a rate of roughly 3.5 millibars for every 30 meters (100 feet). However, because cold air has a higher density than warmer air, the pressure drop can be significantly steeper over it. At 270,000 meters (10^{-6} mb), the pressure is similar to the highest vacuum ever achieved by humans. If cautious acclimatization is not done, the pressure is low enough above 1,500 to 3,000 meters (5,000 to 10,000 feet) to cause mountain sickness and serious physiological issues.

6.1.1. Atmospheric Pressure Measurement

The atmosphere's pressure is measured using a device known as a "barometer." One measure of the weather is the atmospheric pressure. Consequently, barometers are used by meteorologists to forecast brief variations in the weather. An instrument called a barometer is used to measure atmospheric pressure in atmospheres, or bars. One bar, also referred to as one atmosphere or one atm, is the average air pressure at sea level at 15⊠.

An onset of low pressure is indicated by a decrease in atmospheric pressure. Low pressure means that insufficient force or pressure is present to drive the storm or clouds away. So, it's usually cloudy, rainy, or windy outside. An increase in atmospheric pressure signals the departure of clouds, bringing with them cool, dry air and a bright sky.

6.1.2. Atmospheric Pressure Vertical Distribution

The term "vertical distribution of pressure" refers to the columnar distribution of atmospheric pressure. Although the rate of reduction in air pressure may not be the same, it does occur as altitude increases. According to calculations, the average rate of drop in air pressure is 34 millibars for every 300 meters of altitude gain. The pressure drops as one ascends higher in the mountains; this can be attributed to the longer cooking times for rice and the harder breathing at higher elevations. Lower layers are denser than upper layers because the mass of air above the air in a column compresses the air beneath it. As a result, the atmosphere's lower layers experience greater pressure and density.

6.1.3. Atmospheric Pressure Horizontal Distribution

The horizontal distribution of pressure, as it is commonly known, describes how atmospheric pressure is distributed across the surface of the Earth. The isobars allow for the representation of the Earth's atmospheric pressure distribution on maps. "Isobars" are hypothetical lines that join locations with equal pressure values. The air pressure change rate and direction are determined by the isobar spacing. The pressure gradient is another name for the variation in air pressure. Wide spacing denotes mild pressure gradients, whereas closely spaced gradients indicate steep pressure gradients.

There is variation in the horizontal distribution of atmospheric pressure worldwide. It changes in different places and times. The following are the main variables affecting horizontal pressure.

1. Temperature of Air: Because of the uneven distribution of insolation, Earth does not heat up uniformly because the surfaces of the land and the water heat and cool differently. Regional variations in pressure are caused by the inverse relationship between air temperature and air pressure within the atmosphere. In an open system, the air pressure decreases with increasing air temperature and vice versa. It is thought that the pressure variations brought on by the thermal heating are what created the equatorial low-pressure system.
2. Rotation of the Earth: The air is deflected from its original location by the centrifugal force created by the Earth's rotation, which lowers the pressure differential. Earth's rotation creates the high-pressure

belts in the subtropical regions and the low-pressure belts in the subpolar regions. Air also experiences convergence and divergence due to Earth's rotation. Pressure is produced by convergence and divergence, respectively.

3. Presence of Water Vapor: Higher concentrations of water vapor in the atmosphere cause low pressure, whereas lower concentrations cause high pressure. Wintertime on the continents means that the air's ability to hold water vapor is reduced, which leads to the development of high-pressure centers. Because hot air has a greater capacity to hold moisture or water vapor, the continents tend to form low-pressure centers during the summer when temperatures are higher.

   Belts of high or low air pressure define the horizontal distribution of air pressure across the latitudes. However, since they are not always found on Earth in the manner that has been theorized, this is only a theoretical model.

6.1.4. Atmospheric Pressure Factors Influencing

A gas's density, temperature, and pressure are all correlated with one another.

6.1.4.1. The Ideal Gas Law

The ideal gas law, or P = nRT, is an equation that expresses the relationship between pressure, temperature, and density. In this equation, P stands for pressure, n for density, R for the constant of proportionality, and T for temperature. According to this equation, pressure will rise (P) if density stays constant but temperature rises (T), and pressure will rise (P) if temperature rises (n) but density increases (T).

Under a closed container, such as a sealed jar, the relationship between density, temperature, and pressure is simple. But none of these variables can be simply maintained constant in the atmosphere, so the cause-and-effect relationships are actually rather complicated. We can investigate some of the possibilities and see how pressure changes with density, temperature, air movement vertically, and other factors by using the above equation.

6.1.4.2. Pressure and Density Relationships

Density is the mass of matter per unit volume. Liquid densities vary while solid densities remain constant throughout space (whether on Earth, the Moon, or somewhere). The location of the fluids has a significant impact on the density of gases. Gas is free to expand to the extent that ambient pressure permits, hence changes in gas density are easily observed. The density of a gas and its pressure are inversely correlated. The gas is denser at higher pressures. Gas molecules are kept from escaping into space by the force of gravity, which holds the atmosphere to the Earth. At lower altitudes, the molecules of gas in the atmosphere are more closely spaced. At lower altitudes, there are more molecular collisions and higher pressure due to the higher density. At higher altitudes, the pressure drops as a result of the air being less dense.

6.1.4.3. Pressure and Temperature Relationships

Warming the air causes the molecules to become more energetic (as temperature rises, so does their kinetic energy) and accelerate. Since there are more collisions as a result of the molecules' increased speed, more pressure is created. As a result, an increase in gas temperature results in an increase in pressure under the same circumstances. As a result, on warm days, the air pressure will be high, and on cold days, low. Warm air is typically associated with low atmospheric pressure and cool air with high atmospheric pressure, but this is not always the case. A decrease in pressure may coincide with an increase in temperature due to a decrease in density.

6.1.4.4. Dynamic Influences on Pressure

"Dynamic" factors can also affect surface air pressure. To put it another way, air movement, particularly vertical air movement, can affect air pressure. When air gets heated at the surface, it tends to rise upward as air density decreases, creating a vacuum. The surrounding air then fills the vacuum left by the rising air by rushing in to take its place. This causes an air convergence that is visible close to the warm surface. As it ascends in altitude, the rising warm air begins to condense. As the atmosphere's density rises due to condensation, air becomes heavier and causes a divergence in flow. Different rates of air convergence and divergence at the surface and in the upper troposphere are thus related to the air. In general, ascending air causes relatively low pressure at the surface, while descending air causes relatively high pressure.

To put it briefly, variations in temperature, movement, and air density all affect atmospheric pressure. Although it is crucial to comprehend these connections, it is frequently challenging to forecast how changes in one variable will affect the others in a given situation. However, the following helpful generalizations about the elements connected to high- and low-pressure regions close to the surface can be made:

- A dynamic high, or high pressure at the surface, is created by air that is rapidly falling.
- High pressure—a thermal high—is produced at the surface by extremely cold surface conditions.
- A dynamic low, or low pressure at the surface, is caused by strongly rising air.
- A thermal low—a comparatively low-pressure area at the surface—is caused by extremely warm surface conditions.
- Surface pressure conditions are typically caused by a dominating factor from this list.

In actuality, though, there is variation in the pressure distribution throughout the world. They change periodically and from one place to another. We can divide the study of pressure belts into two categories: a horizontal component and a vertical component. This will help us comprehend the distribution of the pressure belts and the factors influencing them.

Weather is influenced by air pressure, especially when it comes to forming or altering atmospheric conditions. It is also one of the most important factors in the creation of precise weather forecasts. Air pressure is the force produced by the weight of the air in the Earth's atmosphere. After the device is used to measure air pressure, it is also referred to as barometric pressure. Despite not being visible, the air weights because it is not empty. Nitrogen, oxygen, argon, carbon dioxide, and other gases are present in trace amounts. The weight of the airborne particles exerts pressure due to Earth's gravitational pull. An instrument used to measure air pressure is the barometer.

6.2. WIND FORMATION AND FACTORS AFFECTING WIND

The natural movement of gases, such as air, about a planet's surface is called wind. There are many different types of winds: local breezes produced by heating land surfaces that last for a few hours, thunderstorm flows that last for tens of minutes, and global winds that result from variations in solar energy absorption between Earth's climate zones. Large-scale atmospheric circulation is primarily caused by the planet's rotation (Coriolis effect) and the difference in temperature between the equator and the poles. Monsoon circulations can be driven by thermal low circulations over terrain and high plateaus within the tropics and subtropics. Local breezes can be defined by the sea breeze/land breeze cycle in coastal areas; mountain and valley breezes can predominate in areas with varied terrain.

The factors that cause winds, the regions in which they occur, their spatial scale, speed, and direction, as well as their effects, are frequently used to classify winds. There are several components to winds, including wind speed, gas density, and energy content, or wind energy. In the field of meteorology, winds are frequently described in terms of their intensity and direction of motion. A "western" or "westerly" wind blows from the west to the east, a "northern" wind blows south, and so on. This convention for directions refers to the direction of the wind. This can occasionally seem counterintuitive. Gusts are brief, extremely fast wind bursts. Squalls are strong winds that last between one and two minutes. Depending on their average strength, long-duration winds go by a variety of names, including breeze, gale, storm, and hurricane.

Planetary wind is the outflow of light chemical elements from a planet's atmosphere into space, whereas solar wind is the motion of gases or charged particles from the Sun across space. Neptune and Saturn have the strongest winds ever recorded on any planet in the Solar System. The idea of wind has been studied in mythology throughout human civilization, impacting historical occurrences, extending the possibilities for travel and conflict, and offering a source of energy for electricity, mechanical labor, and leisure. Sailships travel the oceans of Earth propelled by the wind. Wind is used by powered flight to reduce fuel consumption and increase lift, and it is also used by hot air balloons to make short trips. Aircraft may find themselves in hazardous situations when wind shear is caused by different meteorological phenomena. Trees and man-made structures are susceptible to damage or destruction from strong winds. Through a number of aeolian processes, including the creation of fertile soils like loess and erosion, winds can sculpt landforms. The prevailing winds in large deserts can transport dust great distances from their source region winds associated with dust outbreaks, which are accelerated by rough terrain, have been given regional names in several parts of the world due to their significant impact on those areas. Wildfire spread is influenced by wind as well. Many plant species, as well as populations of flying insects and birds, can survive and spread thanks to the ability of winds to disperse their seeds. Wind affects livestock negatively when combined with cold temperatures. Animals' food stores, hunting, and defensive tactics are all impacted by wind.

6.2.1. Causes

Temperature variations are the primary cause of variations in atmospheric pressure, which in turn produce wind. Winds of different speeds are produced when there is a difference in atmospheric pressure because air moves from the higher to the lower pressure area. The Coriolis Effect will also deflect air on a rotating planet, except the equator. The differential heating between the equator and the poles (difference in absorption of solar energy leading to buoyancy forces) and the planet's rotation are the two main global drivers of large-scale wind patterns or atmospheric circulation. Large-scale winds typically approach geostrophic balance when they are outside of tropical regions and away from the surface's frictional effects. Friction near the Earth's surface makes the wind move more slowly than it otherwise would. Winds in low-pressure areas also blow more inward due to surface friction.

The breakdown and analysis of wind profiles make use of winds that are defined by the balance of physical forces. They help provide qualitative arguments regarding the horizontal and vertical distribution of horizontal winds as well as in streamlining the atmospheric equations of motion. The equilibrium between the Coriolis force and the pressure gradient force produces the geostrophic wind component. It approximates the flow above the atmospheric boundary layer in the midlatitudes and flows parallel to isobars. The difference in geostrophic wind between two atmospheric levels is known as the thermal wind. Only in an atmosphere where there are horizontal temperature gradients does it exist. Air "filling up" cyclones over time is caused by the geostrophic wind component, which is the difference between geostrophic and actual wind. Similar to the geostrophic wind, the gradient wind also incorporates centripetal acceleration, also known as centrifugal force.

6.2.2. Measurement

The direction of the wind is typically indicated by the direction of its origin. A northerly wind, for instance, blows from the north to the south. Weather vanes rotate to show which way the wind is blowing. Windsocks at airports are used to measure wind direction and, depending on their angle of hang, to estimate wind speed. Anemometers, which typically use rotating cups or propellers, measure wind speed. Wind can be measured using the ultrasonic signal propagation speed or the ventilation effect on the heated wire resistance when a high measurement frequency is required, as in research applications. Another kind of anemometer uses pitot tubes, which measure the dynamic pressure and use it to calculate wind speed. Pitot tubes take advantage of the pressure difference between an outer and inner tube that is exposed to the wind.

Globally, sustained wind speeds are recorded at a height of 10 meters (33 feet) and are averaged over a period of 10 minutes. For tropical cyclones, the United States reports winds over a 1-minute average, and for weather observations, it reports winds over a 2-minute average. India usually reports average winds of more than three minutes [Sharad K. Pushpendra K. and Jain. Agarwal and Vijay P. Singh (2007). A one-minute sustained wind usually has a value that is 14% higher than a ten-minute sustained wind, so knowing the wind sampling average is crucial. A wind gust is defined as a brief, high-speed burst of wind that lasts for a few seconds and is defined technically as the maximum wind speed that exceeds the lowest wind speed recorded over ten minutes by 10 knots (5 m/s). A squall is defined as a minute-long or longer increase in wind speed above a specific threshold.

Radiosondes use GPS, radio navigation, or radar tracking of the probe to measure the wind speed aloft. As an alternative, theodolites can be used to visually track the movement of the parent weather balloon position from the ground. SODAR and Doppler lidars and radars, which measure the Doppler shift of electromagnetic radiation scattered or reflected off suspended aerosols or molecules, are examples of remote sensing techniques for wind. Radiometers and radars can also be used to measure the ocean's surface roughness from space or aircraft. Over oceans, wind velocity near the surface can be estimated using ocean roughness. Using geostationary satellite imagery, one can determine the winds at the cloud top by measuring the distances between cloud layers in each image. The study of how wind affects man-made structures, such as buildings, bridges, and other artificial objects, is known as wind engineering.

6.2.3. Models

Airflow data can be obtained from models in both spatial and temporal dimensions. Data interpolation from multiple measurement stations yields spatial information and facilitates horizontal data computation. As an alternative, vertical information can be extracted using profiles, like the logarithmic wind profile. In numerical weather prediction models, temporal information is typically computed by solving the Navier-Stokes equations, which yields global data for GCMs or specific regional data. Friction, Earth's rotation, and radiation differentials are a few of the variables that affect wind field calculations. Although solving the Navier-Stokes equations numerically takes a lot of time, machine-learning techniques can speed up the computation process.

Our knowledge of atmospheric dynamics has greatly expanded thanks to numerical weather prediction models, which are now essential resources for climate research and weather forecasting. These models improve our understanding of the Earth's intricate atmospheric system by allowing scientists to analyze and forecast both global and regional wind patterns by utilizing both spatial and temporal data (Ahlstrom et al., 2011; Hernández et al., 2021).

6.3. GLOBAL WIND SYSTEMS

The uneven heating of the Earth's surface gives rise to large global wind systems. The surface currents in the oceans are therefore propelled by these global wind systems. It is important to realize that wind is essentially the movement of air from an area of high pressure to an area of low pressure to comprehend how global winds form and propel the main ocean currents. Air pressure is just the weight (force) of the air column above a specific location per unit area. Pressure is defined as force per unit area. Therefore, the average temperature of the air above a given location (hot air is lighter than cold air), the composition of the air, and elevation or altitude (higher up means less air above) all affect air pressure. For instance, because a single water molecule has less mass than a single nitrogen or oxygen molecule, air containing a lot of water vapor is less dense than dry air. Air becomes less dense as elevation or altitude rises (Figure 6.1).

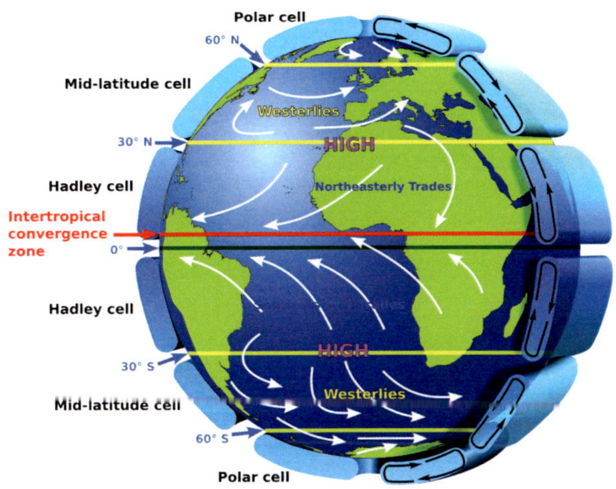

Figure 6.1. Winds are part of Earth's atmospheric circulation.

Source: By Kaidor – Own work based on File: NASA depiction of earth global atmospheric circulation.jpg, CC BY-SA 3.0, https://commons.wikimedia.org/w/index.php?curid=23902538.

Large global wind patterns are also a result of uneven surface heating. For the majority of the year, the sun is nearly directly overhead in areas close to the equator. At the equator, warm air rises and travels toward the poles. The colder air descends at the poles and returns to the equator. But it's not quite that easy. Because of the

Earth's rotation, global winds do not flow directly from north to south or from south to north. In the Northern Hemisphere, all winds seem to curve to the right as they travel. Winds seem to bend left in the southern hemisphere. This is referred to as the Coriolis Effect, which is the apparent change in an object's path caused by the Earth's rotation as it moves across its surface.

The trade winds unite to form a wide zone of light winds that stretches from east to west close to the equator. Because of the light winds, the area is referred to as the "doldrums." The majority of the sun's radiant energy is captured by this belt of air surrounding the equator. The region with the most weather is referred to as the intertropical convergence zone (ITCZ). The latitude of 10°N is the highest point on Earth's mean annual surface temperature. You now know that the Northern Hemisphere is warmer overall and has a larger landmass than the Southern Hemisphere. Warmer, humid air rises at the equator as well, creating a low-pressure area that stretches many kilometers north and south of the equator.

6.3.1. Trade Winds

The permanent east-to-west prevailing winds in the equatorial regions of the Earth are known as the trade winds, or easterlies. The trade winds, which are strongest in the winter and during the warm phase of the Arctic oscillation, primarily originate in the northeast of the Northern Hemisphere and the southeast of the Southern Hemisphere. For centuries, sailors have utilized trade winds as a means of traversing the world's oceans. They made it possible for Europeans to colonize the Americas and for trade routes to be established between the Atlantic and Pacific oceans.

According to meteorology, they serve as the main steering flow for tropical storms that develop over the Pacific, Atlantic, and southern Indian oceans and bring rain to Southeast Asia, Madagascar, East Africa, and North America. When descending air from within the subtropical ridge descends above the surface of the cloud, a trade wind inversion prevents the shallow cumulus clouds from growing larger. More rainfall is to be expected in the nearby landmasses as the trade winds weaken.

Additionally, the trade winds bring phosphate- and nitrate-rich Saharan dust to portions of southwestern and southeast North America, the Caribbean Sea, and all of Latin America. There are occasionally Sahara dust particles visible in Florida sunsets. The movement of Saharan dust across land suppresses rainfall and turns the sky from blue to white, increasing the frequency of red sunsets. Its existence increases the amount of airborne particulates, which has a detrimental effect on air quality.

6.3.1.1. Cause

Surface air flows toward the equator as part of the Hadley cell, while air above it flows toward the poles. The term "doldrums," "near-equatorial trough," "intertropical front," or "Intertropical Convergence Zone" refers to a low-pressure region of calm, mildly variable winds close to the equator [Sverre Petterssen, 1941]. This area of low pressure and wind convergence is sometimes referred to as the monsoon trough when it is found within a monsoon zone. The air starts to descend toward the surface in subtropical high-pressure belts known as subtropical ridges at around 30° in both hemispheres. Since the temperature rises as it descends but the moisture content stays constant, lowering the relative humidity of the air mass, the subsidence (sinking) air is relatively dry. Usually found above a maritime tropical air mass, which is warm and moist, this dry, warm air is referred to as a superior air mass. A temperature inversion is an increase in temperature with height. It is referred to as a trade wind inversion when it takes place within a trade wind regime.

The Coriolis Effect deflects surface air in both hemispheres that flows toward the Equator from these subtropical high-pressure belts toward the west. In the Northern Hemisphere, these winds mostly originate from the northeast, and in the Southern Hemisphere, from the southeast. These winds are known as the northeasterly trade winds in the Northern Hemisphere and the southeasterly trade winds in the Southern Hemisphere because winds are named according to the direction in which they are blowing. The Doldrums is where the trade winds from both hemispheres converge. Air masses warm up over lower latitudes as they pass over tropical regions because of more direct sunlight. As they move northward along the western edge of the subtropical ridge, those that form over land (continental) are hotter and drier than those that form over oceans (maritime). Trade air masses are another name for maritime tropical air masses. There are vast regions of trade winds in every tropical ocean, except for the northern Indian Ocean [John E. Oliver (2005)].S

6.3.1.2. Weather and Biodiversity Effects

When clouds arise above areas under trade wind regimes, they usually consist of a cumulus that reaches a maximum height of 4 kilometers (13,000 feet) before the trade wind inversion prevents it from getting any higher. Trade winds are stronger in the winter than in the summer and originate more from the direction of the poles (northeast in the Northern Hemisphere and southeast in the Southern Hemisphere) during the cold season [James P. Terry (2007)]. For instance, the Guianas, a region of South America at low latitudes, experience their windy season from January to April. The trade winds are stronger in the tropics during warm phases of the Arctic oscillation (AO). Weaker trade winds are a result of the AO's cold phase. Larger regions of rainfall on tropical landmasses, like Central America, when the trade winds are weaker.

The westward-moving trade winds south of the northward-moving subtropical ridge spread northwestward from the Caribbean Sea into southeastern North America (Florida and Gulf Coast) during midsummer in the Northern Hemisphere (July). Red sunsets are more common when dust from the Sahara that is circling the ridge's southern edge moves over land suppresses rainfall and turns the sky white instead of blue. Its existence increases the amount of airborne particulates, which has a detrimental effect on air quality. Despite having some of the cleanest air in North America, the Southeast US is primarily affected by African dust that enters the country through Florida. Since 1970, dry spells in Africa have made dust outbreaks more severe. The dust transport to Florida and the Caribbean varies greatly from year to year. Since the 1970s, dust events have been associated with a decline in the health of coral reefs in Florida and the Caribbean.

Millions of tons of nutrient-rich Saharan dust travel across the Atlantic Ocean each year, replenishing the depleted soils of the Amazon with essential phosphorus and other fertilizers.

6.3.2. Prevailing Westerlies

The predominant winds in the middle

latitudes between 30 and 60° latitude are known as the westerlies, anti-trades, or prevailing westerlies. They blow from the west toward the east. They generally steer extratropical cyclones in this manner, emerging from high-pressure areas in the horse latitudes and trending towards the poles. The increased westerly flow causes tropical cyclones that cross the subtropical ridge axis and enter the westerlies to recur. In the Northern Hemisphere, the winds come primarily from the southwest, and in the Southern Hemisphere, from the northwest.

The westerlies are at their weakest in the summer hemisphere and during periods of higher pressure over the poles, and they are strongest during the winter hemisphere. When land is absent, the westerlies are especially strong, especially in the Southern Hemisphere (also known as "Brave West winds"), because land amplifies the flow pattern and causes the current to become more north-south oriented, slowing the westerlies. This phenomenon affects Chile, Argentina, Tasmania, and New Zealand. In the roaring forties, between 40 and 50° south latitude, the strongest westerly winds in the middle latitudes can be found. Because of the southern hemisphere's enormous oceanic expanse, the westerlies are crucial in bringing warm, equatorial winds and waters to the western coasts of continents.

6.3.2.1. Behavior

Winds in the mid-latitudes would blow away from the subtropical ridge and toward the poles due to solar heating if the Earth were tidally locked to the Sun. On the other hand, poleward winds in the Northern Hemisphere tend to be deflected eastward from the north (to the right) and eastward from the south (to the left) by the Coriolis Effect, which is brought about by Earth's rotation. This explains why winds in the Southern Hemisphere typically originate from the northwest, but in the Northern Hemisphere, they typically originate from the southwest. The westerlies become stronger and consequently warm the mid-latitudes when pressures are lower over the poles. This happens when there is a positive Arctic oscillation and when low pressure is closer to the poles in the winter than it is in the summer. Cold air enters the mid-latitudes when it is negative and there is greater pressure over the poles. This causes a more meridional flow that blows from the pole towards the Equator.

The polar cyclone and the westerlies have varying degrees of intensity throughout the year. The westerlies get stronger as the cyclone reaches its peak intensity in the winter. The Westerlies get weaker in the summer when the cyclone reaches its lowest intensity. When dust plumes from the Gobi desert mix with pollutants and travel great distances downwind, or eastward, into North America, it is an illustration of the effects of the westerlies. Because there is less land in the middle of the Southern Hemisphere, the westerlies can be especially strong and slow the movement of west-to-east winds. Due to the stormy and cloudy weather, the westerlies are commonly referred to as the roaring forties, furious fifties, or shrieking sixties in the Southern Hemisphere, depending on the different latitudes

6.3.2.2. Impact on Ocean Currents

Ocean currents in both hemispheres are driven similarly by persistent winds blowing from west to east on the poleward sides of the subtropical ridges found in the Atlantic and Pacific oceans. Because of the variations in strength between the westerlies in each hemisphere, the currents in the Northern

Hemisphere are weaker than those in the Southern Hemisphere. Currents on an ocean basin's western boundary are stronger than those on its eastern boundary due to the process of western intensification. Warm tropical water is transported poleward by these western ocean currents and ends up in the Polar regions. For centuries, ships traversing both oceans have profited from the ocean currents.

An ocean current that circles Antarctica from west to east is known as the Antarctic Circumpolar Current (ACC), sometimes called the West Wind Drift. At 125 Sverdrups, the ACC is the largest ocean current and the main circulation feature of the Southern Ocean. Strong cyclones of every kind have developed at the base of the Westerlies in the northern hemisphere due to the Gulf Stream, which is a component of the North Atlantic Subtropical Gyre, both in the atmosphere and the ocean. Similar to the Gulf Stream, the Kuroshio, or Black Tide in Japanese, is a powerful western boundary current in the western north Pacific Ocean that has also contributed to the depth of ocean storms in that area.

6.3.3. Polar Easterlies

The dry, cold prevailing winds that circle the high-pressure regions of the polar highs at the North and South Poles are known as polar easterlies in the study of Earth's atmosphere. High-pressure zones are created at the poles when cold air retreats, forcing air to flow equatorward, where it is redirected westward by the Coriolis Effect. The polar easterlies are frequently weak and erratic, in contrast to the westerlies in the middle latitudes and the trade winds in the tropics. Keep in mind that the names of the winds reflect their origins. One of the five main wind zones, or wind belts, that comprise the circulatory system of our atmosphere is the polar easterlies. This specific wind belt reaches the poles from a latitude of roughly 60° north and south.

6.4. CYCLONES, ANTICYCLONES, AND LOCAL WIND SYSTEMS

There are areas of comparatively low pressure in cyclones and high pressure in anticyclones, respectively. They can range in size from the very large semipermanent examples mentioned above to smaller, extremely mobile systems, and they are found over most of the surface of the Earth. The defining circulation patterns of both cyclones and anticyclones are similar. In the Northern Hemisphere, flow around a cyclone (cyclonic circulation) is counterclockwise, and flow around an anticyclone (anticyclonic circulation) is clockwise, according to the geostrophic-wind and gradient-wind models. In the Southern Hemisphere, the directions of circulation are inverted. When friction is present, the motion toward lower pressure that is superimposed creates a "spiraling" effect that moves away from the high-pressure center and toward the low-pressure center.

Extratropical cyclones are cyclones that form outside of the equatorial belt; they can be thought of as large eddies in the broad air currents that generally flow from west to east around middle and higher latitudes in both hemispheres. They play a crucial role in the process that moves Earth's equatorial belt's excess heat from the Sun toward higher latitudes. If the temperatures at these higher latitudes are to remain consistently cool rather than cold, winds from the lower latitudes must carry heat to them since they radiate more heat into space than they receive from the Sun. In the absence of cyclones and anticyclones, air movements would be considerably more restricted north-south, and subtropical winds would have less opportunity to carry heat poleward. In such a scenario, the Polar regions would cool and the lower latitudes would get hotter, intensifying the temperature gradient between them.

Strong temperature gradients in a horizontal direction are especially helpful for the development and formation of cyclones. The Polar Regions and the Equator experience increasing temperature differences, which eventually intensify to the point where new cyclones are produced. The new cyclones lessen the temperature differential as their corresponding warm fronts move poleward and their associated cold fronts sweep equatorward. As a result, the Earth's wind circulation serves as a balance between the heating effects of solar radiation that occur at the Equator and in the Polar regions. Periodically, wind circulation is also destroyed by the action of cyclones, anticyclones, and other wind systems.

Closer to the equator, cyclones with a somewhat different nature typically form in latitudes between 10° and 30°N and S over the oceans. In general, when their winds reach or surpass 74 miles per hour (119 km/h), they are referred to as tropical cyclones. In the western Pacific and China Sea, they are called typhoons; in the Atlantic and Caribbean Seas, they are called hurricanes; and off the coast of Australia, they are called cyclones. These storms have a diameter of 100 to 500 km (60 to 300 miles), which is smaller than that of extratropical cyclones. They are also accompanied by winds that can occasionally reach extremely high levels of violence.

6.4.1. Extratropical Cyclones

Extratropical cyclones are the most common and have the greatest overall impact, affecting the greatest portion of the Earth's surface out of the two categories of large-scale cyclones. Additionally, this category of cyclones is the main contributor to the daily weather variations that occur in middle and high latitudes, making it the main focus of a large portion of contemporary weather forecasting. A group of Scandinavian meteorologists based in Bergen, Norway, planted the seeds between 1912 and 1930 for many of the ideas that are currently held about extratropical cyclones. The Norwegian meteorologist and physicist Vilhelm Bjerknes established the so-called Bergen school, which developed a model for a cyclone that develops as a disturbance along a front—a zone of sharp temperature contrast that separates two contrasting air masses. According to this model, the polar front—the zone that divides warmer tropical air from colder polar air—separates the masses of polar and mid-latitude air on Earth. Since there is a significant temperature gradient in this area, potential energy can be easily extracted and transformed into the kinetic energy linked to extratropical cyclones. A cyclone (also known as a wave, or frontal, cyclone) must form largely along the diagram's lines to access this reservoir. The feature that is most significant before a cyclone develops (cyclogenesis) is a front, which is shown in stage (A) as a thick black line with semicircles or triangles attached in alternation. This front, which is either stationary or moving very slowly, creates a strong horizontal temperature gradient by dividing warm and cold air. It is also sometimes called a baroclinic zone. The beginning of a cyclone is a disturbance along the front, which causes the front to slant in the direction of waves (B; wave appearance). As the disturbance's internal pressure drops, it takes on the characteristics of a cyclone and pushes warm air to travel poleward and cold air to move equatorward, respectively, as indicated by moving frontal boundaries. The triangles represent the cold front, which is advancing cold air, and the semicircles represent the warm front, which is advancing warm air, as shown in the cyclonic circulation stage (C). The cold, dense air rapidly streams equatorward as the cyclone intensifies, producing a cold front that typically has a slope of 1 to 50 and propagates at a speed of 8 to 15 meters per second or more than 18 to 34 miles per hour. In the meantime, warm, less dense air traveling northward passes over the cold air east of the cyclone to form a warm front. This warm front typically has a slope of 1 to 200 and propagates at a much slower rate, typically between 2.5 and 8 meters per second (6 to 18 miles per hour). Because of the two fronts' different rates of propagation, the cold front can pass the warm front and create an occluded front, a further, more intricate frontal structure. A line with alternating triangles and semicircles on the same side is used to depict an occluded front (D). The storm may intensify even more after this occlusion process. However, the storm eventually decays and dissipates (E) due to a process known as cyclolysis, which occurs when the cyclone separates from the warm air and moves toward the Equator.

Such an event usually has a few-day life cycle, and in that time the cyclone may travel several hundred to several thousand kilometers. Dramatic weather changes occur in its wake and path. The diagram shows a typical pattern of weather that could be caused by a cyclone approaching and passing through a region and its fronts. A cross section of the clouds and precipitation that typically occur along line ab is displayed in the occluded-front stage of the cyclogenesis diagram. Stratiform clouds, which rise as the front approaches and may produce rain or snow, are the most common feature of warm frontal weather. Warm fronts bring with them clearing skies and an increase in air temperature. On the other hand, the warmer air might also contain the elements needed to form a thunderstorm or rain shower, a situation that gets better as the cold front gets closer.

A colder air mass moving in, the development of stratocumulus clouds with some residual rain or snow showers, and finally clearing are the telltale signs of a cold front passing through. Even though this is a common occurrence, it's vital to remember that a wide variety of other weather patterns can also happen. For instance, a warm front's stratiform clouds may contain embedded thunderstorms and cumulus formations; the warm sector may be extremely dry with few to no clouds; the weather leading up to the cold front may resemble that of the warm front; or the air after the cold front may be entirely cloud-free. Earth satellite images consistently show cloud patterns spiraling around the cyclone vortex and oriented along fronts.

Any area of low pressure that forms needs the mass in the air column above the Earth's surface to be reduced. The resultant decrease in mass lowers the surface pressure. Three American meteorologists from the Bergen school, Jacob Bjerknes, Jørgen Holmboe, and Carl-Gustaf Rossby, discovered in the late 1930s and early 1940s that transient surface disturbances were accompanied by complementary wave features in the flow in the middle and higher atmospheric layers associated with the jet stream. Regions of mass divergence and convergence accompany these wave features, aiding in the development and direction of surface-pressure fields.

Small-scale cyclones can develop in the center of a single air mass, but extratropical cyclones form and intensify in tandem with fronts. One prominent instance is a group of cyclones that form in polar air streams after a frontal cyclone, and are typically smaller than the frontal variety. These "polar lows," which are most noticeable in subpolar marine environments, are assumed to be the result of middle-tropospheric circulation features being supported as well as heat and moisture being transferred from the warmer water surface into the polar air above. When a mountain blocks the general westerly flow, other small-scale cyclones develop on the lee side of the barrier. These "lee cyclones" have the potential to cause significant dust storms and windstorms below a mountain barrier.

6.4.2. Anticyclones

Anticyclones are typically areas of calm weather, whereas cyclones are usually areas of unfavorable weather. Anticyclones, which are typically larger than cyclones produce dry, stable air that may stretch hundreds of kilometers horizontally and show continuous downward motions. Most of the time, as a cyclone moves away, cold air behind it creates an area over which an actively developing anticyclone forms. Before the following cyclone moves into the region, this anticyclone forms. A cold anticyclone is one such type of anticyclone. Compression of the descending air is a consequence of the downward air motion in an anticyclone, though. The air warms as a result of this compression. Consequently, the anticyclone turns into a warm anticyclone after a few days as the air that makes up the anticyclone at elevations of 2 to 5 km (1 to 3 miles) above the earth tends to get warmer.

Warm anticyclones slow down, causing cyclones to veer off to their side. Anticyclones typically leave the main belt during their transition from cold to warm, trailed by cyclones in middle latitudes. They frequently combine with the so-called subtropical anticyclones, which are quasi-permanent bands of relatively high pressure that are found in both hemispheres around latitudes 20° to 30°. Warm anticyclones can occasionally be found in the belt that is usually inhabited by mid-latitude westerly winds. Atmospheric depressions, or zones of low pressure, are either prevented from moving eastward or redirected to the north or south of the anticyclone, significantly altering the typical cyclone tracks. When an anticyclone prevents the westerly wind from circulating normally, it's referred to as a blocking anticyclone, or blocking high. They usually last for a week or longer, and the presence of a few of these blocking anticyclones can significantly impact a season's overall characteristics. In particular, blocking anticyclones are frequently observed over Alaska, the eastern Atlantic, and Europe.

One could anticipate that an anticyclone's descent and warming would cause the clouds to break up and stop raining. The air can get stagnant close to the anticyclone's center due to light winds. As a result, air pollution may accumulate. For example, the city of Los Angeles frequently experiences poor air quality due to the presence of stationary anticyclones. Winter brings with it a cooling of the earth and the lower atmosphere. As the air cools to its dew point in the stagnant air, fog may form. Under other conditions, clouds may form in regions close to the ground up to a height of roughly 1 km (0.6 miles), and the air trapped in the first kilometer above Earth's surface may absorb moisture from the sea or other moist surfaces. Except for over the continents in the summer, such cloud layers can endure throughout anticyclones, but they hardly ever get thick enough to cause rain. In most cases, precipitation takes the form of light snow or drizzle.

In the summer, anticyclones are typically areas of clear skies and sunny weather; in other seasons, cloudy and foggy conditions are more common, especially over wet ground, snow cover, and the ocean. Surface temperatures during winter anticyclones are typically lower than average, especially when clear skies are present. Periods with little to no rain are caused by anticyclones, and they can last longer when blocking highs are present.

6.4.2.1. Cyclone and Anticyclone Climatology

Tracks, which are specific preferred regions that originate from preferred cyclogenetic and anticyclogenetic regions, are typically where migrating cyclones and anticyclones are distributed. The typical cyclone tracks for both January and July are also shown by the difference between the mean sea-level pressure diagrams for the winter and summer seasons. In the Northern Hemisphere, favorable cyclogenetic zones are located off the east coasts of continents and on the lee side of mountains. After that, cyclones move east or southeast before finally decaying and turning northeast. The tracks' increased northward displacement in July is a reflection of the polar front's summer time northward shift. Though more extreme cases exist, continental cyclones typically intensify at a rate of 0–5 mb (0–05 kPa) per hour or less. On the other hand, explosive development in marine cyclones frequently exceeds 1 mb (0.1 kPa) per hour, especially during the winter.

Before dissipating or combining with a warm anticyclone, anticyclones typically move eastward and equatorward after leaving cold air mass regions. Warm anticyclones move slowly poleward with the warm season, just like cyclones do. In the Southern Hemisphere, where oceans encircle most of the planet, cyclones are dispersed fairly evenly across the different longitudes. The Antarctic continent is typically ringed by several mature or decaying cyclones because cyclones typically form in latitudes between 30° and 40°S and move generally southeastward until reaching maturity in latitudes close to 60° S. The "roaring forties" are the oceanic belt between 40° and 60°S, which is characterized by strong, continuous westerly winds that are part of the circulation north of the main cyclone centers. The westerly winds are only occasionally disrupted by cyclones that are developing and moving southeastward.

6.4.3. Local Winds

6.4.3.1. Scale Classes

The spatial dimensions of organized wind systems range from tens of meters to thousands of kilometers, and their residence times range from a few seconds to several weeks. The typical size and lifespan of a phenomenon are taken into account by the concept of scale. Many attempts have been made to classify different phenomena into scale classes because the atmosphere displays such a wide range of temporal and spatial scales. The planetary scale is the category that includes the biggest and most persistent of these phenomena. These occurrences usually span several days to several weeks and have a size of a few thousand kilometers. Planetary-scale phenomena encompass a variety of phenomena, such as the semi-permanent pressure centers previously mentioned and specific upper-air waves encircling the globe.

The synoptic scale is a second class. Within this class are the migrating cyclones and anticyclones that regulate daily weather variations. They have shorter lifetimes, ranging from a few days to several thousand kilometers, and span smaller distances. The planetary and synoptic scales are occasionally combined to form the large-scale, or macroscale, classification. The dominance of horizontal motions over vertical motions and the central role played by the Coriolis force in determining wind characteristics are characteristics of large-scale wind systems. The westerlies and trade winds are two instances of large-scale wind systems.

A third category comprises phenomena that are even smaller in size and have a shorter lifespan. In this class, the Coriolis force frequently has less of an impact and vertical motions may be just as significant as horizontal movements. This class, referred to as the mesoscale, is distinguished by lifetimes of one day or less and spatial dimensions ranging from ten to a few hundred kilometers. The effect of the Coriolis force in mesoscale phenomena is sometimes overlooked due to the shorter time scale and the possibility of much larger other forces. The thunderstorm and its destructive byproduct, the tornado, are two of the most well-known instances of mesoscale phenomena. The current conversation centers on less powerful, but still frequently observed, wind systems found in relatively particular geographic areas and are therefore frequently referred to as local wind systems.

6.4.3.2. Local Wind Systems

The local wind system known as the "sea and land breeze circulation" is caused by variations in the temperature of the water's surface and the nearby land surface. It is generally found along coastlines next to large bodies of water. The heat capacity of water is higher (i.e., a given temperature change in a volume of water requires more units of heat than the materials on the land surface). The water mixes vertically, the volume slowly warms, and daytime solar radiation reaches several meters into the water. On the other hand, because daytime solar radiation only reaches a few centimeters below the surface, it heats the land surface more quickly. An area of low pressure is produced when the land surface, warmer than the air next to it, transfers more heat to the air mass above it. As a result, a circulation cell that resembles the one in the diagram is induced. It should be noted that the surface flow is known as a sea breeze because it is coming from the water and heading toward the land.

The land cools more quickly at night than the water because the landmass has a lower heat capacity than the latter. Consequently, a zone of relatively higher pressure is created at night by the cooler landmass producing a cooler overlying air mass. As a result, a circulation cell with air motions that are different from those during the day is created. A land breeze is this movement of air from the land to the water. Since the cooling of the atmosphere over land is limited to a shallower layer at night than the heating of the air during the day, the land breeze is usually shallower than the sea breeze. In the absence of a strong, broad-scale wind system, sea and land breezes occur along the coastal regions of oceans or large lakes during intense daytime heating or nighttime cooling. People who live within 10 to 20 km (6 to 12 miles) of the coast frequently experience cooler sea breezes on sunny afternoons, ranging from 19 to 37 km/h (12 to 23 miles/h), which later turn into sultry land breezes in the late evening. A zone of low-level air convergence in the surface flow termination region is one of the characteristics of the sea and land breeze. Cloud formation and local upward motions are frequently caused by this convergence. Therefore, it is not unusual to see clouds lying off the coast at night in areas where both sea breezes and land breezes are present. These clouds are then dissipated by the sea breeze during the day, which forms new clouds and may result in afternoon showers over land.

On the surface of the Earth, the existence of valleys and mountains induces another group of local winds. Differential heating or cooling along mountain slopes causes one subset of these winds, known as mountain winds or breezes. The overlying air moves upslope during the day as a result of solar heating of the sunlit slopes. Another name for these winds is anabatic flow. Cool downslope drainage motion happens at night when the slopes cool and the airflow direction changes. Depending on the topographic configuration, such winds can occur in strong gusts or in relatively gentle gusts. A particular kind of katabatic flow is these winds. Thick fog can develop in an enclosed valley due to the cool air draining into the valley. Until daytime heating reverses the circulation and produces clouds linked to the upslope motion at the mountaintop, fog will continue to persist.

Adiabatic temperature changes that occur as air flows over a mountain induce foehn winds, another subset of katabatic flow. These winds are also referred to as Santa Ana winds in southern California and Chinook winds east of the Rocky Mountains. Adiabatic temperature changes are those that happen in the atmosphere when bundles of air are moved vertically; they happen without any heat being added or taken away. Air expands when it is lifted and moves into an area of lower pressure. Adiabatic cooling, or a drop in temperature, coincides with this expansion. The air contracts and warms adiabatically as it subsides. The heat released during precipitation formation may moderate the air's cooling rate as it ascends on the windward side of the mountain. The air on the leeward side of the mountain, however, adiabatically warms more quickly than it cooled during the windward ascent since it has lost a lot of moisture. If this wind does indeed reach the surface, it will therefore result in warm, dry weather. These winds are typically mild and cause a gradual warming. But occasionally, fierce winds can reach speeds of over 185 km/h (115 mi/h) and raise air temperatures by tens of degrees—and occasionally even over 20°C [36°F]—in a matter of hours.

When the underlying geography is characterized by a cold plateau next to a relatively warm region of lower elevation, other forms of katabatic wind may arise. These circumstances are met in regions where significant ice sheets or frigid, elevated land surfaces abut warmer, vast bodies of water. A thick, cold dome of cool air is created as the air above the chilly plateau cools. Depending on the terrain's incline and the distribution of the background pressure field, this cold air will spread out into the lower elevations unless it is prevented from doing so by background wind conditions. Its speeds will range from mild (a few kilometers per hour) to intense (93 to 185 km [58 to 115 miles] per hour). In Europe, two distinct katabatic wind varieties are highly recognized. The two main winds that blow to the Mediterranean Sea are the mistral, which originates in central and southern France, and the bora, which originates in the highlands of Croatia, Bosnia and Herzegovina, and Montenegro. Strong katabatic winds that originate off the Antarctic Ice Sheet frequently blow northward, producing blizzard-like conditions.

6.4.4. Zonal Surface Winds

The mean sea-level pressure diagrams for January and July show that generally speaking, certain regions can anticipate winds that originate from one dominant direction, which is primarily determined by the existence of significant semi-permanent pressure systems. In maritime environments, these predominant winds have long been recognized due to their impact on large sailing vessels. Near the Equator, there is a general band of low pressure that distinguishes tropical and subtropical regions. Centers of high pressure, which have the potential to stretch poleward into the middle latitudes, surround this band. The tropical wind region lies between these areas of low and high pressure. The trade winds are the largest of these. Trade winds flow westward and slightly toward the equator on the equatorward side of the subtropical high-pressure centers, hence their name, which refers to their beneficial effect on trade ships traversing the subtropical North Atlantic. Declining air is a characteristic of the "root of the trades," which is located on the eastern edge of a subtropical high-pressure center. The easternmost regions of the subtropical Atlantic and Pacific Ocean basins experience extremely warm and dry weather due to this, which is produced above a shallow layer of oceanic stratus clouds. However, as the trade winds move westward, the air mass gets more humid, subsidence decreases, and isolated showers start to appear. These showers are more common on islands with elevated topography features that obstruct the warm, humid airflow. The intertropical convergence zone (ITCZ) is the location where the trade winds from the Northern and Southern hemispheres often converge due to their equatorward flow. Along the ITCZ, there are thunderstorms, showers, and deep convective clouds.

The air turns poleward upon reaching the western extremity of the high-pressure center, and in the middle latitudes, it eventually returns eastward. Now loaded with moist maritime tropical air (mT), the poleward-moving air is warm and contributes to the warm, humid, showery climate that characterizes the Caribbean region, eastern South America, and the western Pacific Island chains. The variable weather typical of the middle latitudes is linked to the westerlies. Extratropical cyclones and anticyclones that are migratory produce periods of rain (sometimes with violent thunderstorms), snow, sleet, or freezing rain interspersed with dry, sunny, and occasionally bitterly cold conditions. These patterns are linked to contrasting warm moist air moving poleward from the tropics and cold dry air moving equatorward from polar latitudes. Additionally, these patterns vary with the season, with winter bringing colder air and stronger cyclones, while spring and summer see a higher frequency of frequent thunderstorms. Additionally, the existence of landmasses and significant mountain features, especially in the Northern Hemisphere, complicates these migrations and the climate that goes along with them. On the equatorward side of the semipermanent subpolar centers of low pressure are located the westerlies. The surface winds shift once more westward over a sizable area of the subpolar latitudes poleward of these centers. Similar to the middle latitudes, these polar easterlies vary significantly due to the presence of large landmasses, most notably in the Northern Hemisphere. Furthermore, there is a seasonal dependency between the wind systems and the corresponding climate. The polar latitudes' wind systems are substantially weaker during the brief summer season. These systems get stronger over the long winter months, alternating between long stretches of dry cold air typical of continental arctic or continental polar air masses and snowy periods.

Mean meridional (north-south) circulation patterns are also connected to these major surface circulation regions and the pressure fields that accompany them. These cells are difficult to find every day at any given longitude, even though their presence can be seen in long-term mean statistics gathered over a hemisphere.

6.4.5. Monsoons

Seasonal variations in atmospheric pressure are particularly strong over continents, as the sea-level atmospheric pressure maps in January and July illustrate. The reason for the greater intensity of these seasonal variations, also referred to as monsoons, is that land surfaces experience greater seasonal temperature variations than bodies of water. Land surfaces frequently quickly alter the temperature and density characteristics of air parcels passing over them because they warm and cool quickly than water bodies.

Mostly in South Asia and some parts of Africa, monsoons blow for roughly six months from the northeast and six months from the southwest; however, similar conditions also exist in Central America and the region bounded by Southeast Asia and Australia. The summer monsoons have a strong tendency to converge, rise, and produce rain. They also have a dominant westerly component. The winter monsoons are primarily oriented towards the east and have a strong propensity to diverge, diminish, and result in drought. The causes of both are the variations in yearly temperature trends over land and sea.

6.4.5.1. Diurnal Variability

In the afternoon, especially on cloudless days, landmasses in monsoon-affected regions warm up extremely quickly; surface air temperatures between 35 and 40°C (95 and 104°F) are not unusual. In these circumstances, warm air is gradually and continuously immersed in the monsoon's humid and hazy atmosphere. Thus, energy from this marked diurnal, or daily, change in terrestrial heating is transferred to the cloud, rain, and diurnal circulation systems over 24 hours. This diurnal shift has a scope that encompasses processes the size of continents as well as coastal sea breezes. The effects of rapid diurnal temperature change are confirmed at continental scales by satellite observations. For instance, during the summer afternoon hours in South Asia's warmer land areas, air from the surrounding areas is drawn into the lower troposphere. Rain and clouds are forming in tandem with this afternoon's heating build-up. On the other hand, an opposite circulation is observed during the early morning hours, which is marked by reduced cloud cover and precipitation.

6.4.5.2. Intra-Annual Variability

On multiple timescales, monsoon rainfall and dry spells alternate. A well-known timescale of this kind is found in intervals of 40–50 or 30–60 days. The Madden-Julian oscillation (MJO) was coined in 1971 and named after two American atmospheric scientists, Paul Julian and Roland Madden. This phenomenon manifests as cyclonic and anticyclonic regions that alternate, intensifying and decreasing precipitation, respectively, and flowing eastward along the Equator in the Pacific and Indian oceans. By increasing moisture during its cyclonic (wet) phase and decreasing convection during its anticyclonic (dry) phase, the MJO can affect monsoonal circulation and rainfall. In monsoon regions, dry and wet spells occur at the surface. Locally, these intervals may alternate by two weeks or more every phase.

6.4.5.3. Interannual Variability

The fluctuations in rainfall caused by the monsoon in Australia and the Indian Ocean seem to be similar to El Niño occurrences. Every two to seven years, there is an El Niño event that causes the ocean temperatures over the central equatorial Pacific Ocean to rise by approximately 3°C (5.4°F). The western equatorial Pacific experiences unusual weather marked by increased rising air motion, convection, and rain. Simultaneously, a counteracting lobe of descending air forms near eastern Australia, Malaysia, and India, resulting in rainfall that is below average. A well-known relationship between El Niño and monsoon rainfall is depicted in the graph. Precipitation data from periods of monsoon rainfall over India that are above and below normal are expressed here as a function of years. Barbs on the histogram that are darker indicate years that are associated with El Niño events. The graph demonstrates how many El Niño years are associated with monsoon rainfall that is below average. For seasonal forecasters, this illustration offers limited guidance because, in many El Niño and La Niña years, monsoon rainfall is almost normal.

Apart from the surface temperatures of the equatorial Pacific Ocean, a multitude of other factors also play a role in the interannual variability of monsoon rainfall. The following summer's insufficient monsoon rainfall in India is linked to the Plateau of Tibet's excessive spring snow and ice cover. Furthermore, there is compelling evidence linking insufficient summer rainfall in Asia with the abundance of snow and ice cover in western Siberia. The Indian Ocean's warmer-than-average sea surface temperatures may also have some role in South Asia's higher-than-average rainfall. Researchers face a difficult task in forecasting monsoon strength due to the complex interplay among these numerous factors. Indian Weather Service data shows a fairly distinct signature on the decadal variability of rainfall in India. An annual running mean that combines average rainfall anomalies—totals that deviate from typical rainfall amounts—at all Indian rain gauge sites represents decadal-scale variability in the graph. Decades with somewhat less rainfall occur after periods of higher-than-normal rainfall (James et al., 2007; John et al., 2005).

CASE STUDY

TROPICAL STORM – TYPHOON RAI

Typhoon Rai (also known locally as "Odette") was a tropical storm that affected the Philippines in December 2021. The Philippines is an emerging country in Southeast Asia. With wind speeds of 195 km/h, Typhoon Rai was a Category 5 tropical storm, which was later names a super-typhoon. In some parts of the country, over 150 cm of rainfall was recorded in just three days.

Typhoon Rai hit Siargao Island in the eastern Philippines at 1.30 pm on Thursday 16th December. By 3.10 pm on Friday 17th December, the typhoon had moved west to Palawan, before traveling out into the South China Sea. Typhoon Rai caused significant economic, social, and environmental impacts.

Impacts

Economic Impacts:

- The overall economic impact of Typhoon Rai is estimated at around $550 million (£435 million)
- Crops and food supplies were washed away, affecting the country's international trade and farmers' incomes
- Many flights were canceled and dozens of ports were closed, affecting business and tourism
- The fishing industry was severely affected with the storm, destroying boats and associated equipment worth over $58 million.

Social Impacts:

- By January 2022, 407 people were killed, 78 people were still missing and nearly 1,150 people had been injured by the storm
- The storm affected 11 of the Philippines' 17 regions. The lives of more than seven million people were disrupted
- 36 million homes were damaged or destroyed. More than half a million people were left homeless over the Christmas period
- There were outbreaks of diarrhea and other diseases due to the lack of sanitation, food, water, shelter, and medication
- Thousands of people were housed in temporary shelters, many without running water or sanitation.

Environmental Impacts:

- Widespread floods damaged and in many cases destroyed homes and businesses in coastal areas
- Farmland was damaged by strong winds and flooding
- Thousands of trees were uprooted leading to a massive release of carbon dioxide and loss of habitat with effects on wildlife
- Protected areas, including Puerto Princesa Subterranean River National Park (PPSRNP) and El Nido-Taytay Managed Resource Protected Area, were damaged
- Major roads were blocked by trees and were impassable.

Responses:

Typhoon Rai rapidly grew into a super-typhoon. Although its path was tracked, the storm exceeded all predictions. On the 21st of December, the Philippines formally declared 'A State of National Calamity' and asked for international help.

The country's military, coast guard, and fire personnel brought aid and assistance to the worst-affected areas. Military aircraft and naval vessels were used to find survivors and distribute supplies.

Charities such as Oxfam, Médecins Sans Frontières (MSF), CAFOD, and ShelterBox raised money for essential supplies, such as medication and shelter. In the longer term, these charities have been helping people get their livelihoods back, for example by repairing fishing boats or distributing rice seeds.

The United Nations initially launched an international aid appeal for $107m to finance the humanitarian relief effort. One month after the storm, it was clear that the situation in the Philippines was more serious than they first thought, so they increased this figure to $169m.

CLASS ACTIVITY

Air Pressure Game:

In this activity, students use their bodies to illustrate how temperature, density, and volume affect air pressure.

Temperature affects how particles move. As air warms up, the particles start to vibrate and bump into each other, increasing the space around each particle. Because each particle uses more space for motion, the air expands and becomes less dense (lighter). A lower density means that there are the same number of air molecules that occupy a larger space (volume). If the temperature of air increases but the volume stays the same, the pressure will increase. The opposite effect happens when the air cools (temperature decreases). As the temperature drops, particles move more slowly, taking up less room. The amount of space the air takes up shrinks, and reduces the air pressure.

- Objectives:
  - Describe the relationships between temperature, density, and volume on air pressure.
  - Explain how air pressure works
- Materials:
  - Outdoor field or large playing area
  - Cones (or similar markers)
- Key Questions:
  - When the area decreased, how did it affect the air density?
  - Does this raise or lower air pressure?
  - Were there more collisions between particles when it was hotter or cooler?
- What to Do:
  - Place cones every 20 feet around the perimeter of the playing field. These cones represent the boundaries of the container the air particles are in.
  - The students are individual air particles.
  - Air particles (students) expand to fill the volume (area in this case) they are in, and they react to changes in temperature. An increase in temperature causes faster movement (running), and a decrease in temperature causes slower movement (walking).
  - The students pretend to be air particles. Now and then, call out changes in the temperature (e.g., hot, warm, cool, icy cold): the students must act accordingly.

- Pause the game and gather the students into the middle. Decrease the size of the playing area by moving the cones in (mimicking a decrease in volume).
- Repeat the game with changes of temperature.
- Decrease the size of the field again.
- Continue to shrink the field until the students can barely move.

- Teacher Tip: The game can be modified into freeze tag, but where the playing area gets progressively smaller. Anyone who goes outside of boundaries becomes auto-frozen.
- Extensions
- How is this model different than how real air particles act with a temperature change?
- Can you tell if you're surrounded by a high-pressure system or a low-pressure system by using your five senses?

SUMMARY

- Atmospheric pressure, force per unit area exerted by an atmospheric column (that is, the entire body of air above the specified area).
- To measure the pressure exerted by the atmosphere, an instrument called a "Barometer" is used. A barometer is a device that measures atmospheric pressure in units known as atmospheres or bars. The average air pressure at sea level at 15°C is equal to one bar, sometimes known as one atmosphere, or 1 atm.
- The distribution of atmospheric pressure in a columnar manner is known as the Vertical Distribution of Pressure. There is a decrease in air pressure with an increase in altitude but may not follow the same rate of reduction.
- In common terms, the distribution of atmospheric pressure over the Earth's surface refers to the Horizontal Distribution of Pressure. The distribution of atmospheric pressure over the Earth can be represented on the maps with the help of the isobars.
- Wind is the natural movement of air or other gases relative to a planet's surface.
- Wind is caused by differences in atmospheric pressure, which are mainly due to temperature differences. When a difference in atmospheric pressure exists, air moves from the higher to the lower pressure area, resulting in winds of various speeds. On a rotating planet, air will also be deflected by the Coriolis Effect, except exactly on the equator.
- Wind direction is usually expressed in terms of the direction from which it originates. Wind speed is measured by anemometers, most commonly using rotating cups or propellers.
- Models can provide spatial and temporal information about airflow. Spatial information can be obtained through the interpolation of data from various measurement stations, allowing for horizontal data calculation. Alternatively, profiles, such as the logarithmic wind profile, can be utilized to derive vertical information.
- Large global wind systems are created by the uneven heating of the Earth's surface. These global wind systems, in turn, drive the oceans' surface currents.
- The trade winds or easterlies are permanent east-to-west prevailing winds that flow in the Earth's equatorial region. The trade winds blow mainly from the northeast in the Northern Hemisphere and the southeast in the Southern Hemisphere, strengthening during the winter and when the Arctic oscillation is in its warm phase.
- The westerlies, anti-trades, or prevailing westerlies, are prevailing winds from the west toward the east in the middle latitudes between 30 and 60° latitude. They originate from the high-pressure areas in the horse latitudes trend toward the poles and steer extratropical cyclones in this general manner.
- In the study of Earth's atmosphere, polar easterlies are the dry, cold prevailing

winds that blow around the high-pressure areas of the polar highs at the North and South Poles.

- Cyclones and anticyclones are regions of relatively low and high pressure, respectively. They occur over most of Earth's surface in a variety of sizes ranging from the very large semipermanent examples described above to smaller, highly mobile systems.

REVIEW QUESTIONS

1. What is an atmospheric pressure system?
2. How pressure systems and wind are related?
3. How does atmospheric pressure control wind direction?
4. Name the two instruments used to measure air pressure.
5. Name the three factors that influence horizontal distribution of air pressure.
6. What is air mass?
7. Which type of cyclones cause heavy loss of life and property?
8. In which latitudes temperate cyclones develop?
9. What are temperate cyclones? How do they differ from tropical cyclones?
10. What is the role of Coriolis force in the deflection of winds?

REFERENCES

1. Ahlstrom, M., Blatchford, J., Davis, M., Duchesne, J., Edelson, D., Focken, U., Lew, D., et al. (2011). Atmospheric pressure. IEEE Power and Energy Magazine, 9(6), 97–107. http://dx.doi.org/10.1109/MPE.2011.942353.
2. Hernández, E. Z., & Fernández Arce, M. (2021). Atmospheric pressure anomalies over earthquakes epicenters. European Journal of Environment and Earth Sciences, 2(4), 21–25. http://dx.doi.org/10.24018/ejgeo.2021.2.4.150.
3. Jain, S. K., Agarwal, P. K., & Singh, V. P. (2007). Hydrology and Water Resources of India. Springer. ISBN 978-1-4020-5179-1.
4. Kauffman, G. B. (1988). Atmospheric pressure demonstration. Journal of Chemical Education, 65(4), 376. http://dx.doi.org/10.1021/ed065p376.2.
5. Oliver, J. E. (2005). Encyclopedia of World Climatology. Springer. ISBN 978-1-4020-3264-6.
6. Petterssen, S. (1941). Introduction to Meteorology. McGraw-Hill Book Company, Inc. ISBN 978-1-4437-2300-8
7. Shulyak, D., Valyavin, G., Kochukhov, O., & Burlakova, T. (2008). Analysis of magnetic pressure effects in atmospheres of CP stars. Proceedings of the International Astronomical Union, 4(S259), 405–406. http://dx.doi.org/10.1017/S1743921309030889.
8. Smith, D., & Norton, J. D. (1995). Reduced atmosphere storage of Chinese chestnuts. HortScience, 30(4), 890A–890. http://dx.doi.org/10.21273/hortsci.30.4.890a.

9. Syage, J. A., Hanold, K. A., Lynn, T. C., Horner, J. A., & Thakur, R. A. (2004). Atmospheric pressure photoionization. Journal of Chromatography A, 1050(2), 137–149. http://dx.doi.org/10.1016/j.chroma.2004.08.033.
10. Terry, J. P. (2007). Tropical Cyclones: Climatology and Impacts in the South Pacific. Springer. ISBN 978-0-387-71542-1.
11. Vidotto, A. A., & Cleary, A. (2020). Stellar wind effects on the atmospheres of close-in giants: A possible reduction in escape instead of increased erosion. Monthly Notices of the Royal Astronomical Society, 494(2), 2417–2428. http://dx.doi.org/10.1093/mnras/staa852.
12. Wang, Y., Deng, J., & Wei, W. (2022). Investigation on the cavity backwater of chute aerators under various atmospheric pressures. Water, 14(9), 1513. http://dx.doi.org/10.3390/w14091513.

CHAPTER 7

Air Masses and Fronts

LEARNING OBJECTIVES

After studying this chapter, you will be able to:

- Write the definition and classification of air masses;
- Examine the formation of air masses;
- Evaluate the basics of weather front.

KEY TERMS FROM THIS CHAPTER

Continental polar	Continental tropical	Cyclones
Cyclonic flow	Humidity	Maritime polar
Maritime	Maritime tropical	Meteorology
Stability	Warmer air	Weather patterns

INTRODUCTION

Large groups of air with comparable humidity and temperature are called air masses. They develop over land or water areas. Air masses can be classified as cold, warm, hot, or continental. A front is the line separating two distinct air masses. Both warm and cold fronts are possible. Warm fronts transfer heated air into cooled air. Warmer air is forced into colder air by cold fronts. Understanding weather patterns requires an understanding of fronts and air masses. A large body of air with comparable moisture and temperature characteristics is called an air mass. Fronts are the areas where two distinct air masses meet. The temperature differential between the two air masses determines the kind of front that forms. A cold front forms when temperatures diverge significantly. When there is little temperature variation, a warm front forms. Comprehending fronts and air masses is crucial for comprehending weather patterns, as they aid in elucidating the reasons behind specific weather phenomena. For instance, there will be a lot of precipitation if a cold front enters a region of warm, humid air. On the other hand, mild weather will occur if a warm front enters a region of cold air. Long-term weather patterns can also be predicted with the help of air masses and fronts.

7.1. DEFINITION AND CLASSIFICATION OF AIR MASSES

In meteorology, an air mass is a large body of air that, at any given altitude, has nearly uniform temperature and humidity conditions. This kind of mass has well-defined boundaries; it can stretch hundreds or thousands of kilometers in a horizontal direction and occasionally even reach the top of the troposphere, which is located roughly 10–18 km (6–11 miles) above the surface of the Earth. When the atmosphere is exposed to a sizable, uniform land or sea surface for an extended period of time, it takes on the temperature and moisture characteristics of that surface, leading to the formation of an air mass. Polar or subtropical latitudes are the starting points of the planet's main air masses. In essence, the middle latitudes are a zone where the polar and tropical air masses interact, change, and mix. Four fundamental source regions are commonly used to categorize air masses based on their latitude. These are Equatorial (warm and extremely moist), Tropical (warm), Polar (cold), and Arctic (extremely cold). The four main air mass types found in the United States are usually maritime tropical, continental tropical, and continental polar.

Typically, during the winter months, continental polar (cP) air forms over vast landmasses like northern Canada and central Asia. It is characterized by the lack of condensation forms and is probably stable. Strong turbulence and heat from the ground cause this kind of air mass to form limited convective stratocumulus clouds, which are accompanied by sporadic light rain or snow showers. Strong continental heating during the summer quickly changes the cP air mass's coolness and dryness as it descends in latitude. Cumulus cloud generation occurs during the day on average, but rain showers are typically prevented by the upper-level stability of the air mass. Over the polar regions of both the Northern and Southern hemispheres, maritime polar (mP) air masses form. Compared to the cP air masses, they often contain a significant amount more moisture. When the air is compelled to ascend mountain slopes or becomes entangled in cyclonic activity, heavy precipitation can occur as they move inland in middle and high latitudes.

Mostly in the summer, arid or desert areas in middle or lower latitudes are the source of the continental Tropical (cT) air mass. Although it is generally very hot, the intense dry convection usually cannot bring it down to the condensation level because of its extremely low moisture content. The cT is the driest air mass of all, supporting the global belt of subtropical deserts. Throughout the year, the most significant air mass that produces rain and holds moisture is the maritime tropical (mT). It travels poleward during the winter and is kept cool by the earth's surface. As a result, it is marked by drizzle, low stratus or stratocumulus clouds, fog, and poor vision. In areas of cyclonic activity, a steep lapse rate above guarantees the presence of intense frontal and convective rains. The properties of the mid-tropic (mT) air mass over oceans and in cyclonic activity zones are essentially unchanged from winter to summer. However, the air mass is highly heated over warm continental regions, resulting in afternoon thunderstorms that are widely dispersed and locally intense rather than fog and low stratus clouds.

7.1.1. Definition

A sizable volume of air in the atmosphere with a largely uniform temperature and moisture content is called an air mass. Air masses can travel thousands of kilometers across the Earth's surface and can travel 16 kilometers (10 miles) into the atmosphere, or the stratosphere, above sea level. Large areas known as source regions, which have constant temperatures and humidity, are where air masses originate. Low wind speeds allow the air to stay still long enough to absorb the heat or cold from the source region. When winds transport air masses, they transfer the source region's weather conditions—hot or cold, dry or wet—to a new location. The air mass may collide with another air mass that has a different temperature and humidity when it reaches a new area. This might produce a powerful storm. Air masses are classified by meteorologists based on their formation location over Earth. Air masses are divided into four categories: tropical, arctic, polar, and equatorial. Arctic air masses are extremely cold and form in the Arctic region. Tropical air masses are relatively warm and form in low-latitude regions. High-latitude areas are where polar air masses form, and they are chilly. Warm, equatorial air masses form close to the equator (Figure 7.1).

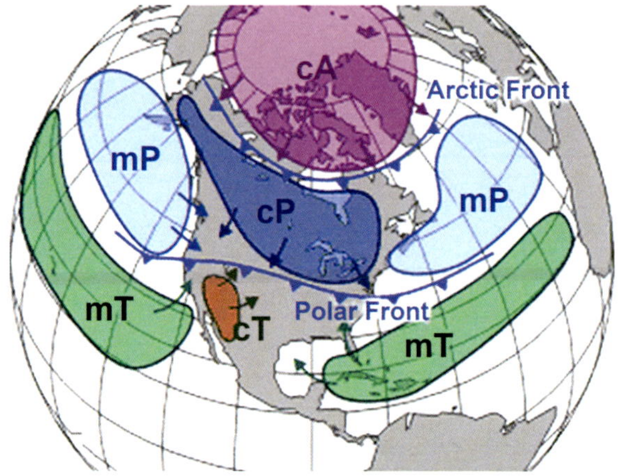

Figure 7.1. Different air masses which affect North America as well as other continents, tend to be separated by frontal boundaries.

Source: https://upload.wikimedia.org/wikipedia/commons/7/76/Airmassesorigin.png.

The fact that air masses form over land or water is another factor used to identify them. Humid maritime air masses form over bodies of water. Dry continental air masses are formed over land. As a result, a cold, dry air mass known as a continental polar air mass forms over northern Canada. A maritime tropical air mass is a warm, humid air mass that forms over the Indian Ocean. On weather maps, air masses are categorized using two or three letters.

- A lowercase letter describes the amount of moisture in the air mass: m for maritime (moist) and c for continental (dry).
- An uppercase letter describes the heat of the air mass: E for equatorial, T for tropical, M for monsoon, P for polar, A for Arctic or Antarctic, and S for superior—a unique situation with dry air formed by a powerful downward motion of the atmosphere.
- A lowercase letter describes the relationship between the air mass

and the earth: k signifies that the air mass is colder than the ground below it, while w describes an air mass that is warmer than the ground below it.

7.1.2. Classification and Notation

An air mass is a volume of air that is specified by its temperature and humidity in meteorology. Numerous hundreds or thousands of square miles are covered by air masses, which adjust to the properties of the surface beneath them. Latitude and the regions of their continental or marine origins are used to classify them. Warmer air masses are referred to as tropical, and colder air masses as polar or arctic. While maritime and monsoon air masses are moist, continental and superior air masses are dry. Weather fronts divide air masses with varying density characteristics, such as moisture content or temperature. Subsurface vegetation and water bodies can rapidly alter the characteristics of an air mass once it has departed from its source region. Classification schemes address both the modification and the characteristics of an air mass.

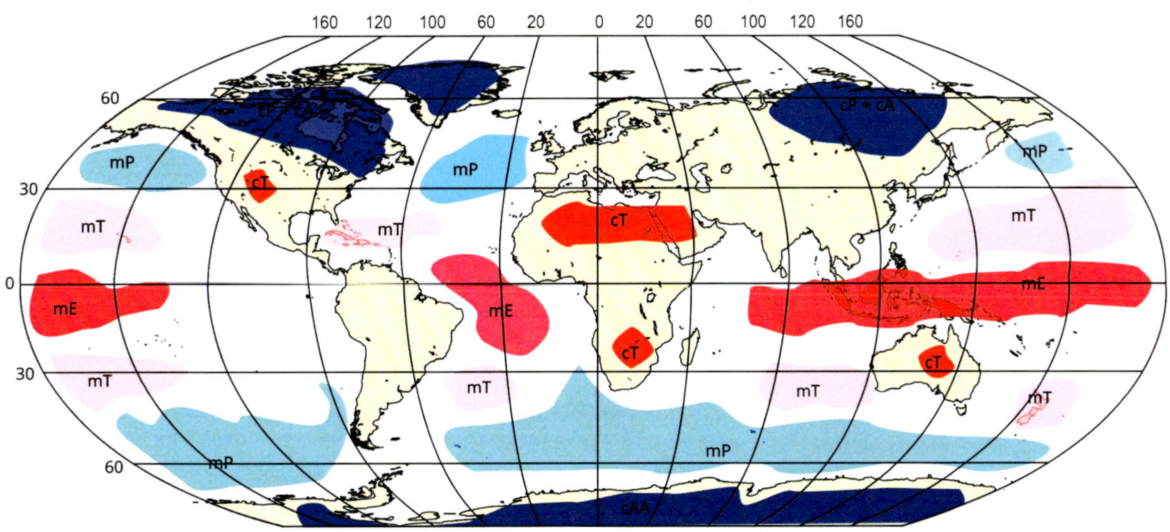

Figure 7.2. Source regions of global air masses.

Source: https://upload.wikimedia.org/wikipedia/commons/thumb/2/2f/Air_masses.svg/1280px-Air_masses.svg.png.

The Bergeron classification is the most widely accepted form of air mass classification, though others have produced more refined versions of this scheme over different regions of the globe. Air mass classification involves three letters. The first letter describes its moisture properties – "c" represents continental air masses (dry), and "m" represents maritime air masses (moist). Its source region follows: "T" stands for Tropical; "P" stands for Polar, "A" stands for Arctic or Antarctic, "M" stands for monsoon, "E" stands for Equatorial, and "S" stands for adiabatically drying and warming air formed by significant downward motion in the atmosphere. For instance, an air mass originating over the desert southwest of the United States in summer may be designated "cT." An air mass originating over northern Siberia in winter may be indicated as "cA" (Figure 7.2). A third letter, either k for an air mass that is colder than the surface below it or w for an air mass that is warmer than the surface below it, can be used to indicate the stability of an air mass. An illustration of this could be a polar air mass, represented by the symbol cPk, blowing over the Gulf Stream. On rare occasions, an apostrophe or degree tick may also be used to indicate that an air mass that is being replaced is colder than the original air mass (usually for polar air masses). For instance, a sequence of fronts across the Pacific might display an air mass with the designation mPk and another with the designation mPk.' An additional convention that makes use of these symbols is the declaration of a change or conversion from one type to another. For example, cA-mPk could represent an Arctic air mass blowing out over the Gulf of Alaska. Another convention denotes the stratification of air masses under specific circumstances. An air mass from the Gulf of Mexico overrunning a polar air mass over the Central United States, for example, could be represented by the notation mT/cP (sometimes using a horizontal line as in fraction notation).

7.1.3. Characteristics

As tropical and equatorial air masses form over lower latitudes, they become hot. Because hot air rises and cold air sinks, tropical air masses have lower pressure. These move poleward along the southern edge of the subtropical ridge and are drier and hotter than those that develop over oceans. They are also known as continental eruptions. Trade air masses are another name for maritime tropical air masses. The Caribbean Sea, southern Gulf of Mexico, and tropical Atlantic east of Florida via the Bahamas are the sources of maritime tropical air masses that impact the United States. The monsoon air masses are unstable and damp. Dry, superior air masses hardly ever touch down. Typically, they are found over maritime tropical air masses, where they create a trade wind inversion by forming a warmer, drier layer on top of the more moderately moist air mass below. Because of their continental source, continental polar air masses (cP) are dry and frigid air masses. Over the interior of Canada, continental polar air masses that impact North America form. The primary source of continental tropical air masses (cT) is low-latitude deserts like the Sahara Desert in northern Africa. CT is a type of tropical air produced by the subtropical ridge over large areas of land. The Arabian Peninsula, the central arid/semi-arid region of Australia, and the deserts in the Southwestern United States are other, less significant sources of cT air masses. Tropical air masses on the continent are incredibly dry and hot. The polar, Antarctic, and Arctic air masses are frigid. Ice and snow-covered ground develop the characteristics of arctic air. Polar air masses are not nearly as cold as the Arctic air. Arctic air can change quickly as it moves equatorward and can be shallow in the summer. Polar air masses are extremely stable, form over land or ocean at higher latitudes, and are typically shallower than arctic air. As polar air becomes more humid over warmer ocean waters, it becomes less stable over the ocean (maritime).

7.1.4. Movement and Fronts

Weather phenomena are primarily caused by weather fronts, which are boundaries dividing two masses of air with different densities. Depending on the type of front, different colored lines and symbols are used to represent it in surface weather analyzes.

The temperature and humidity of the air masses that a front separates are typically different. Cold fronts can occasionally be preceded by squall lines or dry lines, and they can feature narrow bands of thunderstorms and severe weather. Fog and stratiform precipitation often precede warm fronts. Usually, the weather clears up quickly after a front pass through. There is usually a shift in the wind with fronts, but sometimes there is no precipitation or minimal cloud cover. Warm fronts travel poleward, while occluded and cold fronts typically flow from west to east. Cold fronts and cold occlusions move more quickly than warm fronts and warm occlusions due to the higher air density in their wake. Fronts can move more slowly through mountains and warm bodies of water. A front can degenerate into a shear line, or line that divides areas of varying wind velocity when it becomes stationary and the density contrast across the frontal boundary disappears. This occurs most frequently over open waters.

7.1.5. Modification

Numerous techniques can be used to alter air masses. The surface flux of vegetation, like forests, acts to moisten the air mass above it. An air mass can be dramatically altered by heat from warmer waters beneath it over distances as short as 35 to 40 kilometers (22 to 25 miles). For instance, narrow lake-effect snow bands may result from curved cyclonic flow bringing cold air across the comparatively warm water bodies southwest of extratropical cyclones. Large bodies of water, like lakes, efficiently store heat, resulting in significant temperature differences (greater than 13°C or 23°F) between the water's surface and the air above (Jun Inoue et al. 2005). This causes strong localized precipitation to occur in those bands.

7.1.6. Categorization of Air Masses

Meteorologists classify air masses based on where they are formed. Here is the classification.

Typically, there are four types of air masses:

- Arctic: These air masses form in the Arctic region and are very cold.
- Tropical: These air masses form in low-lying latitudes and warm up to a moderate level.
- Polar: These air masses form in the high-latitude region and are cold.
- Equatorial: They start forming over the Equator and are warm.

Besides this classification, there is another way to categorize air masses. Meteorologists classify them based on whether they form over water or land. These distinctions are:

- Maritime: Maritime ones form over the water bodies and are filled with moisture.
- Continental: Whereas, the Continental ones form over the land and are arid.

For instance, an air mass formed over interior Asia is a continental polar air mass that is cold and dry. However, one that is formed over the Indian Ocean is a tropical air mass that is humid and warm.

7.1.6.1. Air Mass Categorization of the USA

- Continental Polar: This type of air mass forms over large land masses like Northern Canada. Mostly stable, these are free of condensation forms as they form over the land. When these air masses get heated or moistened from the ground, they are capable of producing snow or rain.

- Maritime Polar: They form over large water bodies and pick up considerable moisture. They move towards the land in middle and high latitudes. They are responsible for producing heavy rain when forced up mountain slopes or can be caught up in cyclonic activity.
- Continental Tropical: This type of air mass emerges in the lower latitudes over deserts or interiors of land masses. They usually develop in summer. It is subject to great heat, and due to the lack of moisture on the ground, they are typically arid.
- Maritime Tropical: They form over the vast tropical oceans and contain much moisture. They are responsible for producing rain. In case it moves polewards, it is cooled by the ground surface and produces fog. Over the continental areas, they are strongly heated and responsible for convectional rain or thunderstorms.

7.2. FORMATION OF AIR MASSES

A large area of air, typically spanning thousands to millions of square kilometers, is referred to as an air mass. It is a crucial component of the global planetary wind system that is situated in a particular area of the earth's surface and has horizontally uniform physical properties, such as humidity and temperature, for an extended period. It typically covers thousands of kilometers in length and has a subcontinental cost. In terms of humidity, pressure, and temperature, it is typically nearly horizontally uniform. It is a general principle that the changes occur with increasing altitude, independent of air mass or area. In terms of their physical characteristics, it is situated in a particular area on the uniform surface of the earth. It could vertically reach the troposphere's height. This huge amount of air remains horizontally for a considerable amount of time and acquires uniform physical properties, including humidity and temperature. An air mass can also be thought of as a body of air that covers a sizable homogeneous region and is distinguished by the same physical characteristics as the surrounding area, such as heat and moisture. The air mass that maintains continuous contact with the underlying air of any given region achieves the essential equilibrium between humidity and temperature. Such air is frequently referred to as air mass because of its distinct qualities, well-defined equilibrium, and minimal horizontal variation. In addition to the air's moisture content and degree of heat or cold, the vertical distribution of temperature defines the stability of the air mass and controls the condensation process. It is an essential component of the planetary wind system, which links them to either wind belt. In the belt of westerlies, for instance, tropical and polar air masses may coexist. This air mass maintains latitudinal temperature balance by transferring and distributing heat from the tropical to the Polar Regions. The horizontal gradient of moisture and temperature within the air mass is not very strong, at a certain height, the temperature and moisture content are nearly equal or comparable. The horizontal gradient steepens at the edges of air masses.

7.2.1. Origin of Air Mass

The place where air mass originates determines its origin. Its growth begins on a uniformly large area. The term "source region" refers to the area where air masses naturally develop. The temperature, pressure, and moisture content of the source region or area are absorbed by the air masses. The following would be an ideal source region for the air mass to form (Figure 7.3):

- A fairly uniform extensive area;
- Relatively flat and homogenous area with similar types of physical conditions;
- Prevailing calm conditions;
- Preferably high-pressure conditions.

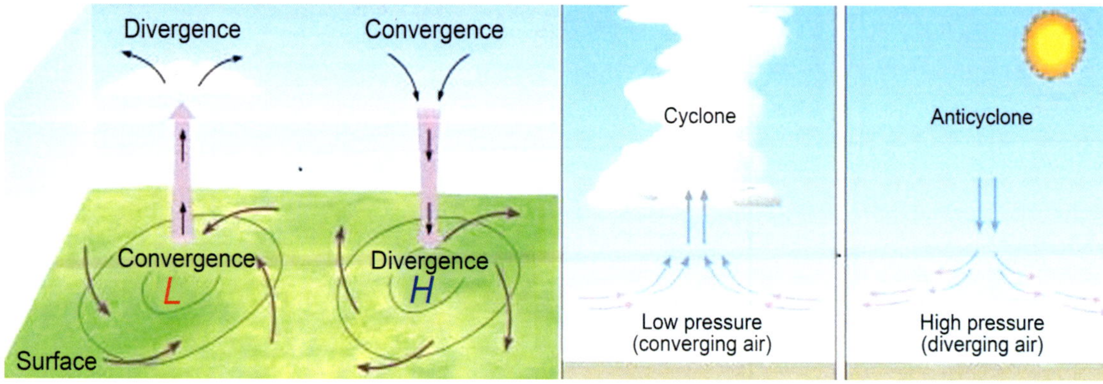

Figure 7.3. Formation of air mass.

Source: https://dspmuranchi.ac.in/pdf/Blog/Air_Masses_Classification.pdf.

Both the thermally induced polar zones and the subtropical vertically air subsiding zone are regions with regular high-pressure zones in the earth's atmosphere. These regions have high-pressure levels accompanied by a low horizontal change in pressure gradient. Since the divergent air circulation is a relatively stable air volume, it is the perfect source region for the air mass to form. Local disturbances such as cyclonic activity predominate in the mid-latitude region. There are essentially none of these things to do. The characteristics of the air mass are determined by the nature and features of the source region, which allows the equilibrium between heat (temperature) and moisture (humidity) with the overlying air mass to be maintained. These areas are typified by calm, light winds and anticyclones that are stationary or moving slowly. The characteristics of the air mass vary with source distance.

7.2.2. Determinant Factors of Air Mass

Large volumes of air with generally uniform horizontal temperature and humidity characteristics are referred to as air masses. The main factors that determine air mass are the source region's atmospheric conditions, surface type, and geographic location. It is essential to comprehend these factors to understand weather patterns and climatic variations. First of all, an air mass's properties are greatly influenced by its geographical origin. Air masses originate over large areas where the air can stay comparatively stable for an extended amount of time, enabling it to take on the properties of the surface beneath it. These areas, which include oceans, deserts, and the interior of continents, are typically large and uniform. For example, air masses that form over tropical oceans are moist and warm, whereas air masses that form over Polar regions are dry and cold. Another important factor to consider is the kind of surface that an air mass forms over. The air mass is directly impacted by the differences in the thermal characteristics and moisture content of land and water surfaces. Because of water evaporation, maritime air masses—which are created over oceans and seas—generally have higher humidity levels. Continental air masses, on the other hand, are usually drier since they originate over vast land areas. The temperature of the air mass is also influenced by the albedo, or reflectivity, of the surface. Darker surfaces, like those found in forests or urban areas, absorb more heat from the sun, resulting in warmer air masses. Surfaces with high albedo, like ice and snow, reflect more solar radiation and cause cooler air masses. (Jeffrey M. et al. 2001).

The stability and predominant wind patterns in the atmosphere also affect the properties of air masses. When the atmosphere is stable, an air mass can stay still and develop uniform characteristics. On the other hand, unstable circumstances may allow the air mass to mix with the surrounding air, changing its properties. Air masses are moved and altered by wind patterns, which are influenced by global circulation systems. For example, air masses are transported from their source regions by the trade winds, westerlies, and polar easterlies, which in turn affect weather patterns far from their origin. Air mass properties can also be influenced by the seasonality and diurnal variations in the source region. Temperature and humidity variations throughout the year have an impact on how air masses form and behave. For instance, a tropical continental air mass in the summer will likely be drier and hotter than one in the winter. Though they are less important than seasonal variations, diurnal variations—the daily cycle of temperature and humidity changes—also contribute to short-term fluctuations in air mass characteristics.

The nature of air masses is primarily determined by three factors. They are:

- Source region;
- Duration of air mass stability;
- Natural changes or modifications with time and space.

7.2.3. Classification of Air Masses

The primary classification of air masses is based on the characteristics of the source region.

They could be:

- Arctic (A);
- Polar (P); or
- Tropical (T).

There are five major source regions of air masses:

- Warm tropical and subtropical oceans;
- The subtropical hot deserts;
- The relatively cold high-latitude oceans;
- The very cold snow-covered continents in high latitudes;
- Permanently ice-covered continents in the Arctic and Antarctica.

Polar air masses are cold, while tropical air masses are warm. The characteristics of the reached areas are altered by the heat transfer processes from the arriving air mass. Whereas a cold air mass makes the area colder than the pre-existing temperature conditions, a warm air mass makes the area warmer than it was previously.

Based on the characteristics of the surface, they are:

- Continental (c);
- Maritime (m).

Apart from the previously mentioned ones, a wide range of secondary types of air masses are classified. For instance, Mediterranean or Equatorial. The two-letter initials may occasionally have a letter k or w appended to them to indicate whether the air is warmer (w) or colder (k) than the air mass's surface of origin. Compared to the warmer, the colder is more stable.

7.2.4. Types of Air Masses Based on Source Region

7.2.4.1. Based on Temperature Conditions

Based on temperature conditions air masses can be listed into two categories as follows:

1. Polar Air Mass (Cold): A cold air mass (polar or arctic) is defined as an air mass that has a temperature below the surface below it, which is typically linked to instability and high atmospheric turbulence. On surface weather maps, high pressure is linked to the centers of cold air masses. A steeper wedge than a warm front is produced when the warmer air is quickly forced upward at a sharp angle by the heavier, denser cold air that is pushing in beneath it. They typically come from one of the following areas:
   - Arctic Ocean – cold and moist;
   - Siberia – cold and dry;
   - Northern Canada – cold and dry;
   - Southern Ocean – cold and moist.
2. Tropical Air Mass (Warm): In general, stable weather conditions are associated with a surface that is not as warm as the mobile warm air mass. On surface weather maps, the cores of extremely warm air masses show up as temporarily located areas of low pressure. Light, warm air moves eastward, resulting in a succession of broad weather bands. These fronts are accompanied by cloud cover and precipitation because of the temperature instability and high humidity. The same air mass can be classified as cold or hot depending on the surface area it travels over and its temperature. They typically come from one of the following areas:
   - Tropical Deserts – warm and dry;
   - Tropical Oceans – warm and moist.

7.2.4.2. Based on the Latitudinal Extent

Different categories can be used to group air masses according to their latitudinal extent and place of origin. Air masses can be broadly categorized into the following groups according to their origin (Table 7.1). Warm air masses are found in tropical regions, whereas cold air masses are found in polar regions. The transfer of heat through conduction or convection allows the air mass to gradually change its properties through migration, movement, or even staying in one place.

After considering the origin of air masses theoretically, a wide variety of air mass types can be listed; however, in practice, some of these types cannot exist or are too weak to be taken into consideration. On weather maps, acronyms are occasionally used to indicate the air mass over a region, especially when the air masses have moved away from their original locations. Practically speaking, air masses are identified upon arrival rather than at the source of origin. Their arrival in a different area causes a noticeable difference, making the clearly defined boundary or zone obvious (Figure 7.4).

Table 7.1. Classification and Characteristics of Air Masses

Air Mass	Symbol	Properties	Region	Temper at ure (° C)	Specific Humidity g/kg
Maritime Tropical	mT	Warm moist	Warm Tropical and Sub-tropical Oceans	24 ° C	17.0
Continental Tropical	cT	Warm dry	The sub-tropic hot deserts	24 ° C	11.0
Maritime Polar	mP	Cool Moist	Warm Tropical and Sub-tropical Oceans	4 ° C	4.4
Continental Polar	cP	Cold dry	The very cold snow covered continents in high latitudes	-11 ° C	1.4
Continental Arctic	cA	Very Cold and dry	Permanently ice covered continents in the Arctic and Antarctica	-46 ° C	0.1

Source: https://dspmuranchi.ac.in/pdf/Blog/Air_Masses_Classification..pdf

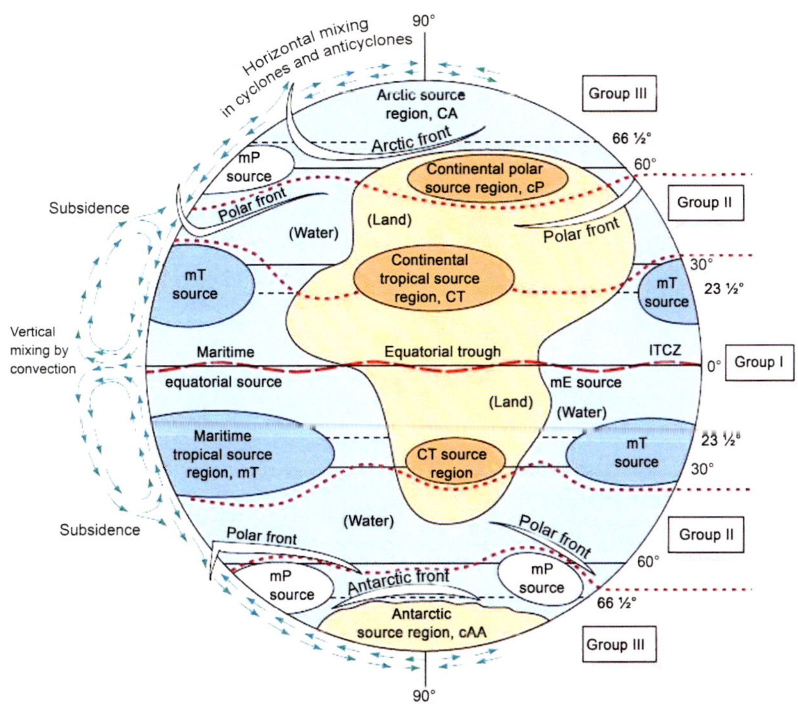

Figure 7.4. A diagrammatic presentation of the air masses on the globe.

Source: https://dspmuranchi.ac.in/pdf/Blog/Air_Masses_Classification.pdf

7.2.4.2.1 Maritime Tropical

The subtropical anticyclones (high-pressure cells) that linger over the oceans year-round at latitudes between 30°N and S are the sources of these air masses. They come from the Atlantic, Pacific, and Mexican Gulf, which are tropical and subtropical oceans. They have a distinct summer and winter character that affects the local climate. Due to the warm water from the seas beneath them, the air is warm, erratic, and humid. During the summer, the air absorbs a lot of moisture from the lower layers because high temperatures increase the air's capacity to absorb moisture. Subsequently, the stable air begins to ascend, ascending from subtropical regions to higher latitudes, and gradually cools down as a result of coming into contact with the frigid sea and land masses they cross. It causes the normal lapse rate (inverse) to decrease, which improves stability and reduces turbulence. The dew point is reached as a result of the lower layer's cooling process, which progressively raises relative humidity. When these two elements come together, fog, drizzle, and eventually rain with the development of stratus clouds are produced. Winters here are moderate, with cloudy, foggy weather. In the summer, this air mass moves over the warmer seas and invades the hotter parts of the continent, exhibiting unstable characteristics and heavy precipitation. The majority of the mid-latitude eastern coastal regions receive the most rainfall from this air mass. Conversely, summertime brings highs in temperature and humidity along with convectional rain and cumulous clouds. (Miller, Samuel T. 10 November 2000)

7.2.4.2.2 Continental Tropical

The tropical and subtropical desert regions of Asia, Australia, and North Africa (Sahara) are the source of these hot, dry, and stable winds. Known locally as Harmattan, these tropical continental dry winds originate in North Africa during the winter and are found in the summertime in the drier western margins of the United States of America. In most cases, these air masses stop at their source. These air mass types are also native to drier regions of Australia. Throughout the year, this air is dry with little to no humidity and stays hot and erratic during the summer. Despite being in an unstable zone, high temperatures and low humidity cause a high condensation level and no rainfall. In contrast, wintertime temperatures are still warm and dry, but due to the presence of descending air in anti-cyclonic zones, the air is stable stratified1. This steady wind is localized because it doesn't move very much. However, as it approaches colder surfaces, its stability rises in the area where it collides with cold air masses. This kind of front happens in the Mediterranean area. Unlike tropical maritime air masses, this one remains mostly dry all year round with little precipitation.

7.2.4.2.3. Continental Polar

These air masses originate from Polar regions such as Antarctica, the Arctic basin, Eurasia, and northern North America, as their name implies. These air masses are extremely cold, dry, and stable. The wind movement is mild and divergent here. The distance from the warmer water bodies (oceans) and the continuous (long-lasting) terrestrial radiation are the main causes of the extremely thick and cold winds. During the summer, these winds are warmer and less stable, with fewer or no anticyclonic winds. Winters are cold, stable, and clear; there is little cloud cover, which causes precipitation to fall as ice or snow.

7.2.4.2.4. Maritime Polar

Between 40° and 60° latitudes are the

oceans from which maritime polar air masses originate. Maritime polar air masses are those continental polar masses that travel over warmer bodies of water, such as oceans, warm up, and absorb moisture (humidity). The air in this area is cool, humid, and erratic. These are the areas that are incapable of remaining still for very long. In this zone, summers are steady, clear, and fair, and winters are marked by high humidity, moderate temperatures, cloudy skies, and sporadic fog and precipitation. One can observe different fronts and air movements worldwide (Figure 7.4).

7.2.4.2.5. Continental Arctic

Large patches of snow and ice, usually close to the poles in both hemispheres, are where these air masses form. When the Poles are extremely cold and receive little insolation, Arctic air masses typically form in the winter months of December through March in the Northern Hemisphere and June through September in the Southern Hemisphere (Figure 7.5).

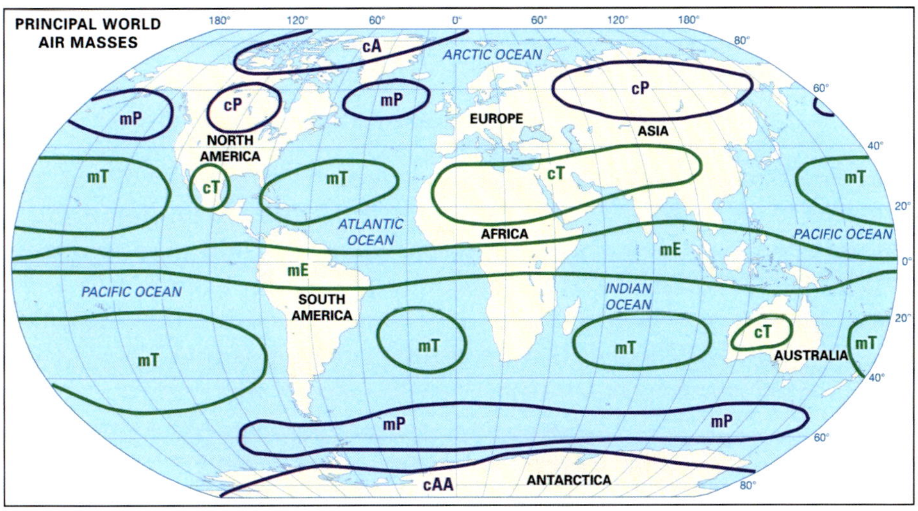

Figure 7.5. Principal air masses on the globe.

Source: https://dspmuranchi.ac.in/pdf/Blog/Air_Masses_Classification.pdf.

7.2.5. Convergence of Air Masses and Fronts

While the various air masses have different origins and destinations, they eventually converge in a low-pressure area (Figure 7.6). The interface zone is where warm and cold air masses collide, producing very little mixing.

The upper layers are essentially unaffected and undisturbed, where the air masses mostly maintain their unique characteristics. The lower atmosphere, close to the earth's surface, is where the instability and turbulence take place. The phenomenon of different air masses meeting and mixing due to convergence causes colder, denser air masses to occupy the space formerly occupied by warmer air masses. The colder, heavier air mass gives way to the warmer, lighter one. Condensation and precipitation are the products of the fronts' ascent and convergence. Temperature and humidity have an impact on how strong the disturbances are during this convergence process. In general, when polar and subtropical air masses converge, strong storms frequently occur.

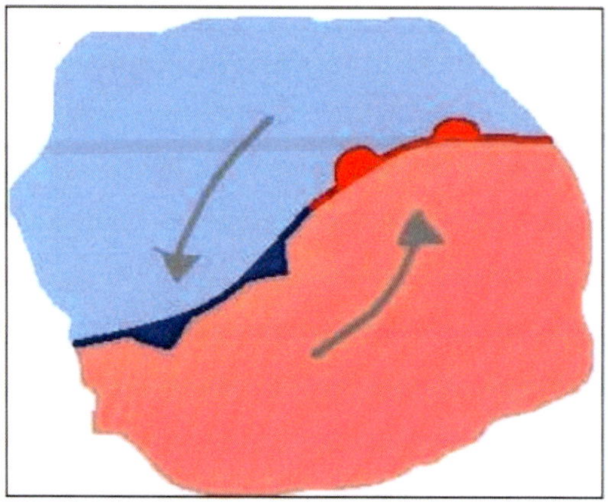

Figure 7.6. Convergence of air masses.

Source: https://dspmuranchi.ac.in/pdf/Blog/Air_Masses_Classification.pdf.

1. Stationary Fronts: A stationary front develops when a frontal system ceases to advance (Figure 7.7). A cold or warm front may stall, resulting in stationary fronts that can stay that way for several days. It could be a warm front or a cold front when a stationary front begins to move again. Frequently, the temperature gradient becomes less pronounced and the front merely disappears. When a front is stationary, the weather is stable and pleasant.

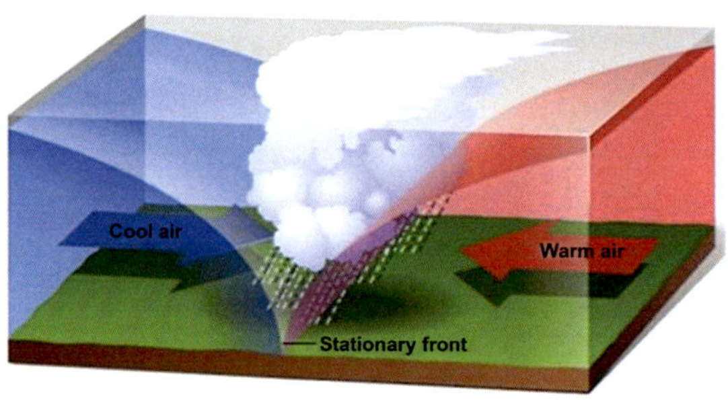

Figure 7.7. Formation of stationary front.

Source: https://dspmuranchi.ac.in/pdf/Blog/Air_Masses_Classification.pdf.

2. Occluded Fronts: As the combined front of the air mass separates from the ground during the later stage of a depression's dissipation, an occluded front forms. Two forms of occlusion exist: warm front type and cold front type (Figure 7.8). When a dense mass of cold air pushes a mass of warm air upward and removes it from the ground, occlusion occurs. The warm sector on the ground decreases and is entirely covered by the cold air mass. Consequently, a long, backward-swinging occluded front—which may or may not be a warm-type

occlusion—forms. An occluded front produces complicated weather that combines elements of both warm and cold frontal systems.

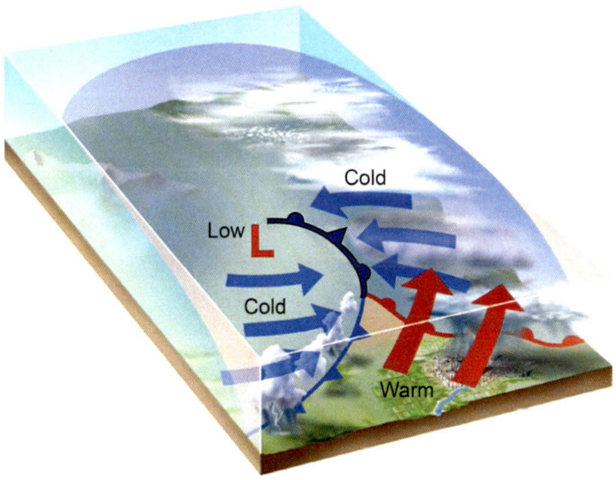

Figure 7.8. Occluded front.

Source: https://dspmuranchi.ac.in/pdf/Blog/Air_Masses_Classification.pdf.

7.2.6. Air Masses Modification

An air mass's characteristics affect the regions in which it occurs. Normally, air masses move from their original locations to other areas that are close by. Two different kinds of air modifications occur in the moving air masses. They are mechanical modifications (b) and thermodynamic modifications (a).

7.2.6.1. Thermodynamic Modification

The heat from the moving air mass transfers to the moved locations, causing thermodynamic modification. At the same time, the moved locations modify the thermodynamics of the modified area by absorbing its features. The effects of thermodynamic processes include the reduction of vertical stability due to below-surface heating. Other include water evaporating into the air mass from below or precipitating into the intermediate layers from the moist air layers above. The degree of modification of an air mass is contingent upon.

- the initial temperature;
- moisture conditions;
- the path taken by air masses;
- the nature of underlying area; and
- time taken or duration of travel.

A mass of air passing over a surface warmer than the ground temperature will inevitably warm from below, steepening the lapse rate and increasing stability as a result. Condensation and precipitation are made more likely by these modifications. A mass of air passing over a colder surface experiences a downward cooling. This circumstance promotes the development of a surface inversion, which raises the air mass's stability. Precipitation and cloud formation are impossible in these circumstances. The polar air masses have a tendency to become more stable as they leave their source regions. On the other hand, the second kind of modification occurs and the tropical air masses become more stable. Changes in thermodynamics result from higher evaporation. The air mass itself may provide the moisture through precipitation from an overlay layer or from the surface the air is traveling over.

7.2.6.2. Mechanical Modification

An air mass may undergo changes due to any of the following reason/s.

- Turbulent mixing caused by eddies or convection.
- Large-scale dynamic effects on in lapse rate; divergence; convergence.
- Subsidence or sinking: in subsidence and latent spreading: movement down from above colder air masses; descent from high elevations to low lands.

- Lifting: over colder air masses; to compensate for horizontal convergence; over elevations of the land.
- Advection of new properties aloft due to the shearing action of the wind.

Through mixing, frequently to a significant height, mechanically induced turbulence resulting from frictional effects at the surface may occur. Air masses are significantly altered as a result of heat and moisture being transported from the surface to different layers of the atmosphere. Significant sections of the atmosphere may move upward or downward due to large-scale divergence or convergences close to or above the surface, which could have an impact on atmospheric stability. A falling air mass becomes more stable through subsidence when it descends on the leeward side of a physical barrier. When an air mass is lifted during a frontal or orographic uplift, it becomes more unstable. Just by examining the names and origins of air masses, one can learn a great deal about the weather conditions they represent. Maritime air masses have a considerable moisture content. As the wind moves across the humid, moisture-laden regions of the warm ocean, it picks up moisture. From oceans to continents, atmospheric moistures are carried by the air mass and wind, which results in precipitation over landmasses. This humid air is cool in the maritime arctic and Polar Regions, where warm, humid weather is produced by the maritime tropical air mass. On the other hand, dry weather is produced by continental air masses. This is because, in terms of moisture, the continents simply cannot compete with the oceans. Winter weather is dry and cold due to the continental arctic and polar air masses, while summer weather is pleasant. The latitudinal heat balance is eliminated by the movement of latent heat by air masses. The majority of migratory atmospheric disturbances, like storms and cyclones, start at the point where two or more air masses collide, and the weather that results from these disturbances is influenced by the properties of the involved air masses (Miller et al., 2000; Roth et al., 2013).

7.3. BASICS OF WEATHER FRONT

In meteorology, a front is the interface or transition zone between two air masses that have different densities and temperatures. The Norwegian meteorologists who gave the front its name during World War I compared the sporadic weather flare-ups along this zone, which occasionally included thunderstorms and electrical activity, to the fighting along Europe's battle lines. Low barometric pressure, also known as a pressure trough, significant variations in wind direction and relative humidity, as well as significant cloud cover and precipitation, are often associated with frontal zones. There are numerous varieties of fronts, primarily based on which way the colder air mass is moving. A cold front is the name given by meteorologists to the leading edge of an advancing mass of relatively cold air. Cold fronts typically move eastward and toward the Equator in middle and high latitudes in both hemispheres, with the most advanced position occurring directly at the ground. The front typically lies 80 to 160 km (50 to 100 miles) behind its surface position at a height of approximately 1.5 km (1 mile); as a result, its slope is 1/50 to 1/100. A cold front typically brings with it thunderstorms and showers. Warm air that has been displaced is forced to rise as the relatively dense cold air undercuts it as it moves forward, sometimes at a speed of 50 to 65 km/h (30 to 40 miles/h). In severe situations, the instability that results could cause a squall line of violent thunderstorms and possibly tornadoes to form parallel to the cold front and roughly 80 km ahead of its surface. Usually, the precipitation ends suddenly after the front moves through. (Ahrens, C. Donald 2007).

The line separating a mass of warm air from a mass of cold air that is retreating is called a warm front. Warm air tends to override rather than displace cold air at constant atmospheric pressure because it is less dense than cold air. A warm front thus tends to move more slowly than a cold front. Compared to cold fronts, its inclination, or slope, is substantially lower. The front typically lies 320 km (200 miles) ahead of its surface position, to the north or northeast in the Northern Hemisphere, at a height of approximately 1.5 km. Precipitation associated with warm fronts is typically far more uniform and widespread than that of cold fronts. Severe ice storms have the potential to form over 100 kilometers (62 miles) ahead of the warm front's surface position in the winter, provided that warm air prevails over cold air at temperatures below freezing. An occluded front occurs when a warm front is surpassed by a cold front that is circling a low-pressure center, as is frequently the case. In high and middle latitudes, cyclonic storms frequently begin as waves or undulations on a frontal boundary separating warm and cold air masses. The surface atmospheric pressure at the center of the storm decreases as the wave gets stronger. The slower-moving cold air beneath the warm front eventually catches up with the cold air behind the cold front as it advances. The cyclone is said to have occluded when the warm air tongue that is in between is forced skyward. The storm's kinetic energy, which is produced by warm air rising and cold air sinking, stops increasing at this point. Occasionally, on a stationary front—a boundary between air masses where the colder air mass has little to no horizontal motion—a wave cyclone forms, bringing with it low atmospheric pressure and stormy weather. For instance, storms that affect the northeastern United States frequently originate from a wave on a stationary front over the southern or southeastern states. The significant thermal contrast between the warm tropical air to the south and the cold polar air to the north provides the storms with significant motion energy. Variations in the location and strength of this polar front boundary between air masses are primarily responsible for the variable weather conditions typical of high and middle latitudes. In both hemispheres, polar fronts are typically found poleward of 30° latitude and can sporadically extend to locations close to the Arctic and Antarctic circles (66°30′ N and S). The large thermal contrast between the air masses drives particularly strong westerly winds in the polar-front jet stream, which is located 10 to 12 km (6 to 7 miles) directly above the front. The formation of cyclonic storm systems (centered on a low-pressure area) along the front is facilitated by the strong thermal contrast within the front.

7.3.1. Characteristics of Fronts

A weather front is a line that divides air masses with different properties, including wind, temperature, humidity, and air density. Along the boundary, disturbed and unstable weather frequently results from these differences. For example, warm fronts are typically preceded by stratiform precipitation and fog, whereas cold fronts can bring bands of thunderstorms and cumulonimbus precipitation, or they can be preceded by squall lines. Dry lines—more subdued variations in humidity—can herald severe weather in the summer. There is usually a shift in the wind with fronts, but sometimes there is no precipitation or minimal cloud cover. Warm fronts can travel in any direction, but they usually move poleward. Cold fronts travel from west to east. Static fronts are halted in their path, and occluded fronts are a hybrid of the two. Because the dense air behind cold fronts and cold occlusions can both lift and push the warmer air, they move more quickly than warm fronts and warm occlusions. Aside from atmospheric conditions, mountains and bodies of water can influence the movement and characteristics of fronts. The front can become little more than a shear line, a line that divides areas of different wind speeds, when the density difference between the air masses has decreased, as it would after flowing out over an evenly warm ocean. This occurs most frequently over open waters.

7.3.2. Bergeron Classification of Air Masses

The Bergeron classification is the most widely accepted form of air mass classification. Air mass classifications are indicated by three letters: Fronts separate air masses of different types or origins, and are located along troughs of lower pressure.

- The first letter describes its moisture properties, with:
  - c used for continental air masses (dry); and
  - m used for maritime air masses (moist).
- The second letter describes the thermal characteristic of its source region:

- o T for tropical;
- o P for polar;
- o A for Arctic or antarctic;
- o M for monsoon;
- o E for equatorial; and
- o S for superior air (dry air formed by a significant upward lift in the atmosphere).

- The third letter designates the stability of the atmosphere; it is labeled:
  - o k if the air mass is colder ("kolder") than the ground below it.
  - o w if the air mass is warmer than the ground below it.

7.3.3. Surface Weather Analysis

A surface weather analysis is a unique kind of weather map that, using data from ground-based weather stations, shows the topography of the weather over a given region at a given time. To locate synoptic scale features like weather fronts, weather maps are produced by identifying, charting, and tracing the values of pertinent quantities like temperature, cloud cover, and sea level pressure onto a geographical map.

Special symbols that display frontal systems, cloud cover, precipitation, and other pertinent data are used in surface weather analyzes. An H, for instance, might stand for a high-pressure region, which suggests clear or pleasant weather. Conversely, an L might stand for low pressure, which usually precedes storms and precipitation. Surface winds originating from high-pressure zones are also produced by low pressure, and vice versa. In addition to frontal zones and other surface boundaries, different symbols are used to show the current weather at different points on the weather map. Furthermore, the location and type of front are determined by areas of precipitation.

7.3.4. Types of Fronts

A front signals a change in the weather when it passes over a region. Weather phenomena like rain, thunderstorms, strong winds, and tornadoes are all brought on by numerous fronts. There could be violent thunderstorms during a cold front. There might be low-stratus clouds at a warm front. After the front moves on, the skies are usually clear. At the surface of the Earth, a weather front is a zone of transition between two distinct air masses. Every air mass has different humidity and temperature properties. At a front, which is the boundary where two distinct air masses converge, turbulence frequently occurs. Storms and clouds can result from the turbulence. Certain fronts merely bring about a shift in temperature rather than clouds and storms. Still, some storm fronts are the catalyst for the biggest storms on Earth. Tropical waves are fronts that form off the coast of Africa in the tropical Atlantic Ocean. If conditions permit, these fronts have the potential to strengthen into hurricanes or tropical storms. Over several days, fronts pass across the surface of the Earth. High winds, like Jet Streams, frequently dictate the direction of motion. Mountains and other landforms can also alter a front's course. The weather surrounding a frontal zone has two distinct meanings in meteorology. When air rises quickly along and over a boundary, causing significant weather changes and heavy precipitation, this is referred to as an anafront. A kata-front is weaker and brings with it less rainfall, less changes in temperature, and less moisture.

7.3.4.1. Cold Front

A closely spaced temperature gradient's

warm side is bordered by and located along a cold front. This temperature gradient is represented by isotherms on surface analysis charts, and since cold fronts frequently line up with a surface trough, isobars can occasionally be used to identify this gradient as well. The cold front is located at the leading edge of the cooler air mass and is indicated on weather maps by a blue line with triangles pointing in the direction that cold air travels. Along with rain, cold fronts can also bring strong thunderstorms. Because cold air is denser than warm air, it can push and lift the warm air ahead of the boundary, causing cold fronts to move at a rate up to twice as fast as warm fronts and produce sharper and more intense changes in the weather. If sufficient humidity is present when the lifted moist warm air condenses, the lifting motion frequently produces a narrow line of showers and thunderstorms. The idea that the less dense, warmer air will eventually wed the denser, colder air is oversimplified because the upward motion is actually a natural reaction of frontogenesis to maintain geostrophic balance on the rotating Earth (Figure 7.9).

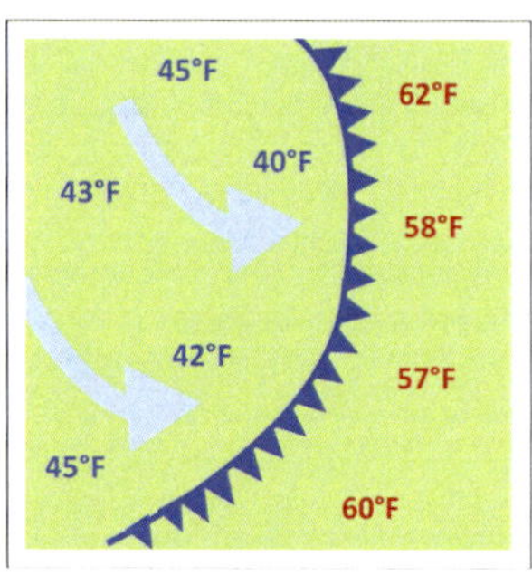

Figure 7.9. The weather map shows a cold front. Air temperatures ahead of the front are warmer than temperatures in the cold air mass behind the front.

Source: https://scied.ucar.edu/sites/default/files/styles/extra_large/public/images/cold_front.webp?itok=8PWD1jO0.

7.3.4.2. Warm Front

As seen in the image to the right (A), a warm front forms when a warm air mass pushes into a cooler air mass. As the warm air mass at the surface rises above the cool air mass, creating clouds and storms, warm fronts frequently bring stormy weather. Since warm air finds it more difficult to push dense, cold air across the surface of the Earth, warm fronts move more slowly than cold fronts. On the east side of low-pressure systems, where warmer air from the south is forced north, warm fronts frequently form. Ahead of a warm front, you can frequently see middle clouds like altostratus and high clouds like cirrus and cirrostratus. The warm air above the cool air is where these clouds form.

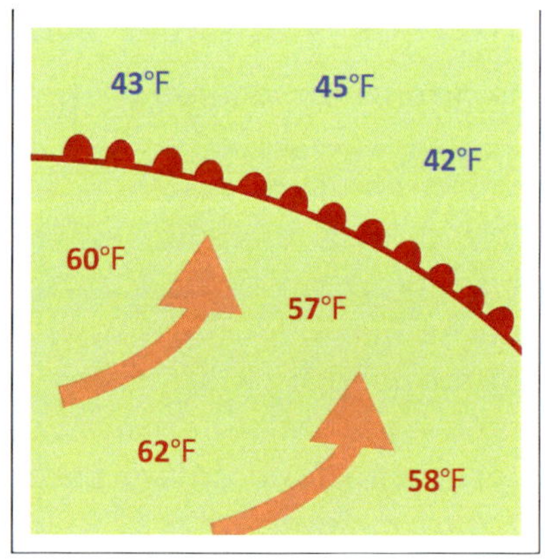

Figure 7.10. The weather map shows a warm front. Air temperatures ahead of the front are cooler than temperatures in the warm air mass behind the front.

Source: https://scied.ucar.edu/sites/default/files/styles/extra_large/public/images/warm_front.webp?itok=tuUO3bCg.

The clouds get thinner as the front moves across a region, increasing the likelihood of rain. If the air is unstable, thunderstorms may form in the vicinity of the warm front. A warm front's surface location is shown on weather maps as a solid red line with red, filled-in semicircles running along it, as shown in map (B) on the right. The semicircles show which way the front is traveling. They are positioned where the front of the line is moving. On the map, you can see that the ground level temperature is lower in front of the front than it is behind (Figure 7.10).

7.3.4.3. Occluded Front

When a cold front passes over a warm front, it creates an occluded front, which typically forms near-developed low-pressure systems, such as cyclones. The triple point, or point of occlusion, is where the cold and warm fronts naturally curve poleward. Though the air mass behind the boundary can be either warm or cold, it is situated within a sharp trough. The air mass that passes over the warm front in a cold occlusion is cooler than the cold air mass that is regressing from the warm front, and it plows under both air masses.

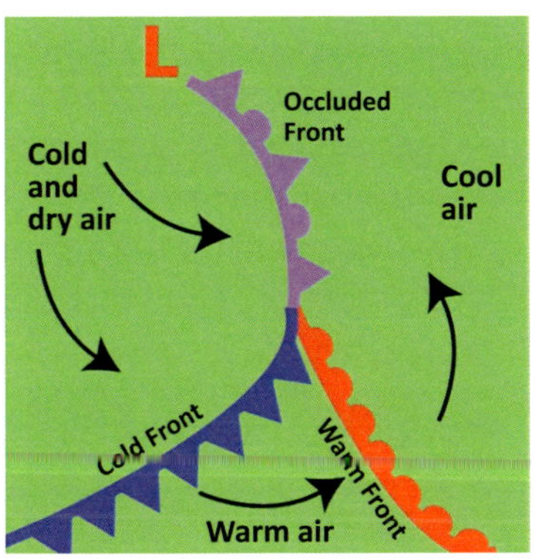

Figure 7.11. On a weather map, an occluded front looks like a purple line with alternating triangles and semicircles pointing in the direction that the front is moving.

Source: https://scied.ucar.edu/sites/default/files/styles/extra_large/public/images/Occluded-front.webp?itok=QctODTLY.

The cold air mass that is eclipsing the warm front in a warm occlusion ride over the colder air while lifting the warm air because it is warmer than the cold air mass that is regressing from the warm front. Depending on the kind of occlusion the front is experiencing, a circulation of air drives cold air drafts downward and warm air upward within the occlusion. The Trowal, or the projection of the warm air tongue above the Earth's surface created during the depression or storm's occlusion process, is linked to clouds and precipitation. On a weather map, an occluded front is shown by a purple line with triangles and half-circles pointing in the direction of travel alternately. An array of red and blue junction lines designates the Trowal (Figure 7.11).

7.3.4.4. Warm Sector

A near-surface air mass situated between the warm and cold fronts, the warm sector is typically observed on the equatorward side of an extratropical cyclone. Warm and humid, this air can support thunderstorms and is prone to convective instability, particularly if it is lifted by the approaching cold front.

7.3.4.5. Stationary Front

A stationary front is a boundary between two air masses that is not strong enough to displace the other and is not moving, also known as stalled. They typically move in waves rather than continuously, staying essentially in one place for extended periods, especially when winds are blowing parallel to one another. Behind the boundary, there is typically a large temperature gradient with more widely spaced isotherm packing. A stationary front can experience a wide range of weather conditions, but clouds and heavy precipitation are typically present. Stationary fronts can become either a warm or cold front if the conditions above change, but they usually fade away after a few days or become shear lines. Weather maps display stationary fronts as alternating red half-circles and blue spikes pointing in opposite directions, signifying no discernible movement. Shearlines are formed when stationary fronts shrink in size and temperature stabilize, eventually degenerating into a narrow zone where wind direction shifts dramatically over a short distance. A shearline is represented by a dashed and red line. Stationary fronts have the potential to bring prolonged periods of light snow or rain (Figure 7.12).

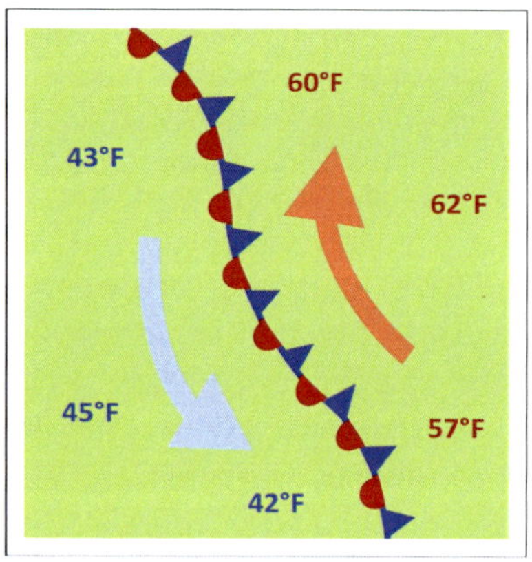

Figure 7.12. A stationary front is represented on a map by triangles pointing in one direction and semicircles pointed in the other direction.

Source: https://scied.ucar.edu/sites/default/files/styles/extra_large/public/images/stationary_front.webp?itok=1e69sCk_.

1. Dry Line: The dry line, which is the boundary between air masses with notable moisture differences rather than temperature differences, is a phenomenon akin to a weather front. Lee troughs form when areas of lowered pressure form downwind of north-south oriented mountain chains, which is caused by an increase in westerlies on the north side of surface highs. Warm, humid air is denser near the surface during the day than dry, higher-temperature air, so it wedges under the drier air like a cold front. The boundary slope reverses and the warm, humid air becomes less dense than the cooler, dry air at higher elevations. Severe weather is possible in the area around the reversal aloft, particularly if a cold front forms an occlusion or triple point. The lee trough, which shows less variance in moisture, is a weaker version of the dry line that is more frequently observed. Diurnal thunderstorms can center on the boundary when moisture collects there during the warm season. Anywhere on Earth that is in between a desert and warm waters can have a dry line. One particularly popular area is the southern plains of the United States, west of the Mississippi River. Normally, the dry line travels west at night and east during the day. On National Weather Service (NWS) surface analyzes, an orange line with scallops pointing into the moist sector represents a dry line. One of the rare surface fronts where the indicated pips do not always correspond to the direction of motion is a dry line.
2. Squall Line: In a pattern where the upper-level jet splits into two streams, organized thunderstorm activity can both reinforce and outrun actively developing cold fronts. The resulting Mesoscale Convective System (MCS) forms at the point of the upper-level split in the wind pattern, which runs southeast into the warm sector parallel to low-level thickness lines. The MCS is referred to as a squall line when the convection is strong, linear, or curved, and the

feature is located at the leading edge of the notable wind shift and pressure rise. Even weaker and less organized thunderstorms produce locally higher pressures and cooler air; additionally, there are outflow boundaries ahead of this kind of activity, which can serve as centers for more thunderstorm activity later in the day. These features, which are located inside surface troughs, are frequently shown on surface analyzes during the warm season throughout the United States. A haboob may occur if squall lines or outflow boundaries form over arid regions. On NWS surface analyzes, squall lines are shown as a pattern of two red.s alternated with a dash labeled SQLN or squall line, and outflow boundaries are shown as troughs labeled outflow boundary.

7.3.5. Precipitation Produced

The main source of significant weather events is fronts. The movement of the cold front or cold occlusion beneath a mass of warmer, moister air causes air to be lifted and condense into clouds, resulting in convective precipitation, which includes showers, thundershowers, heavy rain, and related unstable weather. Roll clouds and tornadoes may develop if there are significant temperature differences between the two involved air masses and intense turbulence due to wind shear and the presence of a strong jet stream. Lee troughs, breezes, outflow boundaries, and occlusions can all cause convection during the warm season if there is sufficient moisture present. When air masses pass over topography like hills and mountains, they produce precipitation known as orographic precipitation. This type of precipitation is most frequently produced when cold fronts move into mountainous regions. It can occasionally happen ahead of warm fronts that are heading eastward over mountainous terrain. On the other hand, warm front-related precipitation, such as light rain or drizzle, is comparatively consistent. Pre-warm-frontal regions are frequently where fog occurs; it can occasionally be dense and extensive. However, since moisture must be present in the air mass being lifted, not all fronts result in precipitation or even clouds.

7.3.6. Weather Patterns Associated with Different Fronts

Understanding and forecasting meteorological phenomena depend heavily on the weather patterns connected to the various front types—cold fronts, warm fronts, stationary fronts, and occluded fronts. These fronts are the boundaries between different air masses that interact to produce a range of weather conditions, including differences in temperature and humidity. When a cold air mass moves forward and pushes aside a warmer air mass, a cold front is created. The warmer air is forced to rise quickly as the colder, denser air presses beneath the lighter, warmer air. This quick ascent causes the warm air to cool and condense, which frequently results in the formation of cumulonimbus clouds. As a result, cold fronts are frequently linked to abrupt changes in the weather, such as thunderstorms, torrential rain, and sporadically severe weather like hail and tornadoes. Following the passage of a cold front, the sky clears, the air gets dryer and more stable, and the temperatures drop dramatically. When a warm air mass passes over a cold air mass that is retreating, a warm front is created. Warm air rises more slowly than cooler, denser air because it is less dense, which causes a more gradual process of lifting and cooling. This more gradual ascent creates a lot of stratiform cloud cover, which usually means long-lasting, consistent precipitation. Overcast skies and light to moderate rain or drizzle are common conditions ahead of a warm front. Temperatures rise and humidity rises as the front moves through, frequently resulting in milder and more stable weather.

When one cold or one warm air mass is insufficiently powerful to push aside the other, a stationary front is created. Days may pass while the boundary between the two air masses stays still, resulting in protracted cloud cover and precipitation. Although stationary front-related weather patterns are usually less severe, they can nevertheless result in prolonged cloud cover, light to moderate rain, and occasionally thunderstorms. Large amounts of rain can result from these fronts, particularly if they linger over the same region for a long period. When a cold front passes over a warm front, lifting the warm air mass completely off the ground, occluded fronts form. Occlusions come in two varieties: warm occlusions and cold occlusions. The air ahead of the warm front is warmer than the overtaking cold front in a warm occlusion, and vice versa in a cold occlusion. Occluded fronts can cause complicated weather patterns that frequently combine elements of both warm and cold fronts. They usually result in widespread cloud cover, precipitation, and the possibility of thunderstorms and moderate to heavy rain. Temperatures typically decrease and the weather stabilizes as the occlusion deepens.

7.3.7. Area of Weather Fronts

Weather fronts, which demarcate the boundaries between various air masses with unique temperature, humidity, and density characteristics, are essential components of meteorology. These fronts have a major impact on weather patterns over wide geographic regions, affecting everything from isolated storms to extensive precipitation events. Usually, cold fronts are linked to constrained but active weather zones. Warm air is forced to rise quickly when a colder, denser air mass enters a warmer region. A line of cumulonimbus clouds and other severe weather phenomena, including thunderstorms, torrential rain, and occasionally hail and tornadoes, are frequently the result of this interaction. A cold front's path can affect a large area, but the area experiencing the strongest weather is typically narrow—often only a few tens of kilometers wide. Clearer and more stable weather conditions follow the front, which is followed by a noticeable drop in temperature and a change in wind direction. On the other hand, warmer fronts are marked by more gradual weather changes over larger areas. A warm air mass rises gradually over the denser, colder air as it moves ahead of the retreating cold air mass. Widespread, consistent precipitation and massive stratiform clouds are produced by this process. Warm front-related weather changes can extend hundreds of kilometers in width, with cloudy skies, light to moderate rain, or drizzle falling before the front. While cold fronts bring abrupt changes in temperature, warmer temperatures, and higher humidity are usually experienced over a longer period.

When neither air mass is powerful enough to push the other side, a stationary front forms, which can result in a standoff that lasts for days. These fronts frequently bring with them protracted cloud cover and precipitation. Sometimes the impact area is quite wide, encompassing whole regions or even several regions. A stationary front may bring persistent rain or snow, cloudy skies, and occasionally thunderstorms. These fronts can linger for long periods, which increases the likelihood of substantial rainfall totals and possible flooding in the impacted areas. When a warm front is completely lifted off the ground by a cold front that passes over it, an occluded front results. This intricate interplay combines the qualities of both warm and cold fronts over a wide region. Precipitation and widespread cloud cover are frequently caused by occluded fronts. An occluded front can affect a region that is several hundred kilometers wide, and it is frequently accompanied by thunderstorms and moderate to heavy rain. Temperatures typically drop and the weather stabilizes as the occlusion deepens (Saravanan et al., 2008; Ahrens et al., 2007).

CASE STUDY

THE GREAT PLAINS BLIZZARD OF 2021:

One of the most illustrative examples of air mass interaction is the Great Plains Blizzard that occurred in February 2021. This event provides a clear demonstration of how contrasting air masses can create severe weather conditions.

In early February 2021, a large continental arctic (cA) air mass moved southward from the Arctic, bringing extremely cold temperatures into the central United States. Simultaneously, a maritime tropical (mT) air mass from the Gulf of Mexico began to push northward. The collision of these two vastly different air masses set the stage for a significant weather event.

The leading edge of the arctic air mass, characterized by temperatures well below freezing, met the warm, moist air of the maritime tropical mass. This interaction occurred over the Great Plains, creating a potent frontal system. As the warm, moist air was forced to rise over the denser, cold arctic air, it cooled and condensed, forming clouds and precipitation.

Development and Impact of the Blizzard:

The Great Plains Blizzard developed rapidly as the frontal system intensified. The stark temperature gradient between the two air masses fueled the formation of a powerful low-pressure system. As the system deepened, it drew in more warm air from the south and cold air from the north, enhancing the temperature contrast and increasing the storm's intensity.

Heavy snowfall, strong winds, and plummeting temperatures were the hallmarks of the blizzard. Snowfall accumulations exceeded 12 inches in many areas, with some locations reporting over 20 inches. The combination of heavy snow and high winds led to whiteout conditions, making travel nearly impossible and resulting in widespread road closures and transportation disruptions.

The blizzard's impact extended beyond immediate weather conditions. The arctic air mass brought record-breaking low temperatures, with some areas experiencing temperatures below –20°F (–29°C). The prolonged cold snap strained energy infrastructure, leading to power outages and heating system failures. The agricultural sector also faced challenges, with livestock at risk from the extreme cold and heavy snow.

Long-Term Implications and Lessons Learned:

The Great Plains Blizzard of 2021 underscores the importance of understanding air mass dynamics and their role in weather forecasting. The ability to predict the movement and interaction of air masses is crucial for anticipating severe weather events and mitigating their impacts. Advances in meteorological technology, such as satellite observations and computer modeling, have improved the accuracy of weather predictions, but challenges remain in forecasting the precise timing and intensity of such events.

Furthermore, this case study highlights the broader implications of air mass interactions in the context of climate variability and change. As global temperatures rise, the characteristics and behaviors of air masses may shift, potentially altering weather patterns and increasing the frequency and severity of extreme weather events. This underscores the need for ongoing research and investment in climate science to better understand and prepare for future changes.

Conclusion:

Air masses are fundamental components of the Earth's atmospheric system, and their interactions significantly influence weather patterns. The Great Plains Blizzard of 2021 serves as a vivid example of how contrasting air masses can create extreme weather conditions with far-reaching impacts. Understanding the formation, movement, and interaction of air masses is essential for accurate weather forecasting and effective response to severe weather events. As our climate continues to change, the study of air masses will remain a critical area of research, helping to build resilience against the increasing challenges posed by extreme weather.

CLASS ACTIVITY

SIMULATING WEATHER FRONTS

- Ø Objective: To understand how different types of weather fronts form and influence weather conditions.
- Ø Materials Needed: Large clear plastic bin, Cold water dyed blue, Warm water dyed red, Ice cubes, Heat source (such as a lamp), Plastic divider (optional), Thermometers, Small paper or plastic boats (optional), Weather front chart/ diagram handouts, Observation worksheets
- Ø Assessment: Have students use thermometers to measure and record the temperature of the water in different parts of the bin during each demonstration. Ask students to place small paper or plastic boats on the water surface to observe how the currents change with different fronts. Provide worksheets for students to record their observations and answer questions about the characteristics and effects of each type of front.
- Ø Extension Activity: Ask students to research a recent weather event involving a significant front (e.g., a cold front leading to a snowstorm) and present their findings to the class. Students can create detailed weather maps showing the progression of fronts and associated weather conditions.

SUMMARY

- Air masses are large bodies of air that have similar temperatures and humidity. They form over areas of land or water. There are four types of air masses: cold, warm, hot, and continental. A front is a boundary between two different air masses. Fronts can be either warm or cold.
- Air masses are commonly classified according to four basic source regions concerning latitude. These are Polar (cold), Arctic (very cold), Equatorial (warm and very moist), and Tropical (warm).
- Continental Polar (cP) air usually forms during the cold period of the year over extensive land areas such as central Asia and northern Canada. It is likely to be stable and is characteristically free of condensation forms.
- Maritime Polar (mP) air masses develop over the polar areas of both the Northern and the Southern hemispheres. They generally contain considerably more moisture than the cP air masses.
- Air masses form over large surfaces with uniform temperatures and humidity, called source regions. Low wind speeds let air remain stationary long enough to take on the features of the source region, such as heat or cold.
- Air masses are also identified based on whether they form over land or water. Maritime air masses form over the water and are humid. Continental air masses form over land and are dry.
- Tropical and equatorial air masses are hot as they develop over lower latitudes. Tropical air masses have lower pressure because hot air rises and cold air sinks. Those that develop over land (continental) are drier and hotter than those that develop over oceans and travel poleward on the southern periphery of the subtropical ridge.
- A weather front is a boundary separating two masses of air of different densities and is the principal cause of meteorological phenomena. In surface weather analyzes, fronts are depicted using various colored lines and symbols, depending on the type of front.
- An air mass is an extensive body of air, usually many thousands to millions of square kilometers of area. It is an integral part of global planetary wind system that is located in a specific region on the surface of the earth and stays for a substantial period of time with horizontally uniform physical characteristics, such as temperature and humidity.
- Air masses are large volumes of air that have relatively uniform temperature and humidity characteristics horizontally. The determinants of air mass are primarily geographical location, surface type, and the atmospheric conditions prevalent in the source region.
- Thermodynamic modification is caused by the transfer of heat from the moving air mass to the moved places and simultaneously gets modified by the moved places by acquiring the characteristics of that area.

REVIEW QUESTIONS

1. How do meteorologists classify and denote different types of air masses, and what criteria are used in this classification system?
2. What mechanisms drive the movement of air masses, and how do these movements create various types of weather fronts?
3. What are the primary categories of air masses, and how do their characteristics differ based on their classification?
4. What factors determine the characteristics of an air mass, and how do geographical location and surface type influence these factors?
5. How air masses classified are based on their source region, and what are the distinctive weather patterns associated with each type?

REFERENCES

1. Ahrens, C. D. (2007). Meteorology Today: An Introduction to Weather, Climate, and the Environment (9th ed.). Cengage Learning. ISBN 978-0-495-01162-0.
2. Byrd, G. (1998). Lake effect snows. University Corporation for Atmospheric Research.
3. Freedman, J. M., & Fitzjarrald, D. R. (2001). Postfrontal airmass modification. Journal of Hydrometeorology, 2(4), 419–437. doi:10.1175/1525-7541(2001)002<0419>2.0.CO;2.
4. Geerts, B. (1998). Lake effect snow. University of Wyoming.
5. Inoue, J., Kawashima, M., Fujiyoshi, Y., & Wakatsuchi, M. (2005). Aircraft observations of air-mass modification over the Sea of Okhotsk during sea-ice growth. Boundary-Layer Meteorology, 117(1), 111–129.
6. Miller, S. T. (2000). Clouds and precipitation. STEC 521 Lecture Notes. University of New Hampshire.
7. Monmonier, M. S. (1999). Air Apparent: How Meteorologists Learned to Map, Predict, and Dramatize Weather. University of Chicago Press. ISBN 0-226-53422-7.
8. Park, C. C. (2001). The Environment: Principles and Applications. Psychology Press. ISBN 978-0-415-21771-2.
9. Roth, D. (2013). Unified Surface Analysis Manual (PDF) (vers. 1 ed.). NOAA / Hydrometeorological Prediction Center / National Hurricane Center.
10. Saravanan, K. J., ed. (2008). Bergeron classification of air masses. weatherfront.blogspot.com.

CHAPTER 8

Climate Classification and Climatic Zones

LEARNING OBJECTIVES

After studying this chapter, you will be able to:

- Understand the history and significance of climate classification;
- Explain traditional climate classification systems;
- Describe Koppen climate classification.

KEY TERMS FROM THIS CHAPTER

Water resource management	Temperature
Meteorological elements	Vegetation
Climate classification	Humidity
Thermal efficiency	Potential
Agriculture	Convectional rainfall
Thunderstorms	Rainfall patterns

INTRODUCTION

Systems of climate classification divide the various climates on Earth into discrete groups according to factors such as temperature, precipitation patterns, and other geographical characteristics. These systems are essential for comprehending and controlling the variability of weather in various geographical areas. Developed by Wladimir Köppen in the early 1900s, the Köppen Climate Classification is still one of the most well-known and widely used systems. Based on average monthly temperature and precipitation, it categorizes climates into five main groups: polar, dry, tropical, temperate, and continental. Every group is further separated into categories, represented by letters, which stand for distinct attributes like temperature ranges and seasonality. For instance, temperate climates, represented by the letter C, have clear seasonal variations, whereas tropical climates, represented by the letter A, are distinguished by year-round warmth. The Trewartha Climate Classification, created by Glenn Thomas Trewartha in the middle of the 20th century, is another noteworthy system. Certain aspects of Köppen's classification are improved by this system, especially the way it differentiates between various temperate climate types and highlights the significance of vegetation and humidity in defining a given climate.

Beyond scholarly curiosity, there are numerous significant uses for climate classifications. They offer a common language and framework for understanding global climate patterns among scientists, decision-makers, and planners. This knowledge is essential for many industries, such as urban planning, agriculture, and water resource management. For instance, understanding the climate of a given area aids farmers in selecting crops and irrigation techniques that are suitable for the climate in the area. In a similar vein, cities can be planned by urban planners to be resilient to climatic shocks like heavy rain or extremely high temperatures. Furthermore, research and mitigation initiatives related to climate change heavily rely on climate classifications. They make it possible for scientists to examine past climate data and forecast future climate conditions depending on various scenarios of greenhouse gas emissions. Policymakers can create plans to adjust to these changes and lessen their effects by knowing how climate classifications may change in response to climate change. The Köppen and Trewartha climate classification systems offer crucial resources for classifying and comprehending the climatic diversity of the planet. They support real-world applications in fields like urban planning and agriculture, facilitate multidisciplinary research, and provide information for policy decisions. These classification schemes are always evolving as our knowledge of climate dynamics grows, so they remain relevant despite the world's climate becoming more complex and dynamic.

8.1. THE HISTORY AND SIGNIFICANCE OF CLIMATE CLASSIFICATION

Climates are categorized according to temperature, precipitation, and other factors. This classification is essential to comprehending global weather patterns. The Köppen Climate Classification is still widely used; it was developed by climatologist Wladimir Köppen. Differentiating between tropical, arid, temperate, and polar climates is helpful for global agricultural, urban planning, and environmental management. Changes made to it throughout its evolution include the Trewartha system, which smoothes out the finer points of climate boundaries and their effects on ecosystems and human activity. Climate classification improves our capacity to forecast climate trends, create policies, and lessen the effects of climate change globally by offering a standardized framework.

8.1.1. Climate Classification: A Historical Perspective

The word "climate," which has Greek roots, refers to the declination of the Sun. This declination's seasonal variation is solely dependent on latitude. Ancient Egyptian and Hellenistic authors were cognizant of the significant impact that latitude has on climate. The efforts of multiple German scientists over the past 200 years are principally responsible for the current differentiation between climatic regions. Alexander von Humboldt plotted the yearly mean temperature on a map of the world in 1817. This map was improved by Wladimir Koeppen (1846–1940), who in 1884 plotted the seasonal temperature range and developed his climate classification. This classification was hierarchical, with major categories subdivided, then subcategories divided again, and so on. It followed the classifications of plants (by Carolus Linnaeus in 1735) and clouds (by Luke Howard in 1802). Koeppen had studied botany at St. Petersburg before going on to earn a Ph. D. in Heidelberg regarding how temperature affects plant growth. His system is based on five sets of temperature limits at its highest level. These originated from his classification of the thermal zones suitable for different types of vegetation in 1884. He made numerous system improvements, most notably in 1918 and 1936 (when he was ninety years old).

Koeppen moved from Germany to Graz, Austria, in 1924 with his son-in-law, the meteorologist Alfred Wegener (1880–1930). Wegener's theory of continental drift caused him to be shunned by German universities. Koeppen in Graz inferred geological climates to bolster the theory. Additionally, he got to know Rudolf Geiger (1894–1981) and worked with him to create the 1936 climate classification system. Subsequent revisions were attributed to Geiger. As he gathered a plethora of data to comprehend "the climate near the ground"—to borrow the title of his book, which was translated into English in 1960—and its fluctuations owing to topography and land use, Geiger founded the field of microclimatology.

In 1928, Swedish-born Tor Bergeron (1891–1971), who received his education in Germany, created a new system for categorizing climates based on their causes. His "genetic" system assigns a location a classification based on the frequencies at which it experiences different types of airmass. Every airmass has a label that indicates the latitude at which its temperature was previously measured as well as the type of surface—continental or marine—that exists there. Although Helmut Landsberg (1906–1985) used this process in Pennsylvania and central Europe in the late 1930s, it is rarely employed today. There are not enough classes offered, and the system needs a lot of data.

In the 1920s, Griffith Taylor (1880–1963) caused political controversy in Australia by adamantly defending the legitimacy of the significant portion of the continent that Koeppen classified as "desert" (BW) or "semi-desert" (BS), which had the unintended consequence of restricting the country's ability to grow. In Western Australia, aspirational politicians banned his geography textbook. Thus, in 1928, he departed the nation, first traveling to Toronto and then Chicago.

A hierarchical classification system was created in the USA in 1931 by Warren Thornthwaite (1892–1963), primarily based on the annual pattern of soil moisture conditions. These were thought to be intricately dependent on the monthly input—rain—and the implicit output—evaporation, which is indicated by temperature—of the system. Research conducted in New Zealand, for instance, revealed that, except low latitudes, Thornthwaite's classification made more sense than Koeppen's. Subsequently, Thornthwaite's seminal "approach toward a rational classification of climate" clarified the link to soil moisture. However, it used an empirical evaporation formula that was replaced in the same year in England by Howard Penman's (1909–1984) physics-based formula. These categorization schemes are all different from what has been done subsequently to address real-world issues. These days, we categorize locations with different climates and identify regions with "similar" climates using intricate statistical techniques.

8.1.2. Significance of Climate Classification

The attempt to organize massive amounts of meteorological data on weather phenomena in various parts of the Earth is called climate classification. The goal of classifying climates is to identify different types of climates and to identify patterns of climate that are similar and different across the globe. The understanding of physical mechanisms influencing climate, as well as the assessment of climatic potentialities and aspects of their utilization in the various regions of the Earth, are the scientific and practical significance of climate classification.

The majority of categorization systems focus on macroclimates, i.e., with the typification of continent-specific climates or the typification of global climates, depending on the interplay of large-scale processes. When interpreting the results of this classification, it is important to remember that processes operating on different scales, ranging from tens to hundreds of kilometers, have a significant impact on the values of specific meteorological elements and shape the microclimate in the region. The range of microclimates, which are the result of physical processes at the tens of meter scale, in turn, reveal mesoclimate. Meso- and microclimates are influenced by factors such as urbanization, water bodies' presence, vegetation type, slope type, and local height. The answers to these queries are thoroughly examined in Local Climate. On the one hand, the classification of climates requires the identification of regions with unique climatic patterns. However, it offers the chance to discover global climate parallels across various regions.

8.2. TRADITIONAL CLIMATE CLASSIFICATION SYSTEMS

An area's climate is the culmination of its external factors, including its soils, vegetation, weather, and so forth.) that have been prevalent there for a very long time. Averages of the climatic components as well as measurements of variability (like extreme values and probabilities) are included in this synthesis. A complicated, abstract idea, climate is based on information about every facet of the Earth's environment. As a result, it is impossible to say that two places on Earth have precisely the same climate.

However, it is easily observed that, over a small portion of the Earth, climates differ within a narrow range and that there are climatic regions that are identifiable and where some degree of consistency in the patterns of climatic elements is evident. Furthermore, when a region's geographic relationships are similar to another, even geographically distant parts of the world have similar climates. The symmetry and structure of the climate environment point to an underlying global regularity and order in the processes that shape the climate (such as air mass, temperature, winds, vegetation, soils, and incoming solar radiation patterns). Even with these underlying patterns, developing a precise and practical climate scheme is a difficult undertaking.

First of all, since climate is a multifaceted concept, deciding which of the numerous observed environmental variables to use as the classification's foundation is not always clear-cut. It is necessary to make this decision based on several theoretical and practical considerations. For instance, employing an excessive number of components increases the likelihood that the classification will have too many categories for ease of interpretation and that many of the categories won't accurately reflect climates. Furthermore, data on a great number of climate-related factors are either unavailable for extensive regions of the globe or have only been gathered recently. The main exceptions are data on temperature, precipitation, vegetation, and soil that have been recorded over longer periods and are more widely available.

The classification's goal (e.g., to explain soil formation processes, account for the distribution of natural vegetation, or categorize climates according to human comfort) also influences the variables chosen. This goal will establish the classification's pertinent variables as well as the threshold values of the variables selected to distinguish between climatic zones. The second challenge stems from the fact that variations in the climatic elements occur gradually throughout the surface of the Earth. Temperature, precipitation, and other climatic factors typically change very slowly over distance, unless there are exceptional circumstances brought on by mountain ranges or coastlines. As a result, when one travels from one area of the Earth's surface to another, climate types often change gradually. Selecting a set of parameters to differentiate between two climatic types is akin to creating a line on a map to separate the climatic region that possesses one type from that possesses the other. It is important to keep in mind that borders between neighboring climatic regions are drawn somewhat arbitrarily through regions of continuous, gradual change, and that the areas defined within these boundaries are far from homogeneous in terms of their climatic characteristics. This is not unlike many other classification decisions that one makes regularly in daily life.

The majority of classification systems define regions that are significant divisions of continents that range in size from hundreds to thousands of kilometers. They are designed for use on a global or continental scale. We could refer to these as macroclimates. There won't just be gradual shifts (from hot to cold, wet to dry, etc. There will be mesoclimates within these regions associated with climatic processes occurring at a scale of tens to hundreds of kilometers that are created by elevation differences, slope aspect, bodies of water, differences in vegetation cover, urban areas, and the like.) across such a region as a result of the geographic gradients of climatic elements over the continent of which the region is a part. Mesoclimates can be further subdivided into multiple microclimates that exist at less than 0.1 km (0.06 mi) in size. Examples of these microclimates include the climatic variations found in forests, crops, and bare soil, as well as at different depths in the soil, on different sides of buildings, and in the plant canopy.

Despite these drawbacks, climate classification is essential for several reasons: it helps to generalize the geographic distribution and interactions among climatic elements; it helps identify combinations of climatic influences that are significant for different phenomena that are dependent on the climate; it encourages the search for the mechanisms that govern the climate; and it serves as a teaching tool by illuminating some of the ways that distant regions of the world differ from and are similar to one's home region.

8.2.1. Approaches to Climatic Classification

Greek classical times are credited with creating the first known climatic classifications. These systems typically separated the Earth into latitudinal zones according to the length of the day and the important latitude parallels of 0°, 23.5°, and 66.5° (the Equator, the Tropics of Cancer and Capricorn, and the Arctic and Antarctic circles, respectively). The first temperature and precipitation maps of the Earth's surface were published in the middle of the 19th century, which paved the way for the development of techniques for simultaneously using both variables to classify climates. This is where modern climate classification got its start (Figure 8.1).

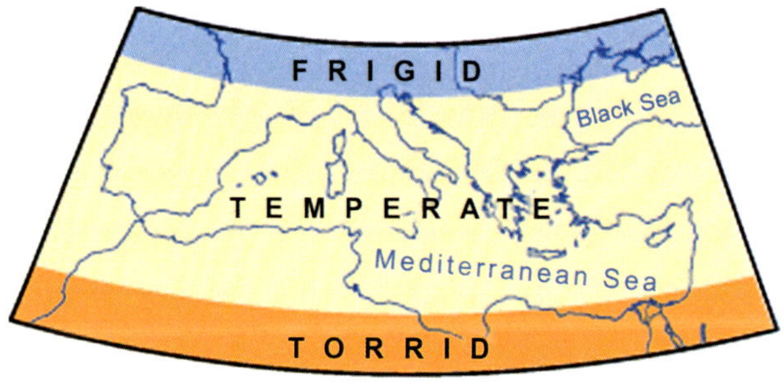

Figure 8.1. Map of climatic zones.

Source: https://www.britannica.com/topic/classification-1703397.

More than a hundred different classification schemes for climate have been developed, but they can all be broadly divided into two categories: genetic and empirical. Based on the type of data used for classification, this distinction is made. Empirical techniques utilize simple quantities (like evaporation) or observed environmental data (like temperature, humidity, and precipitation). The activity and features of all the factors (air masses, circulation systems, fronts, jet streams, solar radiation, topographic effects, and so on) that result in the spatial and temporal patterns of climatic data are the basis for the classification of climate by a genetic method, in contrast. Therefore, genetic methods are (or should be) explanatory, whereas empirical classifications are primarily descriptive of climate. Regretfully, genetic schemes are intrinsically more challenging to execute due to their reliance on complex observations, despite their scientific desirability. Thus, these kinds of schemes are generally less successful and less common. Furthermore, there is no guarantee that the regions delineated by the two categories of classification schemes will coincide; in particular, it is frequently the case that numerous common empirical schemes will group similar climatic forms that arise from distinct climatic processes.

8.2.2. Genetic Classifications

Climates are categorized genetically according to their causes. Three categories of these techniques can be identified: (1) those that rely on the surface energy budget, (2) those that are based on the geographic determinants of climate, and (3) those that are derived from air mass analysis.

Several schemes (mostly developed by German climatologists) are used in the first class to classify climates based on variables like latitudinal temperature control, continentality versus ocean-influenced factors, location with pressure and wind belts, and mountain effects. All of these classifications have one thing in common: they are qualitative, meaning that rather than being determined by applying a strict differentiating formula, climatic regions are assigned based on subjective criteria.

Werner H's 1970 classification serves as an intriguing illustration of a methodology based on the energy balance of Earth's surface. Terjung, a geographicist from America. His technique makes use of data on net solar radiation received at the surface, available energy for evaporating water, and available energy for heating the air and subsurface for over 1,000 locations worldwide. The maximum energy input, the annual range in input, the annual curve's shape, and the number of months with negative magnitudes (energy deficits) are the factors used to categorize the annual patterns. Regions with comparable net radiation climates are mapped, and the combination of attributes for a location is represented by a label made up of multiple letters with clear meanings.

On the other hand, genetic systems that make use of air mass concepts are likely the most widely used. Air masses are big groups of air that have generally the same temperature, humidity, and other characteristics. Within the horizontal. These characteristics and the contrasts between them at fronts can be used to interpret the weather on specific days. Classifications based on air mass have been primarily influenced by two American geographer-climatologists. 1951 saw Arthur N. Based on the combination of air masses present at a specific location throughout the year, Strahler described a qualitative classification. After a few years, in 1968 and 1970, John E. By offering a quantitative framework that classified specific air masses and air mass combinations as "dominant," "subdominant," or "seasonal" at specific locations, Oliver gave this kind of classification a stronger foundation.

8.2.3. Empirical Classifications

The majority of empirical classifications aim to classify climates according to one or more components of the climate system. Although numerous phenomena have been utilized in this manner, natural vegetation is the most significant. Many climatologists believe that natural vegetation serves as a long-term integrator of a region's climate; in essence, vegetation serves as a tool for measuring climate in a similar manner to how a thermometer measures temperature. The fact that many textbooks and other sources refer to climates by the names of vegetation, such as rainforest, taiga, and tundra, indicates how dominant it is.

Of these vegetation-based classifications, the most widely used (though not the first) was created by German botanist and climatologist Wladimir Köppen. His goal was to create formulas that would define climatic boundaries so that they would line up with the newly mapped vegetation zones for the first time in his lifetime. In 1900, Köppen released his initial plan, and in 1918, he revised it. Up to the time of his death in 1940, he kept making changes to his classification scheme. Based on their global experience, other climatologists altered some of Köppen's methodology.

Based on the division of terrestrial climates into five major types—represented by the capital letters A, B, C, D, and E—Köppen's classification was developed. Except B, each of these climate types is determined by temperature parameters. Type B climates are those in which dryness—rather than cold—is the primary factor influencing vegetation. The relationship between the amount of precipitation added to the soil where plants are grown and the amount lost through evaporation defines the state of aridity, which is not solely dependent on precipitation levels. Köppen was compelled to use a formula that defines aridity in terms of a temperature-precipitation index because evaporation is challenging to assess and is not a common measurement at meteorological stations (that is, evaporation is assumed to be controlled by temperature). Dry climates can be further classified into arid (BW) and semiarid (BS) subtypes, with the addition of a third code denoting either warm (h) or cold (k).

The other four major climate types are defined by temperature, as mentioned above. These are further classified, with additional letters being used once more to denote the different subtypes. The seasonality of precipitation distinguishes Type A climates, the warmest, into three categories: Af (no dry season), Am (short dry season), and Aw (winter dry season). The two traditional classifications of Type E climates, which are the coldest, are tundra (ET) and snow/ice (EF). A second letter, f (no dry season), w (winter dry) or s (summer dry), designates the midlatitude C and D climates. A third symbol, one of a, b, c, or d (the last subclass is exclusive to D climates), designates the warmth or coldness of the summer or winter. The highland climate category, or H climate, is occasionally added to climate classification systems to account for elevations above 1,500 meters (about 4,900 feet), despite Köppen's classification not taking into account the uniqueness of highland climate regions. The Köppen-Geiger-Pohl system from 1953 is specified in Table 8.1.

Table 8.1. Classification of Major Climatic Types According to the Modified Köppen-Geiger Scheme

Letter Symbol			
1st	**2nd**	**3rd**	Criterion
A			Temperature of the coolest month 18°C or higher.
	f		Precipitation in the driest month of at least 60 mm.
	m		Precipitation in the driest month less than 60 mm but equal to or greater than 100 – (r/25)1.
	w		Precipitation in driest months less than 60 mm and less than 100 – (r/25).
B[2]			70% or more of annual precipitation falls in the summer half of the year and r less than 20t + 280, or 70% or more of annual precipitation falls in the winter half of the year and r less than 20t, or neither half of the year has 70% or more of annual precipitation and r less than 20t + 1403.
	W		r is less than one-half of the upper limit for classification as a B type (see above).
	S		r is less than the upper limit for classification as a B type but is more than one-half of that amount.
		h	t equal to or greater than 18°C.
		k	t less than 18°C.
C			The temperature of the warmest month greater than or equal to 10°C, and the temperature of the coldest month less than 18°C but greater than –3°C.
	s		Precipitation in the driest month of the summer half of the year is less than 30 mm and less than one-third of the wettest month of the winter half.
	w		Precipitation in the driest month of the winter half of the year less than one-tenth of the amount in the wettest month of the summer half.
	f		Precipitation more evenly distributed throughout the year; criteria for neither s nor w satisfied.
		a	The temperature of the warmest month is 22°C or above.

		b	The temperature of each of the four warmest months is 10°C or above but the warmest month is less than 22°C.
		c	The temperature of one to three months 10°C or above but the warmest month less than 22°C.
D			The temperature of the warmest month greater than or equal to 10°C, and the temperature of the coldest month –3°C or lower.
	s		Same as for type C.
	w		Same as for type C.
	f		Same as for type C.
		a	Same as for type C.
		b	Same as for type C.
		c	Same as for type C.
		d	The temperature of the coldest month less than –38°C (d designation then used instead of a, b, or c).
E			The temperature of the warmest month is less than 10°C.
	T		The temperature of warmest month greater than 0°C but less than 10°C.
	F		The temperature of the warmest month is 0°C or below.
H[4]			Temperature and precipitation characteristics highly dependent on traits of adjacent zones and overall elevation—highland climates may occur at any latitude.

[1]In the formulas above, r is the average annual precipitation total (mm), and t is average annual temperature (°C). All other temperatures are monthly means (°C), and all other precipitation amounts are mean monthly totals (mm).

[2]Any climate that satisfies the criteria for designation as a B type is classified as such, irrespective of its other characteristics.

[3]The summer half of the year is defined as the months April–September for the Northern Hemisphere and October–March for the Southern Hemisphere.

[4]Most modern climate schemes consider the role of altitude. The highland zone has been taken from G.T. Trewartha, An Introduction to Climate, 4th ed. (1968).

There are numerous reasons to criticize the Köppen classification. It has been suggested that extreme occurrences, like a recurring drought or an uncommon cold snap, have an equal impact on regulating vegetation distributions as the average conditions Köppen's scheme is predicated on. It has also been noted that vegetation depends on elements like wind and sunshine that are not included in the classification. Furthermore, it has been argued that because natural vegetation can only adapt slowly to changes in the environment, the vegetation zones that are visible today have partially adapted to earlier climates. The rather weak agreement between the Köppen zones and the observed distribution of vegetation in many parts of the world has been brought up by several critics. Despite these and other drawbacks, the Köppen system is still the most widely used climatic classification system in existence.

The American geographer and climatologist made a significant contribution to the classification of climates. 1948; C. Warren Thornthwaite, 1931. He first specified atmospheric effects on vegetation using a vegetation-based approach that used the derived concepts of temperature efficiency and precipitation effectiveness. The concepts of a moisture index and a thermal efficiency index were kept in his second classification, but the classification criteria were drastically altered. Rather than using vegetation as the climatic integrator, he rejected the idea and tried to classify "rationally" based on the numerical values of these indices. Although his 1948 scheme is found in many climatology texts, it has not become as popular as the Köppen classification system among general readers. This could be due to its complexity and the vast number of climatic regions it defines.

Although vegetation-based climate classifications may be relevant to human activity in terms of what they might reveal about the potential for agriculture and the surrounding environment, they are unable to convey an understanding of what it would be like for people to live in the various climate types. Terjung attempted to categorize climates according to how they affect human comfort in his 1966 scheme. Four physiologically significant parameters are used in the classification: temperature, relative humidity, wind speed, and solar radiation. To express atmospheric conditions as perceived to be extremely hot, oppressive, warm, comfortable, cool, keen, cold, very cold, extremely cold, and ultracold, the first two are combined to create a comfort index. A wind effect index, which combines temperature, wind speed, and solar radiation, expresses the net effect of solar radiation's addition of heat to the human body and wind chill, or the cooling effect of wind on exposed surfaces. To represent how people feel in different regions of the world on an annual basis, these indices are combined for different seasons in different ways. Terjung envisioned his classification being useful as a general analytical tool as well as in the fields of medical geography, climatological education, tourism, housing, and apparel.

Numerous other specialized empirical classifications have been developed. Some theories distinguish between various desert and coastal climates, while others take into consideration the varying rates of rock weathering and soil formation. Still others are predicated on the identification of comparable agricultural climates (Chen et al., 2017; Geiger et al., 1961).

8.3. KOPPEN CLIMATE CLASSIFICATION

Wladimir Koppen, a German climatologist and botanist, developed the plant-based Koppen climate classification system, which is currently in widespread use. An attempt was made to develop a formula to draw climatic boundaries that correspond with vegetation zones or biomes worldwide with the Koppen climate classification system. When Koppen was developing his system of climate classification in 1900, these biomes were being developed and mapped for the first time. In 1918, Koppen released an updated version of his Koppen climate classification system. He would go on to make revisions to the system until he died in 1940.

Numerous attempts have been made throughout history to divide the climates of the Earth into climatic regions. The finer details of Aristotle's division of the Earth into three zones—torrid, temperate, and frigid—have long since been forgotten. Many attempts were made afterward to categorize the climatic regions of Earth, but none had the same impact as the Koppen system. In 1928, Koppen's students Rudolph Geiger and Koppen co-authored a map that included the Koppen climate classification system. Since then, several geographers have made changes to the Koppen climate classification system.

The five major types of climactic zones are classified into five categories by the Koppen climate classification system, also known as the Koppen-Geiger climate classification system. Koppen denoted these types with the letters A, B, C, D, and E. The current Koppen climate classification system is based on the categorization of climactic zones according to temperature, precipitation, and associated vegetation. All climactic zones are defined by temperature, except for B, where vegetation is determined by dryness, which falls under precipitation.

Aridity however, is not determined only by precipitation, and precipitation input in soil also works along with evaporation losses among plants.

The five major climactic zones defined by Koppen as elucidated by Michael Pidwirmy, 2014 are (Figure 8.2; Table 8.2):

1. Tropical moist climates (average temperature above 18°C in all months).
2. Dry climates (deficient precipitation for most of the year).
3. Moist mid-latitude climates with mild winters.
4. Moist mid-latitude climates with cold winters.
5. Polar climates (extremely cold summers and winters).

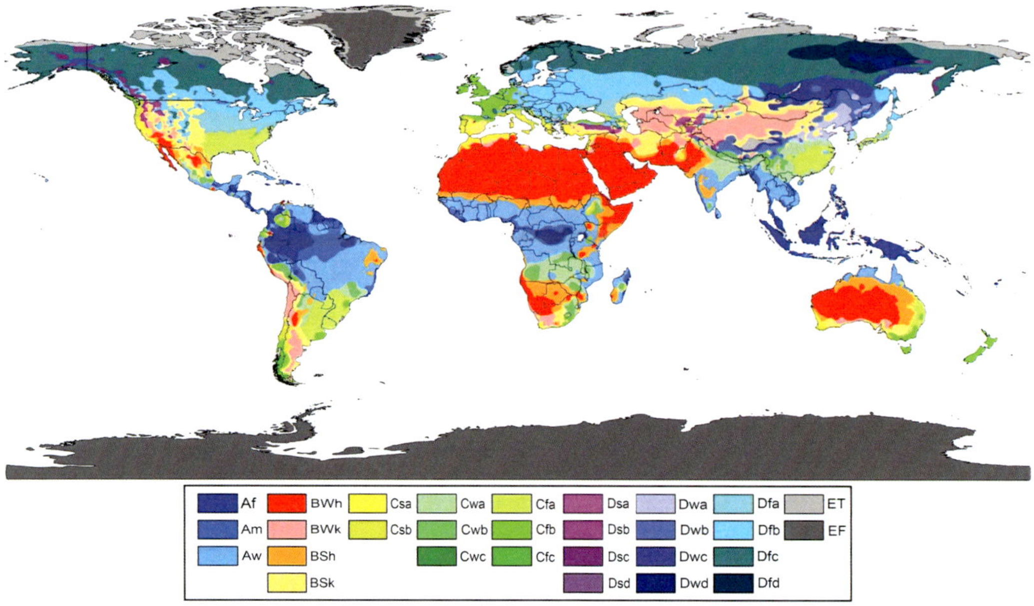

Figure 8.2. Köppen climate classification map.

Source: https://www.britannica.com/science/Koppen-climate-classification.

Table 8.2. Climate Classification Chart

Code	**Group**	**Description**	**Precipitation Type**
Af	Tropical	Tropical rainforest	Rainforest
Am	Tropical	Tropical monsoon	Monsoon
As	Tropical	Tropical dry savannah	Dry, savannah
Aw	Tropical	Tropical wet savannah	Wet, savannah
BSh	Arid	Hot semi-arid – steppe	Steppe
BSk	Arid	Cold semi-arid – steppe	Steppe
BWh	Arid	Hot desert	Desert
BWk	Arid	Cold desert	Desert
Cfa	Temperate	Humid subtropical	No dry season
Cfb	Temperate	Temperate oceanic	No dry season
Cfc	Temperate	Sub polar oceanic	No dry season
Csa	Temperate	Hot summer mediterranean	Dry summer

Csb	Temperate	Warm summer mediterranean	Dry summer
Csc	Temperate	Cool summer mediterranean	Dry summer
Cwa	Temperate	Monsoon humid subtropical	Dry winter
Cwb	Temperate	Sub tropical highland or temperate oceanic with dry winters	Dry winter
Cwc	Temperate	Cold subtropical highland or subpolar oceanic with dry winters.	Dry winter
Dfa	Cold – continental	Hot summer humid	No dry season
Dfb	Cold – continental	Warm summer humid	No dry season
Dfc	Cold – continental	Subarctic climate	No dry season
Dfd	Cold – continental	Extremely cold subarctic	No dry season
Dsa	Cold – continental	Hot dry summer type continental climate	Dry summer
Dsb	Cold – continental	Warm dry summer continental type	Dry summer
Dsc	Cold – continental	Dry summer subarctic type climate	Dry summer
Dwa	Cold – continental	Dry winter	Dry winter
Dwb	Cold – continental	Dry winter	Dry winter
Dwc	Cold – continental	Dry winter	Dry winter
Dwd	Cold – continental	Dry winter	Dry winter
EF	Polar	Icecap climate	Icecap
ET	Polar	Tundra type climate	Tundra

8.3.1 Tropical Moist Climates

Tropical moist climates are found north and south of the equator at latitudes between 15 and 25°. This climactic zone is unique in that its year-round temperatures consistently stay above 18°C. In this climactic zone, annual precipitation is typically greater than 1,500 mm. Three minor climactic types, whose classification is based on the seasonal distribution of rainfall in these climactic zones, also exist within this broad climactic zone. Naturally dense tropical forests make up the areas that fall under these climatic zones. The first is Af, or tropical wet climate, which has year-round precipitation and a tropical temperature. In these areas, monthly temperature variations are less than 3°C. These areas have extremely high surface temperatures and humidity, which leads to the formation of cumulus and cumulonimbus clouds early in the afternoons every day, producing a lot of precipitation. The second climate type, known as Am, is tropical monsoon. The yearly precipitation in these areas is almost identical to that in Af; however, the majority of the precipitation falls between the seventh and ninth warmest months of the year. During the rest of the year, these regions receive less rainfall.

Aw, also known as the tropical wet and dry climate or the savanna climate, is the third subdivision. During the winter, these climatic zones have an extended dry season. Precipitation during the wet season is typically less than 1,000 mm and mostly falls in the summer.

8.3.2 Dry Climates

Precipitation—or rather, the absence of it—is a more significant factor in these climactic zones than temperature. There is more transpiration and evaporation than precipitation overall in these climatic zones. These climatic zones are found in sizable continental regions in the mid-latitudes and are surrounded by mountainous regions. They stretch 20 to 35° latitude north and south from the equator.

This climactic zone is divided into four major sections. The first is the BW climate, also known as the true desert climate or dry arid climate, which makes up around 12% of the planet's land area. This climactic zone contains areas that serve as xerophytic vegetation's habitat. To indicate whether the dry arid zone is found in the sub-tropics or the mid-latitudes, respectively, the letters h and k are suffixed after BW.

The second climate type is called steppe climate, or BS, or dry semi-arid climate. This creates a climate akin to a grassland, which covers roughly 14% of the planet's surface. Because of the inter-tropical convergence zone and mid-latitude cyclones, regions with a dry semi-arid climate, or BS, receive more precipitation than those with a dry arid climate, or BW. Similar to BW zones, the letters h and k are suffixed to indicate where the climactic zone is—in the subtropics or the mid-latitudes, respectively.

8.3.3 Moist Sub-Tropical Mid-Latitude Climates

This climatic zone typically experiences mild winters and warm, humid summers. These climactic zones are found mostly at the eastern and western edges of most continents, extending 30 to 50° latitude north and south from the equator. Convective thunderstorms are common in the summer, and mid-latitude cyclones are occasionally seen in the winter. This type of climactic zone has three subdivisions.

The first is the humid subtropical climate, also known as Cfa, which experiences hot, muggy summers with plenty of thunderstorm activity. The mid-latitude cyclones that produce precipitation during this time, such as those in the southeast of the United States, are responsible for the relatively mild winters. The second type of marine climate are the Cfb types, which are typically found on continents' western coasts. This area experiences hot, dry summers and a generally humid climate. Mid-latitude cyclones contribute to the heavy precipitation that accompanies mild winters. Third is the Mediterranean Climatic Zone, or Cs, where mid-latitude cyclones are primarily responsible for rainfall during the mild winters.

In this climatic zone, summertime precipitation can be incredibly rare. This climatic zone encompasses, for instance, parts of Portland, Oregon, and California (Figure 8.3).

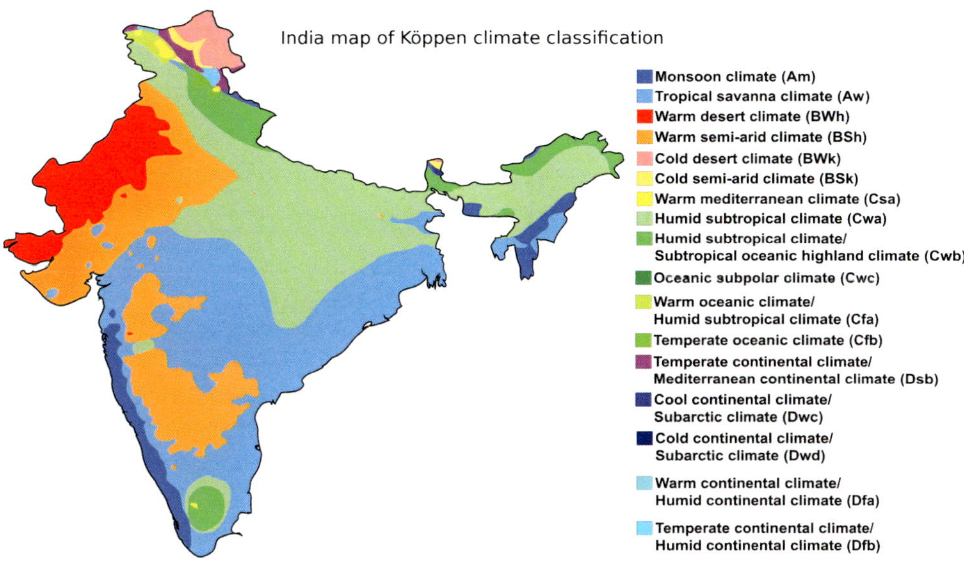

Figure 8.3. India as per the Koppen climate classification system.

Source: World Koppen Classification; https://geographyandyou.com/koppen-climate-classification-system/.

8.3.4 Moist Continental Mid-Latitude Climates

Summers in moist continental mid-latitude climates are warm, with occasional cool spells, and winters are chilly. The moist sub-tropical mid-latitude climates, or C climates, are typically poleward of the regions with moist continental mid-latitude climates. The warmest months typically have average temperatures of over 10°C, while the coldest months can have average temperatures of less than minus 3°C.

Strong winds and snowstorms from the Arctic and Continental Polar air masses can make for extremely cold winters in these areas. This Koppen climate classification has three sub-divisions: Dw, which has dry winters, Ds, which has dry summers, and Df, which has year-round precipitation.

8.3.5.Polar Climates

In Polar Regions, the warmest month has temperatures of less than 10°C. The climate is cold throughout the year. Greenland and Antarctica, as well as the northern coasts of Asia, Europe, and North America, all have polar climates. There are two categories within polar climates.

The first is the Eastern Treaty Zone, or Polar Tundra, where hundreds of meters of permanently frozen soil are found. The majority of the vegetation in this area is composed of woody shrubs, dwarf trees, lichens, and mosses. The second is the Polar Ice Caps, or EF, which have a surface that is covered in snow or ice all year round.

8.3.6. Determining Factors for Koppen Climate Classification

Latitudinal position, solar radiation received, air masses, air pressure (with high and low-pressure zones), wind patterns, heat exchange with the ocean, topographical features (such as mountains), land and sea distribution, altitude, and so on all affect these regions' climactic features.

The Koppen climate classification uses heat and its aftereffects as a primary determinant of global climactic zones. Solar radiation, air masses, and air pressure in terms of high and low-pressure zones have the greatest influence on these climactic zones.

However, the model does not consider weather extremes in a meaningful way, such as temperature extremes, cyclical wind patterns, cloud cover, or the quantity of sunny days, to name a few. For broader climactic zones on Earth, the Koppen climate classification system acts as a general guide. The system, however, represents broad classification categories of global climate and is not responsible for instantaneous or microclimatic aberrations or shifts.

In one of its subsequent modifications, the Koppen-Geiger climate classification that we have previously discussed has been replaced with the Koppen-Trewartha climate classification. This plan considers climatic changes that have occurred and may occur on Earth, taking into account the consistent variability that different land surface types have demonstrated.

With the scheme provided in the Koppen-Trewartha climate classification, it is possible to model simulations of the period between 1900 and 2099. Based on this, the Intergovernmental Panel on Climate Change (IPCC) in its 4th Assessment, projected 16 global climate models continuing into the near future. In this assessment, the average annual mean surface temperature over the Arctic landmasses is expected to witness an increase of 3.1, 4.6, and 5.3°C.

8.3.7. Climatic Region Descriptions

The world's climatic regions are divided into eight groups for the discussion that follows. These climates are divided into groups according to the Köppen classification and the seasonal dominance of air masses.

8.3.7.1 Tropical Wet

- Köppen Classification – Af.
- Dominated by Maritime Tropical air masses all year long.

The somewhat constant daily high temperatures in the 20–30° Celsius range define the tropical wet climate. The average monthly temperature ranges from 24 to 30°C. The monthly temperature ranges over a year are roughly 3°C. The entire year brings fairly consistent precipitation with totals of at least 2000 millimeters.

The region experiencing this climate lies within the effects of the intertropical convergence zone all year long. Convergence and high maritime humidity create cumulus clouds and thunderstorms almost daily (Figures 8.4 and 8.5).

Af – Andagoya, Columbia 5°N, Elevation: 65 m

	Jan.	Feb.	Mar.	Apr.	May	June	July	Aug.	Sept.	Oct.	Nov.	Dec.	Year
Temp. °C	27	27	28	28	27	27	27	27	27	27	27	27	27
Precip. mm	554	519	557	620	655	655	572	574	561	563	563	512	6905

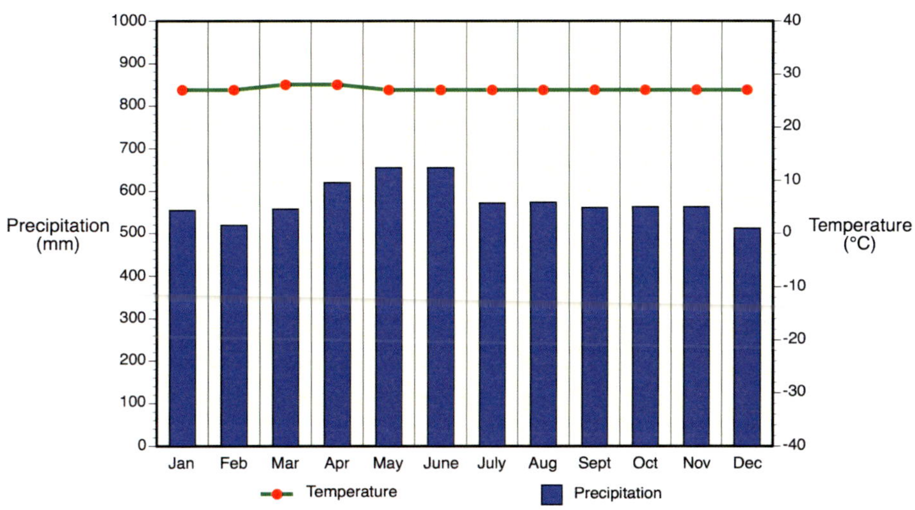

Figure 8.4. Mean monthly temperature and precipitation values for Andagoya, Columbia.

Source: http://www.physicalgeography.net/fundamentals/7v.html.

Af – Iquitos, Peru 4°S, Elevation: 104 m

	Jan.	Feb.	Mar.	Apr.	May	June	July	Aug.	Sept.	Oct.	Nov.	Dec.	Year
Temp. °C	27	27	27	27	26	26	25	27	27	27	27	27	26
Precip. mm	256	276	349	306	271	199	165	157	191	214	244	217	2845

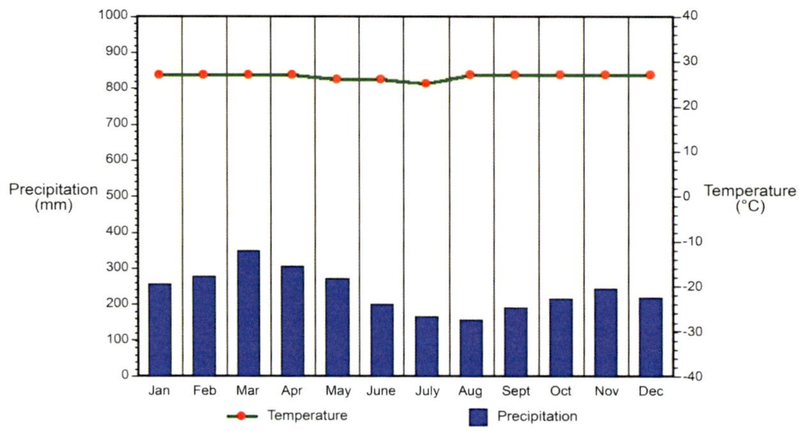

Figure 8.5. Mean monthly temperature and precipitation values for Iquitos, Peru.

Source: http://www.physicalgeography.net/fundamentals/7v.html.

8.3.7.2. Tropical Wet and Dry

- Köppen Classification – Aw and Am.
- Maritime Tropical air masses in the high Sun season and Continental Tropical air masses low Sun season.

This climate has discrete dry and wet seasons. The intertropical convergence zone's migration is the cause of the seasonal pattern of moisture. The high Sun and the existence of the convergence zone coincide with the wet season.

The more stable air that results from the subsidence brought on by the subtropical high zone's presence during the low Sun season is what causes the dry season. This place has a tropical wet climate during the rainy season, which is warm, humid, and prone to frequent thunderstorms. Conditions are essentially semi-desert during the dry season. Rainfall intensity may increase in certain areas due to orographic uplift and monsoon development (Figures 8.6–8.9).

Aw – Calcutta, India 22.5°N, Elevation: 6 m

	Jan.	Feb.	Mar.	Apr.	May	June	July	Aug.	Sept.	Oct.	Nov.	Dec.	Year
Temp. °C	20	23	28	30	31	30	29	29	30	28	24	21	27
Precip. mm	13	24	27	43	121	259	301	306	290	160	35	3	1582

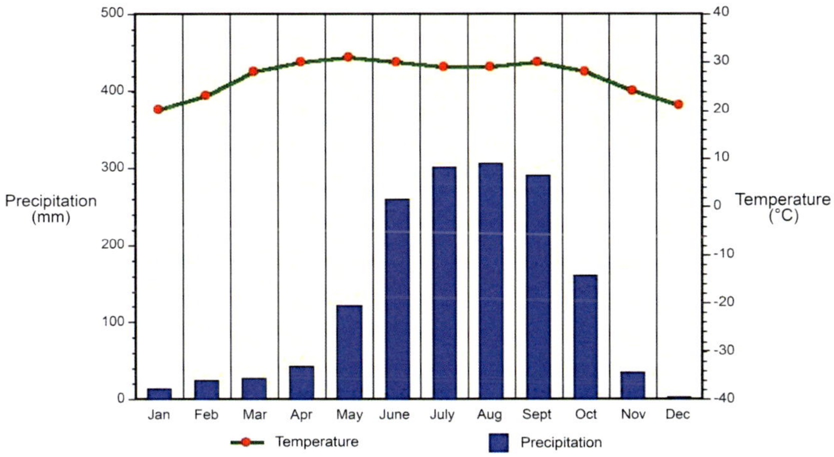

Figure 8.6. Mean monthly temperature and precipitation values for Calcutta, India.

Source: http://www.physicalgeography.net/fundamentals/7v.html.

Aw – Cuiaba, Brazil 13.5°S, Elevation: 165 m

	Jan.	Feb.	Mar.	Apr.	May	June	July	Aug.	Sept.	Oct.	Nov.	Dec.	Year
Temp. °C	27	27	27	27	26	24	24	26	28	28	28	27	27
Precip. mm	216	198	232	116	52	13	9	12	37	130	165	195	1375

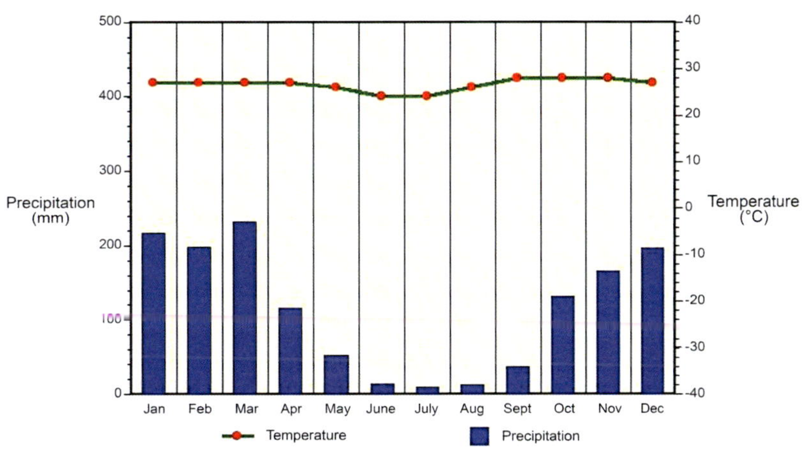

Figure 8.7. Mean monthly temperature and precipitation values for Cuiaba, Brazil.

Source: http://www.physicalgeography.net/fundamentals/7v.html.

Aw – Darwin, Australia 12.5°S, Elevation: 27 m

	Jan.	Feb.	Mar.	Apr.	May	June	July	Aug.	Sept.	Oct.	Nov.	Dec.	Year
Temp. °C	28	28	28	28	27	25	25	26	28	29	29	29	28
Precip. mm	341	338	274	121	9	1	2	5	17	66	156	233	1563

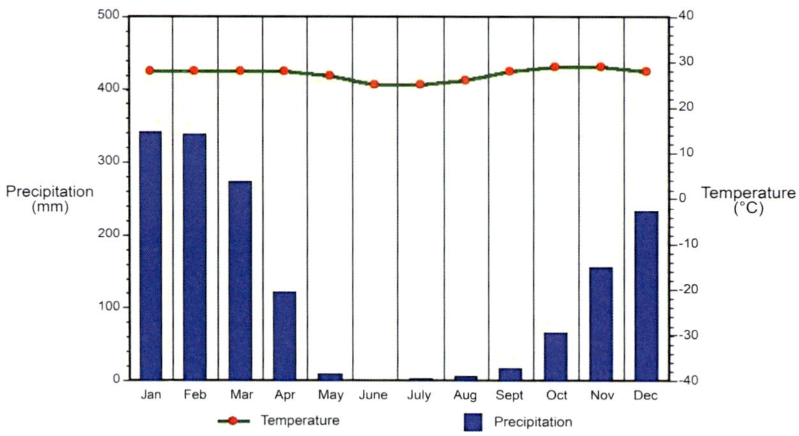

Figure 8.8. Mean monthly temperature and precipitation values for Darwin, Australia.

Source: http://www.physicalgeography.net/fundamentals/7v.html.

Am – Mangalore, India 13°N, Elevation: 22 m

	Jan.	Feb.	Mar.	Apr.	May	June	July	Aug.	Sept.	Oct.	Nov.	Dec.	Year
Temp. °C	27	27	28	29	29	27	26	26	26	27	27	27	27
Precip. mm	5	2	9	40	233	982	1059	577	267	206	71	18	3467

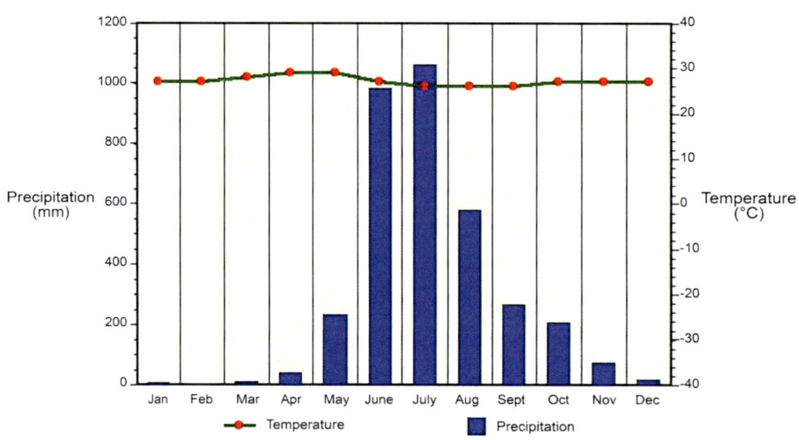

Figure 8.9. Mean monthly temperature and precipitation values for Mangalore, India.

Source: http://www.physicalgeography.net/fundamentals/7v.html.

8.3.7.3. Subtropical Desert and Steppe

- Köppen Classification – BWh and BSh.
- Dominated by Continental Tropical air masses all year.

This climate type covers 12% of all land area on the continents. The heart of the tropical desert climate is found near the tropics of Cancer and Capricorn, usually toward the western side of the continents. Regions with this climate have the following common climatic characteristics:

- low relative humidity and cloud cover.
- low frequency and amount of precipitation.
- high mean annual temperature.
- high monthly temperatures.
- high diurnal temperature ranges.
- high wind velocities.

Because of the subtropical high-pressure zone, upper air stability and subsidence have an impact on the climate of tropical deserts. In interior locations, relative humidity typically ranges from 10 to 30% on average. There is very little precipitation and very little distribution of precipitation over time and space.

BWh – Berbera, Somalia 10.5°N, Elevation: 8 m

	Jan.	Feb.	Mar.	Apr.	May	June	July	Aug.	Sept.	Oct.	Nov.	Dec.	Year
Temp. °C	25	26	27	29	32	37	37	37	34	29	26	26	30
Precip. mm	8	2	5	12	8	1	1	2	1	2	5	5	52

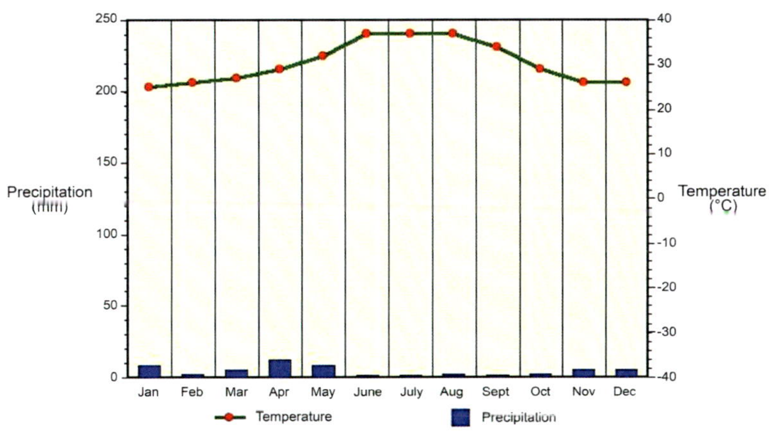

Figure 8.10. Mean monthly temperature and precipitation values for Berbera, Somalia.

Source: http://www.physicalgeography.net/fundamentals/7v.html.

BWh – Alice Springs, Australia 23.5°S, Elevation: 579 m

	Jan.	Feb.	Mar.	Apr.	May	June	July	Aug.	Sept.	Oct.	Nov.	Dec.	Year
Temp. °C	28	28	25	20	15	12	12	14	18	23	26	27	21
Precip. mm	44	34	28	10	15	13	7	8	7	18	29	39	252

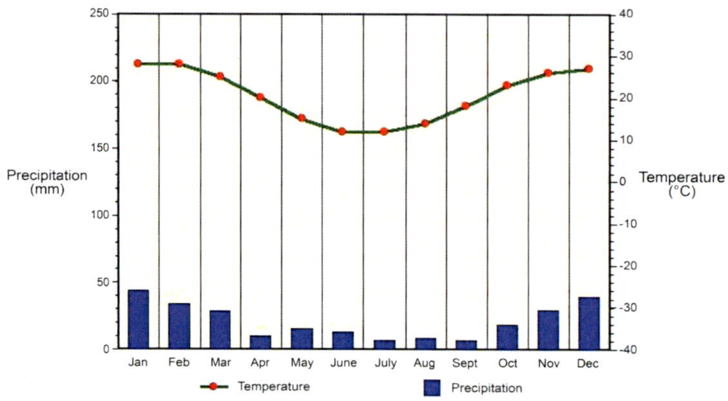

Figure 8.11. Mean monthly temperature and precipitation values for Alice Springs, Australia.

Source: http://www.physicalgeography.net/fundamentals/7v.html.

BSh – Monterrey, Mexico 26°N, Elevation: 512 m

	Jan.	Feb.	Mar.	Apr.	May	June	July	Aug.	Sept.	Oct.	Nov.	Dec.	Year
Temp. °C	14	17	20	23	26	27	28	28	26	22	18	15	22
Precip. mm	18	23	16	29	40	68	62	151	78	26	29	20	606

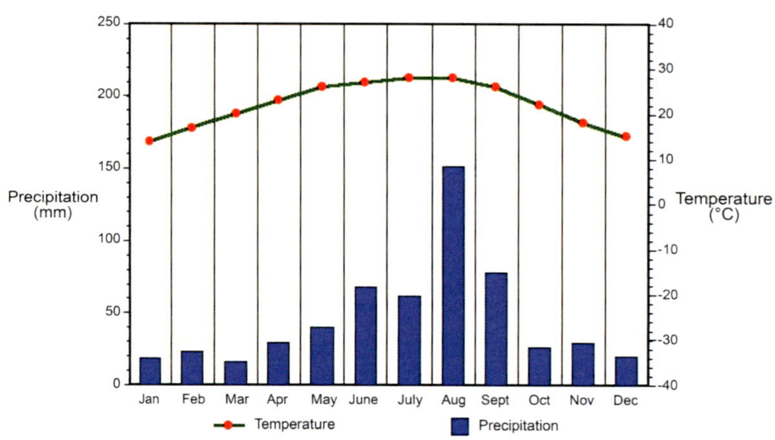

Figure 8.12. Mean monthly temperature and precipitation values for Monterrey, Mexico.

Source: http://www.physicalgeography.net/fundamentals/7v.html.

The daily and yearly variations in temperature are substantial. The tropical desert has the highest monthly average temperatures on Earth. They are in the 29–35°C range. Monthly winter temperatures can be between 15 and 25° colder than summer ones. Extreme daily temperature variations are another feature of this climate. The typical daily temperature range is between 14 and 25°C (Figures 8.10–8.12).

8.3.7.4. Mid-Latitude Desert and Steppe

- Köppen Classification – BWk and BSk.
- Dominated by Continental Tropical air masses during summer and Continental Polar in winter.

This climate type covers 14% of all land area on the continents. Regions with this climate have the following similar climatic characteristics:

- low relative humidity and cloud cover.
- low frequency and amount of precipitation.
- moderate to high annual temperature.
- moderate to high monthly temperatures.

Because of the impact of high elevations and extreme continentality, these climates are dry. A continent's center restricts the amount of moisture that can be drawn from the ocean. Precipitation cannot happen in the absence of moisture. Because of the rain shadow effect, the presence of mountains upwind of these climates can further reduce the availability of moisture. Large areas of mid-latitude deserts are located in the western United States, east of the Caspian Sea, north of the Himalayas, and in a narrow area in southern South America to the east of the Andes. The range of daily and annual temperatures in mid-latitude deserts is larger than that of subtropical deserts. Generally speaking, mid-latitude deserts experience milder summer temperatures than subtropical deserts. Nonetheless, there are some outliers, such as Death Valley, California, which is among the hottest locations on Earth. Wintertime is usually quite chilly.

Significant portions of central Asia and western North America have steppe climates at mid-latitudes. In general, the temperature characteristics of this climate are comparable to those of mid-latitude deserts. Steppe climates in the middle of latitudes do, however, receive somewhat more precipitation than do mid-latitude deserts (Figures 8.13–8.15).

BWk – Lovelock, Nevada, USA 40°N, Elevation: 1211 m

	Jan.	Feb.	Mar.	Apr.	May	June	July	Aug.	Sept.	Oct.	Nov.	Dec.	Year
Temp. °C	0	4	6	10	15	20	24	22	18	12	5	0	11
Precip. mm	16	16	11	14	13	14	4	6	7	13	13	15	143

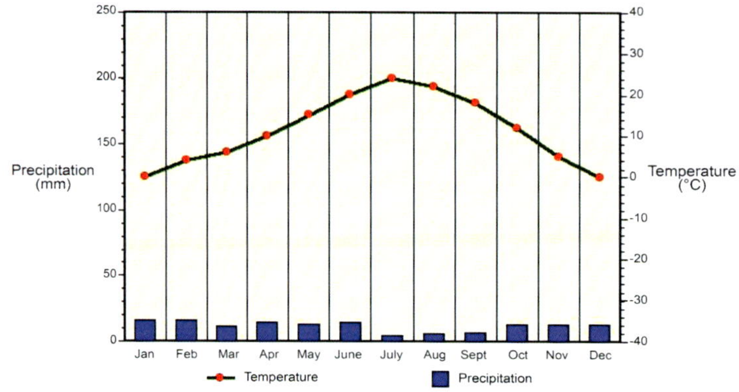

Figure 8.13. Mean monthly temperature and precipitation values for Lovelock, USA.

Source: http://www.physicalgeography.net/fundamentals/7v.html.

BSk – Denver, Colorado, USA 40°N, Elevation: 1611 m

	Jan.	Feb.	Mar.	Apr.	May	June	July	Aug.	Sept.	Oct.	Nov.	Dec.	Year
Temp. °C	–1	1	4	9	14	19	23	22	17	11	4	–1	10
Precip. mm	14	16	34	45	63	43	47	38	28	26	23	15	391

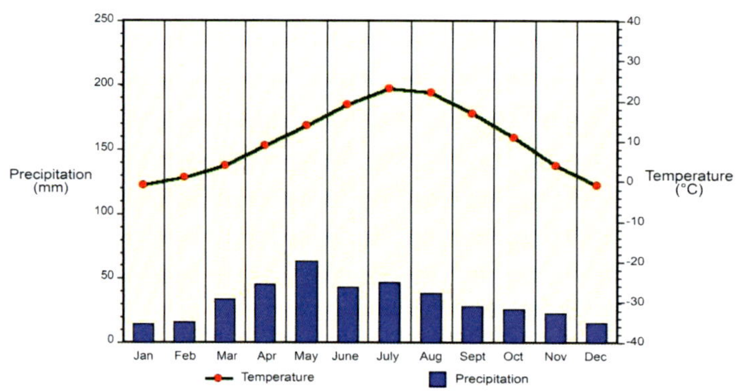

Figure 8.14. Mean monthly temperature and precipitation values for Denver, USA.

Source: http://www.physicalgeography.net/fundamentals/7v.html.

BSk – Williston, North Dakota, USA 47.5°N, Elevation: 579 m

	Jan.	Feb.	Mar.	Apr.	May	June	July	Aug.	Sept.	Oct.	Nov.	Dec.	Year
Temp. °C	–12	–10	–4	6	13	17	22	20	14	8	–2	–8	5
Precip. mm	14	12	18	24	36	84	48	38	28	19	15	13	349

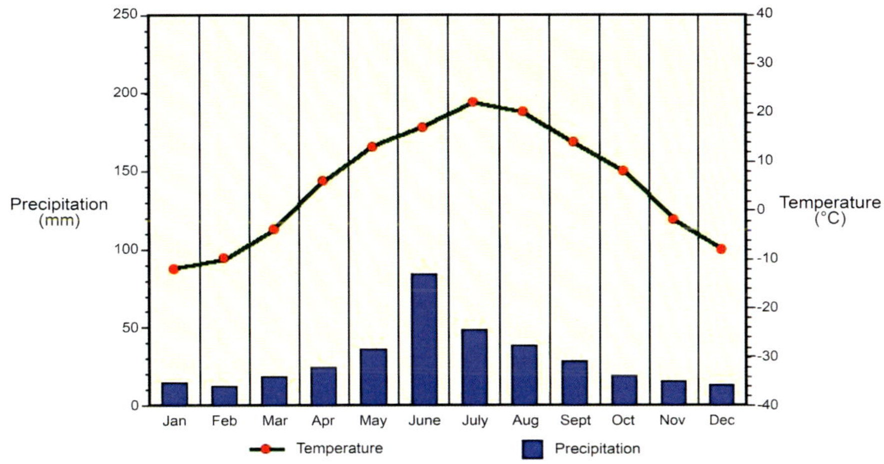

Figure 8.15. Mean monthly temperature and precipitation values for Williston, USA.

Source: http://www.physicalgeography.net/fundamentals/7v.html.

8.3.7.5. Mid-Latitude Wet

- Köppen Classification – Cf and Df.
- Maritime Tropical in summer and Maritime Polar in winter.

The Northern Hemisphere's Mid-Latitude Wet climate is found mostly along the eastern edges of the continents, from 60° North to 25–30° North. This climate is predominant in the Eastern half of the continent, stretches across North America from the Canadian Pacific coast at latitudes above 55° eastward to the Atlantic coast. This climate is found in the Southern Hemisphere in New Zealand, Australia's southeast coast, and the southeast tip of South America.

Maritime tropical air masses dominate the summer weather, causing a lot of thunderstorms due to daytime heating. The tropical regions experience temperatures as high as 29°C, with monthly averages between 21 and 26°C. The humid tropics are not as warm as this. In all regions, frontal weather linked to the mid-latitude cyclone is more common during the winter and dominates the climate of more polar areas.

In this climate, precipitation is spread fairly evenly over the year. The latitude and continental position of the various regions affect the annual totals of precipitation, which can vary greatly. Convectional rainfall is the main precipitation mechanism in the summer and on the equatorial margins. There are 40 to 60 days on average of thunderstorms in the southeast of the United States each year. Thunderstorm frequency rapidly decreases with northward movement. Additionally, hurricanes offer a means of generating precipitation in the climate's more tropical regions (Figures 8.16–8.19).

Cf – New Orleans, USA 30°N, Elevation: 1 m

	Jan.	Feb.	Mar.	Apr.	May	June	July	Aug.	Sept.	Oct.	Nov.	Dec.	Year
Temp. °C	12	13	16	20	24	27	28	28	26	21	16	13	20
Precip. mm	98	101	136	116	111	113	171	136	128	72	85	104	1371

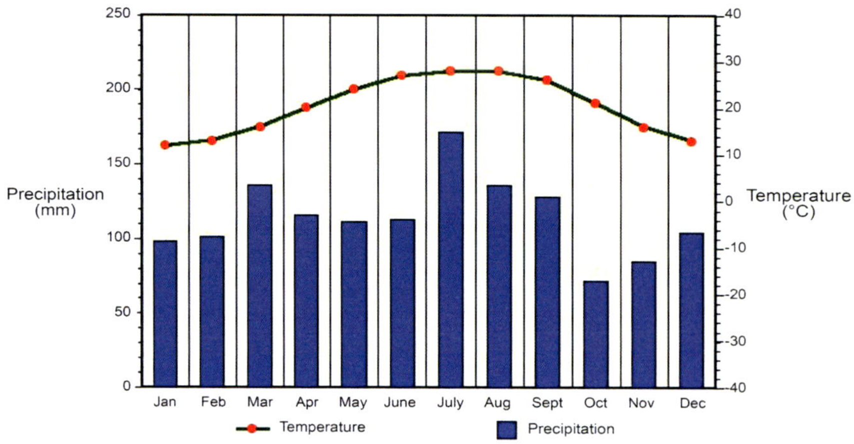

Figure 8.16. Mean monthly temperature and precipitation values for New Orleans, USA.

Source: http://www.physicalgeography.net/fundamentals/7v.html.

Cf – Buenos Aires, Argentina 34.5°S, Elevation: 27 m

	Jan.	Feb.	Mar.	Apr.	May	June	July	Aug.	Sept.	Oct.	Nov.	Dec.	Year
Temp. °C	23	23	21	17	13	9	10	11	13	15	19	22	16
Precip. mm	103	82	122	90	79	68	61	68	80	100	90	83	1026

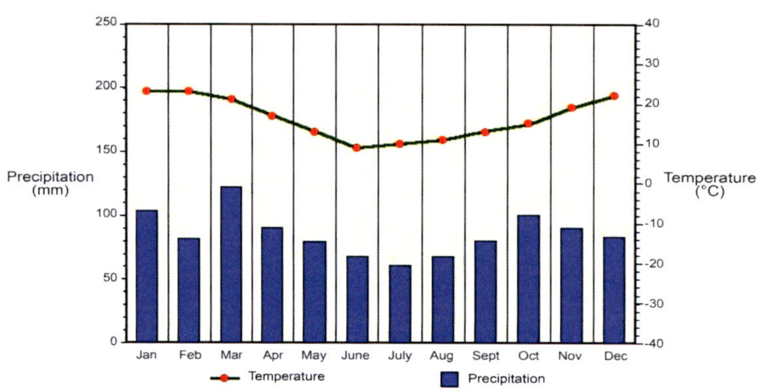

Figure 8.17. Mean monthly temperature and precipitation values for Buenos Aires, Argentina.

Source: http://www.physicalgeography.net/fundamentals/7v.html.

Cf – London, England 51.5°N, Elevation: 5 m

	Jan.	Feb.	Mar.	Apr.	May	June	July	Aug.	Sept.	Oct.	Nov.	Dec.	Year
Temp. °C	4	4	7	9	12	16	18	17	15	11	7	5	10
Precip. mm	54	40	37	38	46	46	56	59	50	57	64	48	595

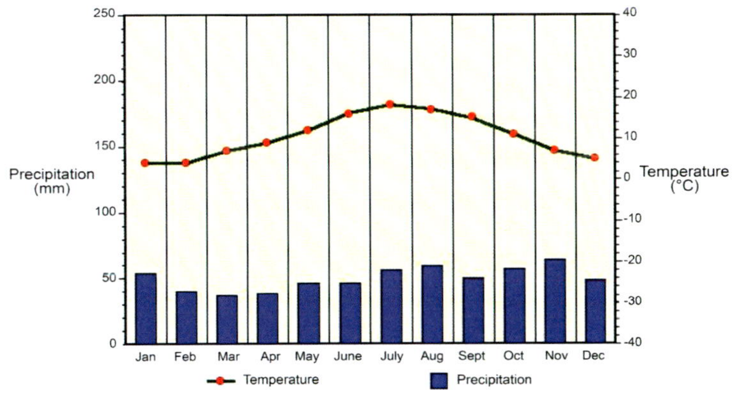

Figure 8.18. Mean monthly temperature and precipitation values for London, England.

Source: http://www.physicalgeography.net/fundamentals/7v.html.

Df – Winnipeg, Canada 50°N, Elevation: 240 m

	Jan.	Feb.	Mar.	Apr.	May	June	July	Aug.	Sept.	Oct.	Nov.	Dec.	Year
Temp. °C	–18	–16	–8	3	11	17	20	19	13	6	–5	–13	3
Precip. mm	26	21	27	30	50	81	69	70	55	37	29	22	517

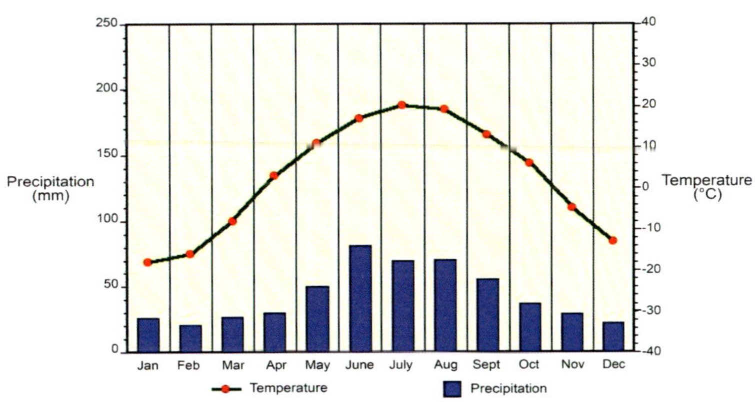

Figure 8.19. Mean monthly temperature and precipitation values for Winnipeg, Canada.

Source: http://www.physicalgeography.net/fundamentals/7v.html.

8.3.7.6. Mid-Latitude Winter-Dry

- Köppen Classification – Cw and Dw.
- Maritime Tropical air masses in summer and Continental Polar air masses in winter.

Strong seasonal patterns in precipitation and temperature define this climate. The interior of the continents at mid-latitudes is typically home to the Mid-Latitude winter dry climate. The continentality of this location results in a wide range of annual temperatures.

Summertime in this climate is characterized by Maritime Tropical air masses and sporadic Continental Tropical air masses from the nearby deserts. Summer convectional storms bring intense heat and humidity to the region. During the winter, continental polar air masses predominate, with sporadic outbreaks of maritime polar air. Cold, dry weather is often linked to continental polar air masses. The summer months see the most precipitation due to thunderstorm activity. In the winter, the mid-latitude cyclone produces less precipitation (Figures 8.20 and 8.21).

Dfa – Omaha, Nebraska, USA 41°N, Elevation: 298 m

	Jan.	Feb.	Mar.	Apr.	May	June	July	Aug.	Sept.	Oct.	Nov.	Dec.	Year
Temp. °C	–5	–3	3	11	17	23	26	25	19	13	4	–2	11
Precip. mm	21	24	37	65	88	115	86	101	67	44	32	20	700

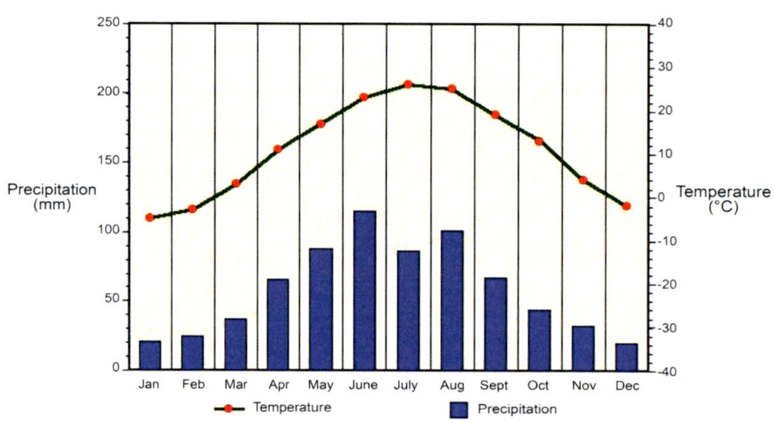

Figure 8.20. Mean monthly temperature and precipitation values for Omaha, USA.

Source: http://www.physicalgeography.net/fundamentals/7v.html.

Dw – Calgary, Canada 51°N, Elevation: 1140 m

	Jan.	Feb.	Mar.	Apr.	May	June	July	Aug.	Sept.	Oct.	Nov.	Dec.	Year
Temp. °C	–10	–9	–4	4	10	13	17	15	11	5	–2	–7	4
Precip. mm	17	20	26	35	52	88	58	59	35	23	16	15	444

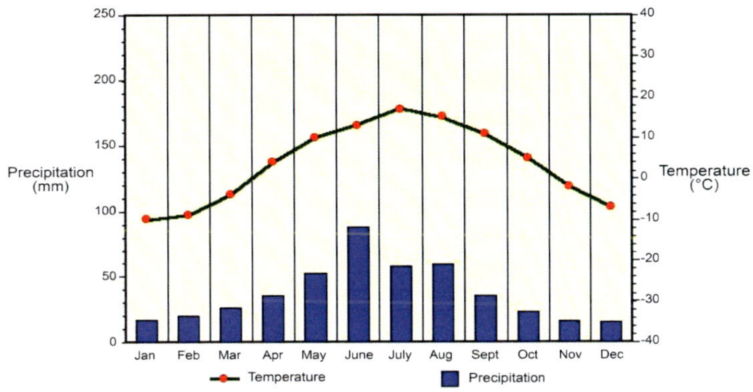

Figure 8.21. Mean monthly temperature and precipitation values for Calgary, Canada.

Source: http://www.physicalgeography.net/fundamentals/7v.html.

8.3.7.7. Mid-Latitude Summer-Dry

- Köppen Classification – Cs.
- Summer weather is dominated by Continental Tropical air, while in the winter, Maritime Polar air masses are frequent.

Between 30 and 40° latitude on the western edge of the continents is where the Mid-Latitude Summer-Dry climate is found. This climate typically does not extend very far across the continents. This type of weather is frequently referred to as Mediterranean.

In this climate, precipitation primarily falls during the winter months due to mid-latitude cyclones. Stable subtropical highs provide dry, warm weather to these regions during the summer (Figures 8.22–8.24).

Cs – Santiago, Chile 33.5°S, Elevation: 512 m

	Jan.	Feb.	Mar.	Apr.	May	June	July	Aug.	Sept.	Oct.	Nov.	Dec.	Year
Temp. °C	19	19	17	13	11	8	8	9	11	13	16	19	14
Precip. mm	3	3	5	13	64	84	76	56	30	13	8	5	360

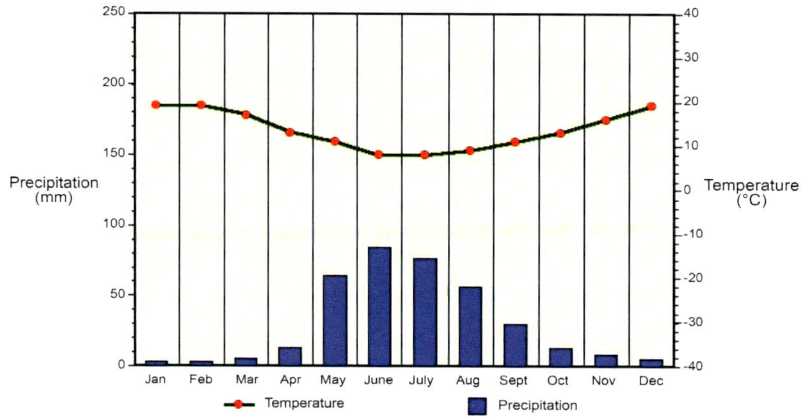

Figure 8.22. Mean monthly temperature and precipitation values for Santiago, Chile.

Source: http://www.physicalgeography.net/fundamentals/7v.html.

Cs – Los Angeles, USA 34°N, Elevation: 37 m

	Jan.	Feb.	Mar.	Apr.	May	June	July	Aug.	Sept.	Oct.	Nov.	Dec.	Year
Temp. °C	13	14	15	17	18	20	23	23	22	18	17	15	18
Precip. mm	78	85	57	30	4	2	0	1	6	10	27	73	373

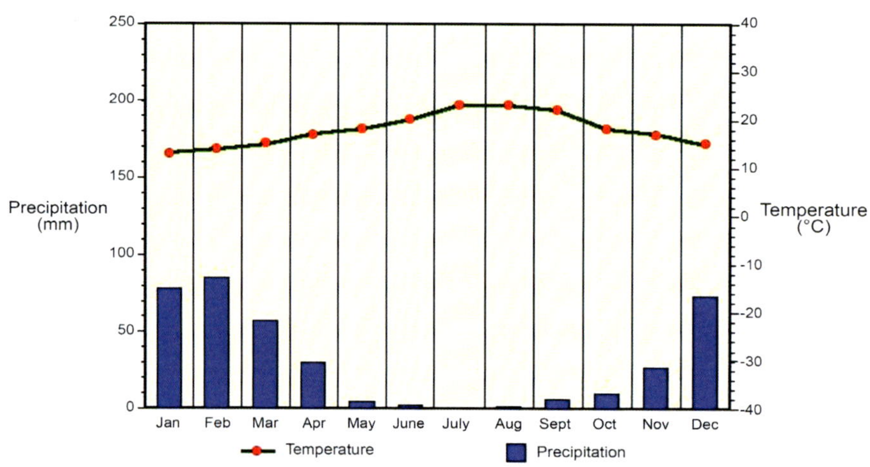

Figure 8.23. Mean monthly temperature and precipitation values for Los Angeles, USA.

Source: http://www.physicalgeography.net/fundamentals/7v.html.

Cs – Rome, Italy 42°N, Elevation: 131 m

	Jan.	Feb.	Mar.	Apr.	May	June	July	Aug.	Sept.	Oct.	Nov.	Dec.	Year
Temp. °C	8	8	10	13	17	22	24	24	21	16	12	9	15
Precip. mm	76	88	77	72	63	48	14	22	70	128	116	106	881

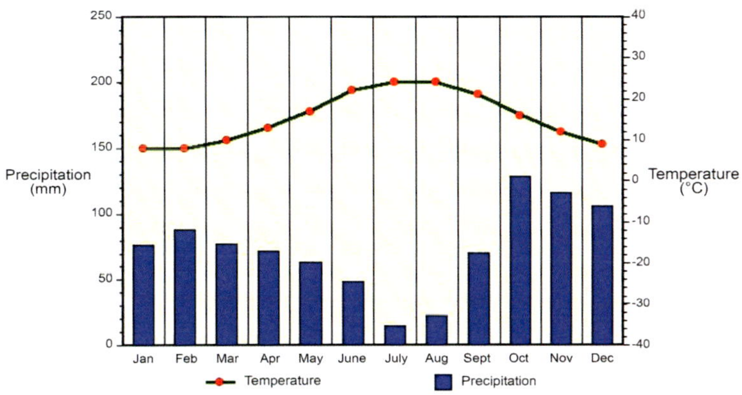

Figure 8.24. Mean monthly temperature and precipitation values for Rome, Italy.

Source: http://www.physicalgeography.net/fundamentals/7v.html.

8.3.7.8. Polar Tundra

- Köppen Classification – ET.
- Maritime Polar in summer and Continental Polar or Arctic in winter.

The polar tundra climate is typified by summer rainfall patterns, chilly summers, and frigid winters. The Arctic coasts of Europe and Asia, Iceland, North America, Greenland's coast, and the islands of McQuarrie, Kerguelen, and South Georgia in the Southern Hemisphere all have this type of climate. In most places, the annual precipitation is less than 250 mm, with summer being the season with the highest amounts (Figures 0.25 and 8.26).

ET – Isachsen, Canada 79°N, Elevation: 35 m

	Jan.	Feb.	Mar.	Apr.	May	June	July	Aug.	Sept.	Oct.	Nov.	Dec.	Year
Temp. °C	–35	–37	–35	–24	–12	0	4	1	–8	–19	–28	–32	–19
Precip. mm	2	2	1	4	8	3	22	23	18	10	4	2	98

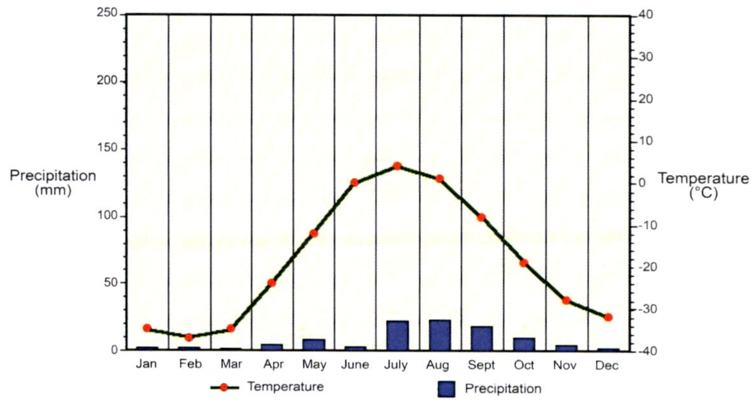

Figure 8.25. Mean monthly temperature and precipitation values for Isachen, Canada.

Source: http://www.physicalgeography.net/fundamentals/7v.html.

ET – Barrow, Alaska, USA 72°N, Elevation: 9 m

	Jan.	Feb.	Mar.	Apr.	May	June	July	Aug.	Sept.	Oct.	Nov.	Dec.	Year
Temp. °C	–25	–28	–26	–19	–7	1	4	3	–1	–10	–19	–24	–13
Precip. mm	5	4	4	4	4	8	22	25	16	12	6	4	113

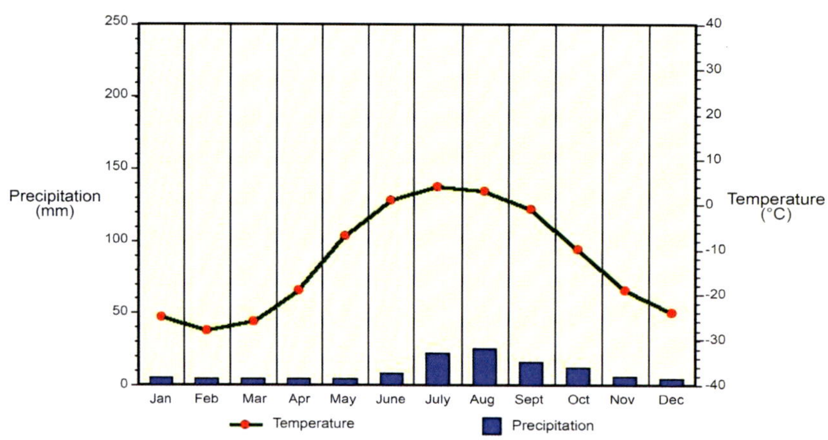

Figure 8.26. Mean monthly temperature and precipitation values for Barrow, USA.

Source: http://www.physicalgeography.net/fundamentals/7v.html.

8.3.7.9. Polar Ice Cap

- Köppen Classification – EF.
- Continental Arctic and Continental Polar air masses dominate.

High latitudes over continents, such as Greenland and Antarctica, are home to polar ice cap climates. Much of the planet is covered by this type of climate. There is no solar radiation for half the year. Due to the long days and relatively transparent atmosphere, available insolation is fairly high during the summer. Nonetheless, up to 90% of the insolation is reflected into space by the albedo of snow-covered surfaces. The majority of the time, monthly average temperatures are below freezing. Most of the time, there are steady winds and sufficient velocity to create blizzard conditions (Figures 8.27 and 8.28).

EF – Eismitte, Greenland 71°N, Elevation: 2953 m

	Jan.	Feb.	Mar.	Apr.	May	June	July	Aug.	Sept.	Oct.	Nov.	Dec.	Year
Temp. °C	–42	–47	–40	–32	–24	–17	–12	–11	–11	–36	–43	–38	–29
Precip. mm	15	5	8	5	3	3	3	10	8	13	13	25	111

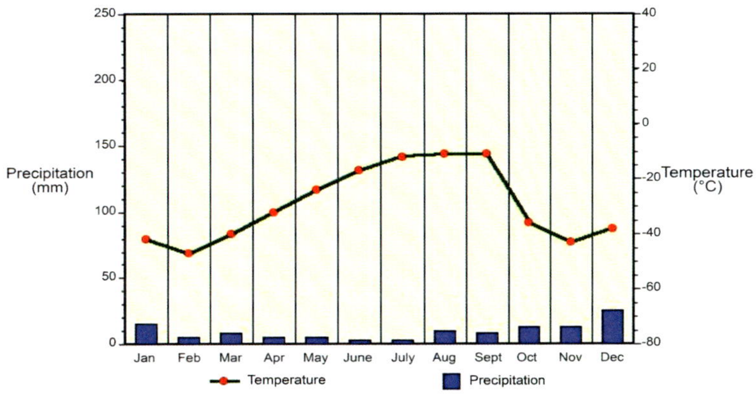

Figure 8.27. Mean monthly temperature and precipitation values for Eismitte, Greenland.

Source: http://www.physicalgeography.net/fundamentals/7v.html.

EF – Plateau Station, Antarctica 79°S, Elevation: 3625 m

	Jan.	Feb.	Mar.	Apr.	May	June	July	Aug.	Sept.	Oct.	Nov.	Dec.	Year
Temp. °C	–34	–44	–57	–66	–66	–69	–68	–71	–65	–60	–44	–32	–56
Precip. mm	NA	NA	NA	NA	NA	NA	NA	NA	NA	NA	NA	NA	NA

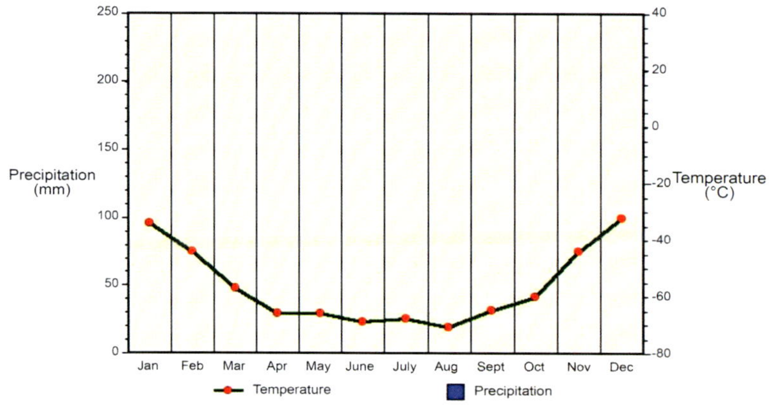

Figure 8.28. Mean monthly temperature and precipitation values for Plateau Station, Antarctica.

Source: http://www.physicalgeography.net/fundamentals/7v.html.

8.3.8. Merits

There are numerous merits of the Koppen climate classification system, out of which some are mentioned below:

- The Koppen climate structure is quantitative along with its division.
- The system is one that is quite easy to understand.
- One can measure the Koppen climate classification system easily as well.
- The Koppen climate classification system coincides with the vegetation pattern that is found in any region.
- The Koppen climate classification system has importance about two major factors, which are temperature and precipitation.
- The Koppen climate classification system is very vast, so much so that it actually covers all the types of climate patterns as well as conditions that are found on the Earth.

8.3.9. Demerits

You must be aware of the saying – Every coin has two sides. It is true in terms of concepts relating to the Koppen climate classification system as well. And so, here are a few of the demerits that you must be aware of:

- The concept is quite vast and has various classifications which might get difficult to memorize at times.
- The Koppen climate classification system is majorly based on the average values. This pertains to the fact that one needs average values of temperature and precipitation to make use of it. You may already know how precipitation is much more of an aspect that is estimated rather than measured. So, such

average values are, on occasions, difficult when it comes to comparisons and yielding the desired results.

- The Koppen climate classification system ignores various features. These include the intensity of the precipitation, the number of rainy days, cloud coverage, variations in temperature or precipitation, and much more.
- In this system of Koppen climate classification, Koppen also ignored the air masses and their significance in the climate formation that contribute a lot in climatology (Köppen et al., 1884; Kottek et al., 2006).

CASE STUDY

CLIMATE

The Sahel, a semi-arid region spanning across Africa just south of the Sahara Desert, faces significant challenges due to its sensitive climate conditions.

Background:

Initially, the Sahel experienced moderate rainfall, supporting pastoralism and agriculture. However, in the mid-20th century, the region entered a prolonged period of severe droughts, exacerbated by unsustainable land use practices and increasing population pressures.

Impacts:

1. Ecological Impact: The prolonged droughts led to desertification, where fertile land turned into barren desert, threatening biodiversity and ecosystem services.
2. Human Impact: The droughts triggered food insecurity and forced migration, as traditional livelihoods became unsustainable. Pastoralists and farmers struggled to adapt to changing climatic conditions.

Response and Adaptation:

1. International Aid and Cooperation: The international community responded with aid and development projects aimed at improving water management, promoting sustainable agriculture, and enhancing resilience among vulnerable populations.
2. Local Initiatives: Local communities implemented traditional knowledge and modern techniques to mitigate desertification, such as reforestation and water conservation measures.

Conclusion:

While the Sahel continues to face challenges from climate change, efforts in adaptation and mitigation are ongoing. The case study of the Sahel underscores the importance of climate resilience and international cooperation in addressing the impacts of climate change on vulnerable regions.

This case study provides a comprehensive overview of the complexities involved in climate change adaptation and resilience building, highlighting the interconnectedness of environmental, social, and economic factors in vulnerable regions like the Sahel.

CLASS ACTIVITY

In a class activity focused on climatic zones, students explore various climate classifications like the Köppen and Trewartha systems. They analyze global maps, identifying regions based on temperature, precipitation patterns, and geographical factors. Discussions delve into the implications of different climatic zones on ecosystems, agriculture, and human activities. Through this hands-on approach, students gain insights into the diversity of climates worldwide and the importance of climate classification in understanding environmental challenges and planning sustainable development. Such activities foster critical thinking and awareness about climate variability and its impacts on a global scale.

SUMMARY

- Climate classification systems categorize the Earth's diverse climates into distinct types based on temperature, precipitation patterns, and other geographical factors.
- Climate classification is aimed at defining climatic types and to find similarity and distinctions in climatic patterns as observed in different regions of the world.
- The climate of an area is the synthesis of the environmental conditions (soils, vegetation, weather, etc.) that have prevailed there over a long period.
- Climate is a multidimensional concept, and it is not an obvious decision as to which of the many observed environmental variables should be selected as the basis of the classification.
- The factors that influence the climactic characteristics of these regions include latitudinal position and amount of solar radiation received, air masses, air pressure (with high and low-pressure zones), wind patterns, heat exchange with the ocean, topographical features (such as mountains), land and sea distribution, and altitude.
- The tropical desert climate is influenced by upper air stability and subsidence which is the result of the presence of the subtropical high pressure zone.

REVIEW QUESTIONS

1. What is the importance of climate classification?
2. Who discovered climate classification?
3. What is the Koppen climate classification system?
4. When did Köppen classify climates?
5. What are the major classifications in Thornthwaite climate classification?
6. What is the difference between the Köppen and Thornthwaite systems?
7. How did Köppen classify climates?

REFERENCES

1. Chen, D., & Chen, H. W. (2013). Using the Köppen classification to quantify climate variation and change: An example for 1901–2010. Environmental Development, 6, 69–79. https://doi.org/10.1016/j.envdev.2013.03.007.
2. Geiger, R. (1961). Überarbeitete Neuausgabe von Geiger, R.: Köppen-Geiger / Klima der Erde (Wandkarte 1:16 Mill.). Klett-Perthes, Gotha.
3. Köppen, W. (1884). Die Wärmezonen der Erde, nach der Dauer der heissen, gemässigten und kalten Zeit und nach der Wirkung der Wärme auf die organische Welt betrachtet [The thermal zones of the earth according to the duration of

hot, moderate and cold periods and to the impact of heat on the organic world)]. Meteorologische Zeitschrift, 20(3). Translated by Volken, E., & Brönnimann, S. (published 2011): 351–360.

4. Kottek, M., Grieser, J., Beck, C., Rudolf, B., & Rubel, F. (2006). World map of the Köppen-Geiger climate classification updated. Meteorologische Zeitschrift, 15(3), 259–263. https://doi.org/10.1127/0941-2948/2006/0130.
5. McKnight, T. L., & Hess, D. (2000). Climate zones and types. In Physical Geography: A Landscape Appreciation (pp. 296). Upper Saddle River, NJ: Prentice Hall. ISBN 978-0-13-020263-5.
6. Peel, M. C., Finlayson, B. L., & McMahon, T. A. (2007). Updated world map of the Köppen–Geiger climate classification. Hydrology and Earth System Sciences, 11(5), 1633–1644. https://doi.org/10.5194/hess-11-1633-2007.

CHAPTER 9

Climate Change and Variability

LEARNING OBJECTIVES

After studying this chapter, you will be able to:

- Explain the natural causes of climate change;
- Focus on anthropogenic causes of climate change;
- Define the impact of climate change on environment and society.

KEY TERMS FROM THIS CHAPTER

Adaptation	Biodiversity loss	Carbon emissions
Carbon footprint	Climate models	Climate policy
Deforestation	Emission reductions	Extreme weather
Fossil fuels	Global warming	Greenhouse gases
Heatwaves	Mitigation	Ocean acidification
Paris Agreement	Renewable energy	Resilience
Sea level rise	Sustainable development	

INTRODUCTION

Climate change is a long-term alteration in the average weather patterns that have come to define Earth's local, regional, and global climates. These changes have been particularly evident since the late 19th century, primarily due to human activities, notably the burning of fossil fuels which increases heat-trapping greenhouse gas levels in the Earth's atmosphere. This phenomenon is not only reshaping natural landscapes and ecosystems but also impacting human societies and economies in profound ways.

The science of climate change is rooted in an understanding of how greenhouse gases such as carbon dioxide (CO_2), methane (CH_4), and nitrous oxide (N_2O) accumulate in the atmosphere and trap heat that would otherwise escape into space. This process, known as the greenhouse effect, is natural and essential for life on Earth; however, human activities have intensified this effect, leading to global warming. The Intergovernmental Panel on Climate Change (IPCC) has documented that the Earth's average surface temperature has risen by about 1.2°C since the late 19th century, with the past few decades being the warmest on record.

The impacts of climate change are widespread and varied. Rising global temperatures are causing glaciers and polar ice caps to melt, contributing to sea level rise and increased coastal erosion. Changes in precipitation patterns are leading to more frequent and severe droughts, floods, and storms, disrupting agriculture, water supply, and infrastructure. Ecosystems are being altered as species migrate to new areas in response to changing conditions, often leading to mismatches between plants and pollinators and disruptions in food chains. Human health is also at risk, with increased heatwaves, the spread of vector-borne diseases, and food and water insecurity posing significant challenges.

Addressing climate change requires a comprehensive approach involving mitigation and adaptation strategies. Mitigation efforts focus on reducing greenhouse gas emissions through the adoption of renewable energy sources, enhancing energy efficiency, and protecting and restoring forests which act as carbon sinks. Adaptation involves making adjustments in social, economic, and environmental practices to minimize the damage caused by climate change and to take advantage of any potential benefits. International agreements like the Paris Agreement aim to unite countries in the effort to limit global warming to well below 2°C above pre-industrial levels, with an aspirational goal of limiting the increase to 1.5°C.

9.1. NATURAL CAUSES OF CLIMATE CHANGE

Natural causes of climate change refer to variations in the Earth's climate that are driven by natural processes, independent of human activities. These natural factors have influenced climate patterns over geological time scales and include:

9.1.1. Solar Radiation Variations

Solar radiation, or sunlight, encompasses electromagnetic waves emitted by the Sun. It includes visible light, ultraviolet (UV) light, and infrared (IR) radiation. The amount of solar radiation received by Earth is known as the solar constant, approximately 1361 watts per square meter (W/m^2) at the top of the atmosphere. This energy drives the Earth's weather systems, ocean currents, and ultimately, the climate. (Dansgaard, W.et al. (1993))

1. Solar Radiation Components:
   i. Direct Solar Radiation: This is the sunlight that directly reaches the Earth's surface without being scattered or reflected.
   ii. Diffuse Solar Radiation: This radiation reaches the Earth's surface after being scattered by molecules and particles in the atmosphere.
   iii. Reflected Solar Radiation: A portion of the incoming solar radiation is reflected into space by clouds, aerosols, and the Earth's surface.
2. Factors Influencing Solar Radiation: Several factors determine the amount of solar radiation reaching the Earth's surface, including:
   i. Solar Angle: The angle at which sunlight strikes the Earth varies with latitude, time of day, and season, affecting the intensity of solar radiation.
   ii. Atmospheric Composition: Gases like ozone, water vapor, and carbon dioxide absorb and scatter solar radiation.
   iii. Cloud Cover: Clouds reflect and absorb solar radiation, influencing the amount reaching the surface.
   iv. Albedo: The reflectivity of the Earth's surface varies with land cover types, such as ice, water, and vegetation, impacting the absorbed and reflected solar energy.
3. Historical Perspective on Solar Radiation Variations: Understanding past variations in solar radiation is crucial for interpreting current climate trends. Historical records and proxy data (like ice cores, tree rings, and sediment layers) provide insights into solar radiation changes over millennia.
   i. Solar Cycles: The Sun's activity follows an approximately 11-year cycle, known as the solar cycle, characterized by fluctuations in the number

of sunspots, solar flares, and radiation levels. These cycles have minor but notable impacts on Earth's climate. For example, during periods of high solar activity (solar maxima), increased solar radiation can lead to slight warming, whereas low activity periods (solar minima) may induce cooling.

ii. Long-Term Variations: Beyond the regular solar cycles, the Sun exhibits longer-term variations in activity, including:

a. Maunder Minimum (1645–1715): A prolonged period of low solar activity, coinciding with the "Little Ice Age," a period of cooler global temperatures.

b. Medieval Warm Period (950–1250): A time of relatively high solar activity associated with warmer temperatures in certain regions.

4. Mechanisms of Solar Radiation Influence on Climate: The influence of solar radiation on climate operates through various mechanisms, including direct heating of the Earth's surface and atmosphere and indirect effects on atmospheric circulation and cloud formation.

i. Direct Heating: Solar radiation directly heats the Earth's surface, leading to increased temperatures. This heating is unevenly distributed due to the Earth's curvature, leading to temperature gradients that drive atmospheric and oceanic circulation patterns.

ii. Atmospheric Circulation: Variations in solar radiation affect atmospheric circulation by altering temperature and pressure gradients. For example:

a. Hadley Cells: Enhanced solar radiation can intensify the Hadley cell circulation, affecting tropical and subtropical climates.

b. Jet Streams: Changes in solar radiation influence the position and strength of jet streams, impacting weather patterns and storm tracks.

iii. Cloud Formation and Albedo: Solar radiation influences cloud formation through its effect on surface and atmospheric temperatures. Clouds, in turn, affect the Earth's albedo (reflectivity). Increased cloud cover generally reflects more solar radiation back into space, leading to cooling, while decreased cloud cover allows more solar radiation to reach the surface, causing warming.

5. Recent Variations in Solar Radiation and Climate Change: In recent decades, human activities have significantly altered the Earth's atmosphere and surface, modifying the natural processes that govern solar radiation and climate.

i. Anthropogenic Influences:

a. Greenhouse Gas Emissions: The burning of fossil fuels has increased concentrations of greenhouse gases, such as carbon dioxide and methane, in the atmosphere. These gases trap outgoing infrared radiation, leading to the greenhouse effect and global warming.

b. Aerosols: Industrial activities release aerosols into the atmosphere, which can scatter and reflect solar radiation, causing a cooling effect. However, aerosols also contribute to cloud formation, complicating their overall impact on climate.

c. Land Use Changes: Deforestation, urbanization, and agriculture alter the Earth's surface albedo, affecting the amount of absorbed and reflected solar radiation.

6. Satellite Observations: Satellite data over the past few decades provide precise measurements of solar radiation and its variations. These observations indicate that while there have been minor fluctuations in solar radiation due to solar cycles, the overall trend of global warming cannot be attributed solely to solar variations.

7. The Role of Solar Radiation in Climate Models: Climate models incorporate solar radiation variations to simulate past, present, and future climate scenarios. These models use complex algorithms to account for the interactions between solar radiation, greenhouse gases, aerosols, clouds, and surface albedo.

   i. Model Projections: Climate models project that while solar radiation variations will continue to influence climate patterns, the dominant driver of future climate change will be anthropogenic greenhouse gas emissions. Scenarios with high emissions predict significant global warming, while low-emission scenarios suggest more moderate temperature increases.

8. Mitigation and Adaptation Strategies: Understanding solar radiation variations and their impact on climate is essential for developing effective mitigation and adaptation strategies.

   i. Mitigation:

      a. Reducing Greenhouse Gas Emissions: Transitioning to renewable energy sources, improving energy efficiency, and reforestation are critical steps to reduce greenhouse gas emissions and mitigate climate change.

      b. Aerosol Management: Regulating industrial emissions of aerosols can help manage their impact on solar radiation and climate.

   ii. Adaptation:

      a. Building Resilience: Enhancing infrastructure and developing adaptive strategies for agriculture, water resources, and coastal management can help communities cope with climate impacts.

      b. Monitoring and Research: Continued monitoring of solar radiation and climate variables, along with research on solar-climate interactions, will improve our understanding and response to climate change.

9.1.2. Volcanic Activity

Volcanic activity has long been a critical factor in Earth's climatic and environmental changes. The influence of volcanic eruptions on climate is multifaceted, involving both immediate and long-term effects. This essay explores the complex relationship between volcanic activity and climate change, covering the mechanisms through which volcanic eruptions impact the climate, historical instances of volcanic-induced climate variations, and the role of volcanic activity in the context of contemporary climate change. (Markson, R. (1978))

9.1.2.1. Mechanisms of Volcanic Influence on Climate

Volcanic eruptions can affect the climate through various mechanisms, including the injection of aerosols into the stratosphere, the release of greenhouse gases, and the alteration of atmospheric chemistry.

1. Aerosols and Solar Radiation: One of the primary ways volcanic eruptions influence the climate is by injecting large quantities of sulfur dioxide (SO_2) into the stratosphere. When SO_2 reacts with water vapor, it forms sulfate aerosols, which can reflect and scatter solar radiation. This phenomenon reduces the amount of sunlight reaching the Earth's surface, leading to a cooling effect known as "volcanic winter." The degree of cooling depends on the eruption's magnitude and the amount of sulfur dioxide released.

   For example, the 1991 eruption of Mount Pinatubo in the Philippines released approximately 20 million tons of sulfur dioxide into the stratosphere, causing a global temperature drop of about 0.5°C over the following two years. The aerosols formed a dense layer in the stratosphere, significantly reducing solar radiation and illustrating the potent cooling effect of volcanic aerosols.
2. Greenhouse Gases: Volcanic eruptions also release significant amounts of carbon dioxide (CO_2), a greenhouse gas. However, the contribution of volcanic CO_2 to the global carbon cycle is relatively small compared to anthropogenic emissions. On average, volcanoes emit about 130–380 million metric tons of CO_2 annually, whereas human activities release approximately 35–40 billion metric tons per year. Despite this, volcanic CO_2 can contribute to long-term climate change, especially over geological timescales.
3. Atmospheric Chemistry and Ozone Depletion: Volcanic eruptions can alter atmospheric chemistry by releasing halogen gases, such as chlorine and bromine, which can deplete the ozone layer. The depletion of stratospheric ozone leads to an increase in ultraviolet (UV) radiation reaching the Earth's surface, with implications for both climate and biological systems. For instance, the 1982 eruption of El Chichón in Mexico significantly increased stratospheric chlorine levels, contributing to ozone depletion.

9.1.2.2. Historical Instances of Volcanic-Induced Climate Change

The impact of volcanic activity on climate has been evident throughout history, with several notable eruptions causing significant climatic and environmental changes.

1. The Little Ice Age: The Little Ice Age (LIA), a period of cooler climate from the 14th to the 19th centuries, is often attributed to a combination of volcanic activity and reduced solar radiation. Several major eruptions during this period, including the 1257 eruption of Samalas (Indonesia), the 1452–1453 eruption of Kuwae (Vanuatu), and the 1600 eruption of Huaynaputina (Peru), injected large quantities of aerosols into the stratosphere, contributing to global cooling.
2. The Year Without a Summer (1816): One of the well-documented instances of volcanic-induced climate change was the "Year without a Summer" in 1816. The 1815 eruption of Mount Tambora in Indonesia was one of the largest eruptions in recorded history, ejecting an estimated 160 cubic kilometers of material into the atmosphere. The resulting sulfate aerosols caused a dramatic decrease in global temperatures, leading to widespread crop failures and food shortages across the Northern Hemisphere.
3. The Impact of Krakatoa (1883): The 1883 eruption of Krakatoa in Indonesia also had significant climatic effects. The explosion generated a massive amount of volcanic ash and aerosols, leading to vivid sunsets and a global temperature drop of about 0.3°C for several years. This eruption highlighted the far-reaching impacts of volcanic activity on global climate patterns.

9.1.2.3. Volcanic Activity in the Context of Contemporary Climate Change

In the context of contemporary climate change, volcanic activity plays a nuanced role. While individual eruptions can cause temporary cooling, the overall impact of volcanic CO_2 emissions is minor compared to anthropogenic sources. However, understanding volcanic activity is crucial for accurately modeling climate systems and predicting future climate scenarios.

1. Short-Term Cooling Effects: Volcanic eruptions continue to cause short-term cooling effects. For instance, the 2010 eruption of Eyjafjallajökull in Iceland released a significant amount of ash and aerosols, temporarily affecting climate patterns and air travel. Although this eruption was relatively small, it underscored the potential for volcanic activity to disrupt climate and human activities.
2. Long-Term Contributions to Climate Models: Incorporating volcanic activity into climate models is essential for accurate predictions. Volcanic aerosols can affect climate feedback mechanisms, such as cloud formation and albedo changes. Climate models must account for the sporadic nature of volcanic eruptions and their varying magnitudes to improve the accuracy of climate projections.

9.1.2.4. The Role of Volcanic Activity in Geoengineering

Given the cooling effect of volcanic aerosols, there is growing interest in geoengineering strategies that mimic volcanic eruptions to mitigate global warming. Stratospheric aerosol injection (SAI) involves the deliberate release of sulfur dioxide or other reflective particles into the stratosphere to reduce solar radiation and lower global temperatures.

- Potential Benefits: SAI could provide a temporary solution to global warming by mimicking the natural cooling effects of volcanic eruptions. This approach could buy time for reducing greenhouse gas emissions and implementing sustainable energy solutions.
- Risks and Uncertainties: However, SAI carries significant risks and uncertainties. The long-term impacts on weather patterns, precipitation, and ecosystems are not well understood. There are also concerns about the uneven distribution of cooling effects, which could exacerbate regional climate disparities. Moreover, relying on SAI could detract from efforts to reduce greenhouse gas emissions, creating a moral hazard.

9.1.3. Orbital Variations (Milankovitch Cycles)

Orbital variations, also known as Milankovitch Cycles, play a crucial role in Earth's long-term climate changes. These cycles, named after the Serbian mathematician Milutin Milankovitch, describe the collective effects of changes in Earth's movements on its climate over thousands of years. This essay delves into the science behind Milankovitch Cycles, their impact on historical climate patterns, their interplay with other climatic factors, and their relevance in the context of contemporary climate change.

9.1.3.1. The Science of Milankovitch Cycles

Milankovitch Cycles comprise three primary types of orbital variations: eccentricity, axial tilt (obliquity), and precession. Each cycle influences the distribution and intensity of solar radiation received by Earth, thereby affecting long-term climate patterns.

1. Eccentricity: Eccentricity refers to the shape of Earth's orbit around the Sun, which varies from nearly circular to more elliptical over a cycle of about 100,000 years. When the orbit is more elliptical, the distance between Earth and the Sun changes more dramatically over a year, leading to variations in the amount of solar energy Earth receives. The difference in solar energy received can amplify or moderate seasonal changes, influencing glacial and interglacial periods.
2. Axial Tilt (Obliquity): Axial tilt, or obliquity, refers to the angle between Earth's rotational axis and its orbital plane. This tilt varies between approximately 22.1° and 24.5° over a cycle of about 41,000 years. Changes in axial tilt affect the severity of seasons; greater tilt angles enhance the contrast between summer and winter, while smaller tilt angles result in milder seasons. Variations in axial tilt are crucial for the distribution of solar energy between the equator and the poles, significantly impacting ice sheet formation and melting.
3. Precession: It refers to the wobble in Earth's rotational axis, which alters the timing of the seasons relative to Earth's position in its orbit around the Sun. This cycle

occurs over roughly 26,000 years. Precession influences the intensity and distribution of seasons, particularly in conjunction with eccentricity and obliquity, and can either amplify or diminish seasonal contrasts.

9.1.3.2. Historical Climate Patterns and Milankovitch Cycles

Milankovitch Cycles have been instrumental in shaping Earth's climate over geological timescales. Evidence from ice cores, marine sediments, and other geological records demonstrates the significant impact of these cycles on past climate events.

1. Glacial and Interglacial Cycles: The most prominent manifestation of Milankovitch Cycles is the alternation between glacial (cold) and interglacial (warm) periods during the Quaternary period, the last 2.6 million years of Earth's history. These cycles are closely linked to variations in eccentricity, axial tilt, and precession, which together modulate the amount and distribution of solar radiation.

   For example, during periods of high eccentricity combined with favorable axial tilt and precession, solar radiation is reduced in summer at high northern latitudes, allowing ice sheets to grow and leading to glacial periods. Conversely, during periods of low eccentricity and less pronounced precession and axial tilt, increased summer radiation at high latitudes leads to ice sheet melting and interglacial periods.

2. The Last Glacial Maximum: The Last Glacial Maximum (LGM), occurring around 20,000 years ago, was a period when ice sheets covered significant portions of North America, Europe, and Asia. The LGM was driven by a combination of low axial tilt and specific precessional alignment, which reduced summer insolation in the Northern Hemisphere, promoting ice sheet growth. The subsequent warming and ice sheet retreat, leading to the current Holocene interglacial period, were influenced by increasing solar radiation due to changes in precession and axial tilt.

9.1.3.3. Interplay with Other Climatic Factors

While Milankovitch Cycles provide a framework for understanding long-term climate changes, their effects are modulated by various other factors, including greenhouse gas concentrations, ocean circulation patterns, and tectonic activity.

1. Greenhouse Gases: The concentration of greenhouse gases, such as carbon dioxide (CO_2) and methane (CH_4), amplifies the climate response to Milankovitch Cycles. During glacial periods, lower greenhouse gas concentrations contribute to cooler global temperatures. Conversely, higher concentrations during interglacial periods enhance warming. Feedback mechanisms, such as the release of CO_2 from oceans during warming phases, further amplify the effects of Milankovitch Cycles.
2. Ocean Circulation: It patterns, such as the thermohaline circulation, play a crucial role in distributing heat around the planet. Changes in these patterns can either enhance or mitigate the climatic effects of

Milankovitch Cycles. For example, during periods of significant ice sheet melting, freshwater input into the North Atlantic can disrupt the thermohaline circulation, potentially leading to abrupt climate changes.

3. Tectonic Activity: This, including the movement of continents and volcanic eruptions, can influence climate by altering atmospheric composition and ocean currents. For instance, the uplift of mountain ranges affects global wind patterns and weather systems, while volcanic eruptions inject aerosols and greenhouse gases into the atmosphere, impacting climate.

9.1.3.4. Milankovitch Cycles and Contemporary Climate Change

In the context of contemporary climate change, driven predominantly by anthropogenic factors, Milankovitch Cycles remain relevant but are often overshadowed by more immediate influences such as greenhouse gas emissions and land-use changes.

1. Current Orbital Configuration: Currently, Earth is in a phase of low eccentricity, with a moderate axial tilt and a precessional alignment that results in relatively stable insolation patterns. This configuration would typically suggest a gradual cooling trend, potentially leading toward another glacial period over tens of thousands of years. However, the rapid increase in greenhouse gas concentrations due to human activities is overriding these natural cycles, leading to unprecedented warming.
2. Future Climate Projections: These must account for the interplay between Milankovitch Cycles and anthropogenic influences. While Milankovitch Cycles operate on timescales of tens to hundreds of thousands of years, human activities are inducing rapid changes over decades to centuries. Understanding the long-term baseline provided by Milankovitch Cycles helps contextualize the magnitude and pace of current climate changes.

9.1.3.5. Implications for Climate Policy and Research

The study of Milankovitch Cycles offers valuable insights for climate policy and research, emphasizing the importance of long-term climate dynamics and the need for comprehensive climate models.

1. Integrating Long-Term Climate Dynamics: Climate models that integrate Milankovitch Cycles and other natural forcings provide a more robust framework for understanding future climate scenarios. These models can help differentiate between natural variability and anthropogenic impacts, aiding in the development of more effective mitigation and adaptation strategies.
2. Emphasizing Sustainable Practices: Recognizing the interplay between natural and human-induced climate drivers underscores the urgency of reducing greenhouse gas emissions and adopting sustainable practices. While Milankovitch Cycles will continue to influence Earth's climate over millennia, human actions in the coming decades will determine the trajectory of global temperatures and environmental stability.

9.1.4. Oceanic Circulation

Oceanic circulation plays a critical role in regulating Earth's climate. The vast and dynamic ocean currents redistribute heat, influence weather patterns, and impact the global carbon cycle. Understanding the interplay between oceanic circulation and climate change is vital for predicting future climatic conditions and developing effective mitigation strategies. This essay explores the mechanisms of oceanic circulation, its role in historical and contemporary climate patterns, the impacts of climate change on oceanic circulation, and the broader implications for global climate. (Paillard, D. (1998))

9.1.4.1. Mechanisms of Oceanic Circulation

Oceanic circulation is driven by a combination of wind patterns, the Coriolis Effect, differences in water density, and the configuration of continents. The two main components of oceanic circulation are surface currents and deep-water currents, which together form the global conveyor belt, also known as thermohaline circulation.

1. Surface Currents: These are primarily driven by wind patterns and are influenced by the Coriolis Effect, which causes moving fluids to turn to the right in the Northern Hemisphere and the left in the Southern Hemisphere. Major surface currents include the Gulf Stream in the North Atlantic, the Kuroshio Current in the North Pacific, and the Antarctic Circumpolar Current.

   These currents play a crucial role in redistributing heat from the equator to the poles. For instance, the Gulf Stream transports warm water from the Gulf of Mexico to the North Atlantic, moderating the climate of Western Europe.

2. Deep-Water Currents: These are driven by differences in water density, which are affected by temperature (thermal) and salinity (haline) variations. This process, known as thermohaline circulation, involves the sinking of cold, salty water in Polar regions and the upwelling of warmer water elsewhere.

   The formation of deep water primarily occurs in the North Atlantic and around Antarctica. In these regions, cold temperatures and high salinity increase water density, causing it to sink and flow along the ocean floor. This deep-water flow is a key component of the global conveyor belt, connecting surface and deep currents in a continuous loop.

9.1.4.2. Historical Role of Oceanic Circulation in Climate

Oceanic circulation has been a major driver of climatic changes throughout Earth's history, influencing periods of glaciation and warming.

1. The Last Glacial Maximum: During the Last Glacial Maximum (LGM), approximately 20,000 years ago, changes in oceanic circulation played a critical role in sustaining the extensive ice sheets. Reduced thermohaline circulation in the North Atlantic, due to lower temperatures and freshwater input from melting ice, led to reduced heat transport to higher latitudes. This contributed to the persistence of cold conditions and extensive glaciation.

2. The Younger Dryas: The Younger Dryas, a period of abrupt cooling around 12,800 years ago, is another example of the significant impact of oceanic circulation on climate. It is believed that a massive influx of freshwater from the melting Laurentide Ice Sheet disrupted thermohaline circulation in the North Atlantic, causing a dramatic reduction in heat transport and triggering a sudden drop in temperatures in the Northern Hemisphere.
3. The Medieval Warm Period and the Little Ice Age: The Medieval Warm Period (approx. 950–1250 AD) and the Little Ice Age (approx. 1300–1850 AD) were also influenced by changes in oceanic circulation. Enhanced thermohaline circulation during the Medieval Warm Period likely contributed to warmer conditions in Europe and the North Atlantic region. Conversely, a slowdown in circulation during the Little Ice Age may have contributed to cooler temperatures and increased glaciation.

9.1.4.3. Contemporary Oceanic Circulation and Climate Change

In the context of contemporary climate change, driven primarily by anthropogenic factors, oceanic circulation continues to play a pivotal role. However, the relationship between oceanic circulation and climate is becoming increasingly complex due to the rapid pace of environmental changes.

1. Global Warming and Thermohaline Circulation: Global warming is impacting thermohaline circulation in several ways. Rising temperatures are leading to increased melting of polar ice and glaciers, resulting in an influx of freshwater into the oceans. This reduces the salinity and density of seawater, potentially slowing down or altering thermohaline circulation.

   One major concern is the potential slowdown of the Atlantic Meridional Overturning Circulation (AMOC), a critical component of thermohaline circulation. The AMOC transports warm water from the tropics to the North Atlantic, where it cools, sinks, and returns southward as a deep current. A slowdown in the AMOC could lead to significant climatic changes, including cooler temperatures in Europe and North America, altered weather patterns, and disruptions to marine ecosystems.
2. Ocean Heat Uptake: Oceans absorb a significant portion of the excess heat generated by anthropogenic greenhouse gas emissions. This heat uptake plays a crucial role in moderating global temperatures but also leads to several adverse effects. Increased ocean temperatures contribute to thermal expansion, a major factor in sea level rise. Warmer oceans also affect marine ecosystems, including coral reefs, which are highly sensitive to temperature changes.
3. Changes in Surface Currents: Climate change is also affecting surface currents. Changes in wind patterns, driven by shifts in atmospheric circulation, can alter the intensity and direction of major ocean currents. For example, the strengthening of trade winds in the Pacific Ocean has intensified the Pacific Ocean's subtropical gyres, impacting heat distribution and marine ecosystems.

9.1.4.4. Feedback Mechanisms and Climate Interactions

Oceanic circulation interacts with other components of the climate system, creating feedback mechanisms that can either amplify or mitigate climate change.

1. Carbon Cycle Feedbacks: Oceans are a major component of the global carbon cycle, acting as both a carbon sink and a source. The biological pump, involving the uptake of carbon by phytoplankton and its subsequent transfer to deep waters through the sinking of organic matter, plays a crucial role in sequestering carbon.

   Changes in oceanic circulation can impact the efficiency of this carbon sink. For instance, a slowdown in thermohaline circulation could reduce the transport of carbon-rich deep water to the surface, affecting the overall carbon balance. Additionally, warming oceans reduce the solubility of CO_2, potentially decreasing the ocean's capacity to absorb carbon from the atmosphere.
2. Sea Ice and Albedo Feedback: The melting of sea ice due to global warming creates a feedback loop that influences oceanic circulation and climate. Reduced sea ice extent decreases the albedo (reflectivity) of the Earth's surface, leading to increased absorption of solar radiation and further warming. This warming accelerates the melting of sea ice and glaciers, contributing to freshwater input into the oceans and potentially disrupting thermohaline circulation.
3. Marine Ecosystem Feedbacks: Changes in oceanic circulation can also impact marine ecosystems, which in turn affect climate processes. For example, shifts in upwelling patterns can alter nutrient distribution, affecting primary productivity and the biological pump's efficiency. Additionally, changes in the distribution of marine species, such as phytoplankton and zooplankton, can influence biogeochemical cycles and the ocean's capacity to sequester carbon.

9.1.4.5. Regional Impacts of Changes in Oceanic Circulation

The impacts of changes in oceanic circulation are not uniform across the globe. Different regions experience varying effects based on their specific oceanographic and climatic conditions.

1. North Atlantic: The North Atlantic is particularly sensitive to changes in thermohaline circulation. A slowdown in the AMOC could lead to cooler temperatures in Europe and the North Atlantic region, increased storm activity, and disruptions to marine ecosystems and fisheries. This region's economy and livelihoods, which are closely tied to the ocean, could face significant challenges.
2. Pacific Ocean: Changes in the Pacific Ocean's circulation, such as variations in the El Niño-Southern Oscillation (ENSO), have far-reaching impacts on global climate patterns. ENSO influences weather patterns, precipitation, and temperature extremes in various regions. Climate change is expected to affect the frequency and intensity of ENSO events,

with potential consequences for agriculture, water resources, and disaster management.

3. Arctic and Antarctic: The Polar regions are experiencing some of the most rapid and pronounced effects of climate change. Melting ice in the Arctic and changes in Antarctic ice shelves are influencing global sea levels and oceanic circulation. The influx of freshwater from melting ice is affecting salinity and density gradients, with potential implications for thermohaline circulation and global climate patterns.

9.1.4.6. Future Projections and Uncertainties

Predicting the future behavior of oceanic circulation in a changing climate involves considerable uncertainties due to the complex interactions between atmospheric, oceanic, and cryospheric processes.

1. Climate Models and Projections: Climate models are essential tools for projecting future changes in oceanic circulation and their impacts on climate. These models incorporate various physical and biogeochemical processes to simulate interactions between the atmosphere and oceans. However, uncertainties in model projections arise from a limited understanding of key processes, such as ice sheet dynamics, deep-water formation, and feedback mechanisms.

   Despite these uncertainties, most models agree that significant changes in oceanic circulation are likely under continued global warming. Projections suggest potential slowdowns in major currents, such as the AMOC, with associated impacts on regional and global climates.

2. Research and Monitoring: Ongoing research and monitoring are crucial for improving our understanding of oceanic circulation and its response to climate change. Observational programs, such as the Argo network of autonomous floats, provide valuable data on ocean temperature, salinity, and currents. Satellite observations also offer insights into sea surface temperatures, sea level changes, and ice cover.

   Enhanced monitoring and research efforts are needed to refine climate models, reduce uncertainties, and improve projections. Understanding the mechanisms driving changes in oceanic circulation and their feedback on the climate system is essential for developing effective climate policies and adaptation strategies.

9.1.4.7. Implications for Climate Policy and Adaptation

The intricate relationship between oceanic circulation and climate change has significant implications for climate policy and adaptation efforts.

1. Mitigation Strategies: Reducing greenhouse gas emissions remains the most critical strategy for mitigating climate change and its impacts on oceanic circulation. Limiting global warming to well below 2°C, as outlined in the Paris Agreement, is essential for minimizing disruptions to oceanic currents and associated climatic changes.

2. Adaptation Measures: Adaptation measures are necessary to address the impacts of changes in oceanic circulation on coastal communities, marine ecosystems, and global climate. These measures include:
   - Coastal Protection: Enhancing coastal defenses and implementing sustainable land-use practices to protect against sea level rise and increased storm activity.
   - Marine Conservation: Establishing marine protected areas and promoting sustainable fisheries management to safeguard marine ecosystems and biodiversity.
   - Climate-Resilient Infrastructure: Developing infrastructure that can withstand changing climate conditions, including extreme weather events and shifting ocean currents.
   - Community Preparedness: Building the resilience of coastal communities through education, early warning systems, and disaster preparedness plans.
3. International Collaboration: Addressing the global challenge of climate change and its impact on oceanic circulation requires international collaboration. Efforts to reduce emissions, protect marine environments, and enhance climate resilience must be coordinated across nations and regions. International agreements, such as the Paris Agreement and the United Nations Sustainable Development Goals, provide frameworks for collective action.

9.1.5. Plate Tectonics

Plate tectonics, the scientific theory explaining the movement of Earth's lithospheric plates, is a fundamental aspect of geological sciences. The movements of these plates shape the Earth's surface, form mountains, trigger volcanic activity, and influence oceanic and atmospheric circulation patterns. This essay delves into the mechanisms of plate tectonics, explores their role in historical climate changes, examines their interplay with contemporary climate change, and discusses the implications for future climate projections and policy. (Richerson, P. J., Boyd, R. & Bettinger, R. L. (2001))

9.1.5.1. Mechanisms of Plate Tectonics

The theory of plate tectonics describes the large-scale movements of Earth's lithosphere, which is divided into several rigid plates that float on the semi-fluid asthenosphere beneath. These plates interact at their boundaries, leading to various geological phenomena.

1. Types of Plate Boundaries:
   i. Divergent Boundaries: At divergent boundaries, plates move apart from each other. This movement results in the formation of a new crust as magma rises from the mantle. Mid-ocean ridges, such as the Mid-Atlantic Ridge, are prominent examples of divergent boundaries. The creation of new oceanic crust at these ridges plays a significant role in seafloor

spreading and the global distribution of continents.

ii. Convergent Boundaries: At convergent boundaries, plates move towards each other. This interaction can result in subduction, where one plate is forced beneath another, leading to volcanic activity and the formation of mountain ranges. The Andes in South America and the Himalayas in Asia are examples of mountain ranges formed by convergent boundary interactions.

iii. Transform Boundaries: At transform boundaries, plates slide past each other horizontally. This lateral movement can cause earthquakes along faults, such as the San Andreas Fault in California.

2. Mechanisms Driving Plate Movements: Plate movements are driven by several mechanisms, including:
   i. Mantle Convection: The flow of mantle material due to heat from the Earth's interior creates convection currents, which drag the plates along the surface.
   ii. Ridge Push: The formation of new crust at mid-ocean ridges creates a force that pushes plates away from the ridge.
   iii. Slab Pull: The subduction of a dense oceanic plate into the mantle creates a pulling force that drags the rest of the plate with it.

9.1.5.2. Historical Role of Plate Tectonics in Climate Change

Plate tectonics have played a crucial role in shaping Earth's climate over geological timescales. The movement of continents, the formation of mountain ranges, and volcanic activity associated with plate tectonics have all contributed to long-term climate changes.

1. Continental Drift and Climate: The position and movement of continents significantly influence global climate patterns. Over millions of years, the drifting of continents has altered oceanic and atmospheric circulation, leading to changes in climate.
   - Supercontinents: The assembly and breakup of supercontinents, such as Pangaea, have had profound climatic impacts. For instance, the formation of Pangaea around 300 million years ago led to extensive arid conditions in the interior of the supercontinent, while its breakup around 200 million years ago facilitated the development of more diverse and humid climates.
   - Polar Positioning: The positioning of continents near the poles can lead to glaciation. For example, the presence of Antarctica over the South Pole has contributed to the formation of the Antarctic ice sheet, which plays a crucial role in Earth's albedo and global climate regulation.
2. Mountain Building and Climate: The formation of mountain ranges through tectonic processes impacts climate by altering atmospheric

circulation and precipitation patterns.

- Orographic Effect: Mountains force air to rise, cool, and condense, leading to increased precipitation on the windward side and arid conditions on the leeward side. This effect can create distinct climatic regions, such as the wet Pacific Northwest and the dry Great Basin in the United States.
- Carbon Cycle: The weathering of silicate rocks in mountains consumes atmospheric CO_2, acting as a long-term carbon sink. Mountain-building events, such as the uplift of the Himalayas, have been linked to global cooling periods due to enhanced weathering rates and CO_2 drawdown.

9.1.5.3. Volcanism and Climate

Volcanic activity associated with plate tectonics has both short-term and long-term impacts on climate:

- Volcanic Eruptions: Explosive volcanic eruptions can inject large quantities of aerosols and ash into the stratosphere, reflecting sunlight and causing temporary global cooling. Notable examples include the eruptions of Mount Tambora in 1815 and Mount Pinatubo in 1991, which led to significant but short-lived drops in global temperatures.
- Long-Term Volcanism: Prolonged periods of volcanic activity, such as the formation of large igneous provinces, can release substantial amounts of CO_2 and other greenhouse gases, contributing to long-term warming. The Siberian Traps eruptions around 250 million years ago are believed to have played a role in the Permian-Triassic extinction event due to their climatic impacts.

9.1.5.4. Plate Tectonics and Contemporary Climate Change

While plate tectonics operate on geological timescales, their interactions with contemporary climate change remain relevant. Human activities, such as fossil fuel combustion and land-use changes, are driving rapid climate change, but tectonic processes continue to influence climate in several ways.

Ø Carbon Sequestration and Geological Storage: "Plate tectonics play a role in the long-term carbon cycle, influencing the storage and release of carbon.

- Carbonate Sedimentation: The formation of carbonate rocks through the deposition of marine organisms sequesters carbon over geological timescales. Tectonic uplift and erosion of these rocks can release CO_2 back into the atmosphere, affecting long-term climate dynamics.
- Subduction Zones: The subduction of oceanic plates carries carbon-rich sediments into the mantle, where it can be stored for millions of years. This process acts as a long-term carbon sink, balancing the global carbon cycle.
- Geological Carbon Storage: Modern efforts to mitigate climate change include the injection of CO_2 into deep

geological formations for long-term storage. Understanding tectonic processes is crucial for assessing the stability and security of these storage sites.

- Ø Volcanic Emissions and Climate: Contemporary volcanic activity continues to influence climate by releasing greenhouse gases and aerosols.
  - Greenhouse Gas Emissions: Active volcanoes emit CO_2, methane (CH_4), and water vapor, contributing to the greenhouse effect. While volcanic emissions are relatively small compared to anthropogenic sources, they still play a role in the global carbon cycle.
  - Aerosol Effects: Volcanic eruptions can have a cooling effect by injecting sulfate aerosols into the stratosphere, which reflect sunlight. The cooling effect of volcanic aerosols can temporarily offset warming from greenhouse gas emissions, but the long-term impact depends on the frequency and magnitude of eruptions.

9.1.5.5. Feedback Mechanisms and Climate Interactions

The interactions between plate tectonics and climate involve complex feedback mechanisms that can amplify or mitigate climatic changes.

- Ø Weathering Feedback: The weathering of silicate rocks, enhanced by tectonic uplift, acts as a negative feedback mechanism for climate change.
  - CO_2 Drawdown: Increased weathering rates during periods of mountain building consume atmospheric CO_2, leading to global cooling. This feedback mechanism has been proposed as a driver of long-term climate stability, helping to prevent runaway greenhouse or icehouse conditions.
  - Erosion and Sedimentation: Tectonic uplift also influences erosion and sedimentation rates, which can impact the carbon cycle. Enhanced erosion can increase the burial of organic carbon in sediments, acting as a long-term carbon sink.
- Ø Sea Level Changes: Tectonic processes influence sea level changes, which in turn affect climate and coastal ecosystems.
  - Mid-Ocean Ridge Activity: Changes in the rate of seafloor spreading at mid-ocean ridges can alter sea levels. Faster spreading rates lead to higher ridges and increased sea levels, while slower spreading rates result in lower ridges and decreased sea levels. These changes can influence ocean circulation patterns and climate.
  - Tectonic Subsidence and Uplift: Local tectonic movements, such as subsidence or uplift, can impact relative sea levels and coastal environments. Understanding these processes is essential for assessing the impacts of

sea level rise due to climate change.

Ø Ocean Circulation: Plate tectonics shape the configuration of ocean basins and influence ocean circulation patterns.

- Gateway Events: The opening and closing of oceanic gateways, such as the formation of the Isthmus of Panama, can alter global ocean circulation. These events impact the distribution of heat and nutrients in the oceans, influencing climate and marine ecosystems.
- Plate Movements: The movement of tectonic plates can change the positions of continents and ocean currents, affecting the global climate system. For example, the northward drift of India and the collision with Asia led to the uplift of the Himalayas, which altered atmospheric circulation and monsoon patterns.

9.1.5.6. Regional Impacts of Plate Tectonics on Climate

The impacts of plate tectonics on climate are region-specific, depending on local geological and climatic conditions.

1. North America:
   - Rocky Mountains: The formation of the Rocky Mountains has influenced regional climate patterns by creating a rain shadow effect, leading to arid conditions in the Great Basin and the development of distinct climatic regions.
   - San Andreas Fault: The transform boundary along the San Andreas Fault generates significant seismic activity, which can impact infrastructure and local climates through changes in topography and water resources.
2. Asia:
   - Himalayas: The uplift of the Himalayas has had profound climatic impacts, including the intensification of the Asian monsoon system and the creation of unique climatic zones. The mountains also act as a barrier, influencing the distribution of precipitation and temperature across the region.
   - Sundaland: The subduction zones around Southeast Asia contribute to volcanic activity and the formation of island arcs, which impact regional climate and biodiversity.
3. Africa:
   - East African Rift: The East African Rift is a region of divergent tectonic activity, leading to the formation of rift valleys and volcanic activity. These geological processes influence local climate and water resources, affecting agriculture and human settlements.
   - Sahara Desert: The tectonic history of North Africa, including the uplift of the Atlas Mountains, has contributed to the arid conditions of the Sahara Desert. Changes

in tectonic activity could influence future climatic conditions in the region.

9.1.5.7. Future Projections and Uncertainties

Predicting the future impacts of plate tectonics on climate involves considerable uncertainties due to the complex interactions between geological and climatic processes.

1. Climate Models and Projections: Climate models incorporate various geological processes to simulate future climate scenarios. However, uncertainties in tectonic activity, volcanic emissions, and feedback mechanisms pose challenges for accurate projections.
    - Geological Inputs: Incorporating accurate geological data, such as tectonic movements and volcanic activity, into climate models is essential for improving projections. Enhanced monitoring and research efforts are needed to refine these inputs.
    - Feedback Mechanisms: Understanding the feedback mechanisms between tectonics and climate, such as weathering rates and carbon cycle dynamics, is crucial for reducing uncertainties in climate projections.
2. Research and Monitoring: Ongoing research and monitoring are vital for improving our understanding of the interactions between plate tectonics and climate change.
    - Geological Surveys: Conducting geological surveys and mapping tectonic features can provide valuable data on tectonic activity and its climatic impacts. This information is essential for assessing the risks and impacts of tectonic processes on climate and society.
    - Volcanic Monitoring: Monitoring volcanic activity and emissions is crucial for understanding their impacts on climate. Enhanced monitoring networks and remote sensing technologies can provide real-time data on volcanic eruptions and their atmospheric effects.

9.1.5.8. Implications for Climate Policy and Adaptation

The interactions between plate tectonics and climate change have significant implications for climate policy and adaptation efforts.

1. Mitigation Strategies: Efforts to mitigate climate change should consider the role of geological processes in the global carbon cycle.
    - Carbon Sequestration: Geological carbon storage, such as carbon capture and storage (CCS) technologies, can play a role in mitigating climate change. Understanding tectonic processes is essential for assessing the stability and security of storage sites.
    - Volcanic Risk Management: Managing the risks associated with volcanic activity, including monitoring and early warning systems, can help mitigate the impacts of volcanic emissions on climate and society.

2. Adaptation Measures: Adaptation measures should address the impacts of tectonic processes on climate and society.
    - Coastal Protection: Enhancing coastal defenses and implementing sustainable land-use practices can help protect against sea level rise and tectonic subsidence. Understanding local tectonic activity is crucial for effective coastal management.
    - Disaster Preparedness: Building the resilience of communities to seismic and volcanic hazards through education, early warning systems, and disaster preparedness plans is essential for reducing vulnerability to tectonic events.
3. International Collaboration: Addressing the global impacts of plate tectonics on climate change requires international collaboration.
    - Research Networks: International research networks and collaborations can enhance our understanding of the interactions between tectonics and climate. Sharing data and expertise can improve climate models and projections.
    - Policy Frameworks: International agreements and frameworks, such as the Paris Agreement, should incorporate geological processes and risks into climate policy. Coordinated efforts are needed to address the global impacts of tectonic activity on climate and society.

9.2. ANTHROPOGENIC CAUSES OF CLIMATE CHANGE

Climate change, driven by human activities, has emerged as one of the most pressing global challenges of the 21st century. The term "anthropogenic" refers to changes or effects that result from human activities. In the context of climate change, anthropogenic causes primarily stem from the emission of greenhouse gases (GHGs) and other pollutants into the atmosphere. This essay explores the various anthropogenic causes of climate change, their historical context, current impacts, future projections, and implications for global policy and adaptation strategies.

9.2.1. Historical Context

Human influence on climate can be traced back thousands of years, but the Industrial Revolution marked a significant turning point. The burning of fossil fuels, deforestation, industrial activities, and changes in land use have accelerated since the 19th century, leading to substantial increases in atmospheric concentrations of GHGs.

- Industrial Revolution and Fossil Fuel Consumption: The Industrial Revolution, beginning in the late 18th century, saw a rapid increase in the use of fossil fuels—coal, oil, and natural gas—for energy production, transportation, and industrial processes. This marked the beginning of large-scale emissions of carbon dioxide (CO_2) and methane (CH_4) into the atmosphere.
- Land Use Changes and Deforestation: The expansion of agriculture, urbanization, and infrastructure development has led to significant changes in land use patterns worldwide. Deforestation, particularly in tropical regions, releases stored carbon dioxide and reduces the Earth's capacity to absorb CO_2 through photosynthesis.
- Technological Advancements and Population Growth: Technological advancements, population growth, and increased consumption patterns have further intensified anthropogenic impacts on the climate system. The demand for energy, food, and resources has driven emissions from agriculture, livestock production, and waste management.

9.2.2. Anthropogenic Greenhouse Gas Emissions

Greenhouse gases are substances that trap heat in the Earth's atmosphere, leading to the greenhouse effect—a natural phenomenon essential for maintaining Earth's temperature.

However, human activities have significantly increased GHG concentrations, enhancing the greenhouse effect and causing global warming.

- Carbon Dioxide (CO_2): Carbon dioxide is the primary greenhouse gas emitted through human activities, primarily from burning fossil fuels (coal, oil, natural gas) for energy production, transportation, and industrial processes. Deforestation and land use changes also contribute to CO_2 emissions by reducing the Earth's capacity to absorb carbon through photosynthesis.
- Methane (CH_4): Methane is another potent greenhouse gas emitted from agriculture (rice paddies, livestock), landfills, natural gas production, and fossil fuel combustion. While methane has a shorter atmospheric lifetime than CO_2, it is significantly more effective at trapping heat, contributing to global warming.
- Nitrous Oxide (N_2O): Nitrous oxide is emitted from agricultural practices (fertilizer use), industrial processes, and combustion of biomass and fossil fuels. It is a powerful greenhouse gas and also contributes to ozone depletion in the stratosphere.
- Fluorinated Gases: Fluorinated gases, such as hydrofluorocarbons (HFCs), perfluorocarbons (PFCs), and sulfur hexafluoride (SF_6), are synthetic compounds used in refrigeration, air conditioning, insulation, and semiconductor manufacturing. These gases have high global warming potentials and can persist in the atmosphere for long periods.

9.2.3. Mechanisms and Processes

Anthropogenic activities impact climate through various interconnected mechanisms and processes, altering the Earth's energy balance and atmospheric composition.

- Greenhouse Effect and Radiative Forcing: The greenhouse effect occurs when greenhouse gases absorb and re-emit infrared radiation, trapping heat in the atmosphere. Human-induced increases in GHG concentrations enhance this effect, leading to a positive radiative forcing that warms the planet.
- Aerosols and Particulate Matter: While GHGs predominantly cause warming, aerosols and particulate matter from human activities can have both cooling and warming effects. Sulfate aerosols, for example, reflect sunlight and have a cooling effect on climate, partially offsetting warming from GHGs. However, their atmospheric lifetime is short compared to GHGs.
- Land Use Changes and Albedo: Changes in land use, such as deforestation and urbanization, alter the Earth's surface albedo—the amount of sunlight reflected into space. Darker surfaces absorb more solar radiation, leading to localized warming, while lighter surfaces reflect more radiation, contributing to cooling.

9.2.4. Current Impacts of Anthropogenic Climate Change

The consequences of anthropogenic climate change are increasingly evident and widespread, affecting natural ecosystems, human societies, and economies globally.

- Temperature Rise and Heatwaves: Global temperatures have risen steadily due to anthropogenic emissions, leading to more frequent and intense heatwaves. Extreme heat events pose risks to human health, agriculture, and infrastructure, particularly in urban areas.
- Changes in Precipitation Patterns: Climate change is altering regional precipitation patterns, leading to increased frequency of droughts, floods, and storms. These changes impact water availability, food security, and the livelihoods of millions of people worldwide.
- Sea Level Rise and Coastal Erosion: Rising temperatures cause thermal expansion of seawater and the melting of polar ice caps and glaciers, contributing to global sea level rise. Coastal communities are increasingly vulnerable to flooding, erosion, and saltwater intrusion.
- Ocean Acidification: Increased atmospheric CO_2 levels are absorbed by the oceans, leading to ocean acidification—a reduction in pH levels. This process threatens marine ecosystems, including coral reefs, shellfish, and fish populations, with cascading impacts on coastal economies and food security.
- Biodiversity Loss and Ecosystem Disruption: Climate change exacerbates habitat loss, fragmentation, and species extinction, disrupting ecosystems and reducing biodiversity. Vulnerable species, such as polar bears, corals, and migratory birds, face increasing threats to their survival.

9.2.5. Future Projections and Climate Scenarios

Climate models project a range of future scenarios based on different levels of greenhouse gas emissions, socio-economic factors, and policy interventions.

- Intergovernmental Panel on Climate Change (IPCC) Scenarios: The IPCC's Representative Concentration Pathways (RCPs) outline different emission trajectories and their corresponding impacts on global temperature rise, sea level rise, and other climate indicators. These scenarios help policymakers assess the potential risks and benefits of mitigation and adaptation measures.
- Temperature Projections: Under high-emission scenarios (RCP8.5), global temperatures are projected to increase by 2.6°C to 4.8°C by the end of the 21st century compared to pre-industrial levels. Lower-emission scenarios (RCP2.6) aim to limit warming to below 2°C, consistent with the goals of the Paris Agreement.
- Sea Level Rise: Sea levels are expected to continue rising throughout the 21st century and beyond, with projections ranging from 0.3 meters to 1.1 meters under different scenarios. Coastal adaptation measures will be crucial for protecting vulnerable communities and ecosystems.
- Regional Impacts: Future climate impacts will vary regionally, with some areas experiencing more severe droughts, heatwaves, and extreme weather events. Vulnerable regions, including small island states, coastal cities, and arctic ecosystems, face disproportionate risks from climate change.

9.2.6. Implications for Global Policy and Adaptation Strategies

Addressing anthropogenic climate change requires coordinated global efforts to reduce emissions, adapt to impacts, and build climate resilience.

- International Agreements and Commitments: The Paris Agreement, adopted in 2015, aims to limit global warming to well below 2°C above pre-industrial levels, with efforts to pursue a 1.5°C target. Countries have committed to nationally determined contributions (NDCs) to reduce emissions and enhance climate resilience.
- Mitigation Strategies: Mitigation strategies focus on reducing greenhouse gas emissions through energy transition, renewable energy deployment, energy efficiency improvements, and sustainable land use practices. Carbon pricing mechanisms, such as carbon taxes and emissions trading systems, incentivize emission reductions.
- Adaptation Measures: Adaptation measures seek to reduce vulnerability and build resilience to climate impacts, including investments in climate-resilient infrastructure, disaster risk reduction, water management, agricultural adaptation, and ecosystem restoration.
- Technology and Innovation: Advancements in clean energy technologies, carbon capture and storage (CCS), and climate-smart agriculture are essential for achieving climate goals and transitioning to a low-carbon economy. Research and development funding support innovation and deployment of climate solutions (Dansgaard et al., 1993; Haug et al., 2003).

9.3. IMPACT OF CLIMATE CHANGE ON THE ENVIRONMENT AND SOCIETY

Climate change impacts the environment and society in numerous ways, and these impacts are increasingly being felt around the globe. Here are some of the key impacts:

9.3.1. Social and Economic Impact of Climate Change

Climate change, driven primarily by human activities such as the burning of fossil fuels and deforestation, is altering global weather patterns and exacerbating natural disasters. These changes have significant social and economic implications, affecting livelihoods, public health, food security, and economic development. This essay will delve into the multifaceted impacts of climate change on society and economies across different regions of the world. (Zachos, J.et al., (2001))

9.3.1.1. Social Impacts

1. Health Impacts: Climate change affects human health directly and indirectly through various pathways. Heatwaves, exacerbated by rising temperatures, increase the incidence of heat-related illnesses such as heatstroke and dehydration, particularly in vulnerable populations such as the elderly and outdoor workers. Additionally, extreme weather events such as hurricanes, floods, and wildfires can cause injuries, deaths, and mental health issues such as post-traumatic stress disorder (PTSD) among survivors.
2. Food Security: Changes in temperature, precipitation patterns, and extreme weather events affect agricultural productivity and food availability. Droughts and floods can lead to crop failures, reduced yields, and food shortages, particularly in regions with already precarious food security. This situation often exacerbates poverty and malnutrition, particularly among rural communities dependent on agriculture for their livelihoods.
3. Water Resources: Climate change impacts water availability and quality, affecting communities' access to safe drinking water and sanitation. Changes in precipitation patterns can lead to droughts or floods, disrupting water supplies and sanitation

infrastructure. In coastal areas, sea level rise and saltwater intrusion can contaminate freshwater sources, posing health risks and exacerbating water scarcity.

4. Migration and Displacement: Climate change contributes to forced migration and displacement, as people seek to escape the impacts of extreme weather events, droughts, or rising sea levels. This phenomenon, known as climate-induced migration, places additional pressure on urban areas and host communities, leading to social tensions and challenges in providing adequate housing, healthcare, and livelihood opportunities for displaced populations.
5. Vulnerable Populations: Vulnerable groups such as low-income communities, indigenous peoples, women, and children are disproportionately affected by climate change. They often have limited resources and adaptive capacities to cope with its impacts, exacerbating existing social inequalities and barriers to development. Addressing the needs of these populations is crucial for building resilience and ensuring equitable adaptation measures.

9.3.1.2. Economic Impacts

1. Agriculture and Livelihoods: Agriculture is highly sensitive to climate variability and change, impacting crop yields, livestock productivity, and fisheries. Extreme weather events, such as droughts and floods, can destroy crops, damage infrastructure, and disrupt agricultural supply chains, leading to income losses and food insecurity for farmers and rural communities. Changes in temperature and precipitation patterns also affect agricultural practices and crop suitability, necessitating adaptation strategies such as crop diversification or improved irrigation systems.
2. Infrastructure and Property Damage: Climate change poses risks to infrastructure such as roads, bridges, buildings, and utilities, particularly in vulnerable coastal and low-lying areas. Sea level rise, storm surges, and extreme weather events can cause infrastructure damage, disrupt transportation networks, and increase maintenance costs. The economic impacts extend to private property damage, insurance liabilities, and the costs of rebuilding or relocating affected communities.
3. Tourism and Recreation: Climate change affects tourism and recreation industries, which depend on stable weather conditions and natural landscapes. Rising temperatures, coral bleaching, and extreme weather events can degrade tourist attractions such as beaches, national parks, and cultural heritage sites. This can reduce visitor numbers, harm local economies dependent on tourism revenue, and undermine livelihoods in tourist-dependent communities.
4. Energy Supply and Demand: Climate change influences energy supply and demand dynamics, affecting energy production, distribution, and consumption patterns. Shifts in temperature patterns may alter heating and cooling needs, while extreme weather events can disrupt energy infrastructure, leading to power outages and supply shortages. Renewable energy sources such as solar and wind power offer opportunities for mitigating greenhouse gas emissions

and enhancing energy security but require investment in infrastructure and technological advancements.

5. Healthcare Costs and Productivity: Climate change impacts human health, resulting in increased healthcare costs associated with treating heat-related illnesses, respiratory diseases, and injuries from extreme weather events. Lost productivity due to illness, absenteeism, or displacement further strains economies and undermines socio-economic development. Addressing climate-related health risks through adaptation measures and public health interventions is essential for reducing economic burdens and improving community resilience.

9.3.2. Policy Responses and Adaptation Strategies

1. Mitigation Strategies: Mitigating climate change requires reducing greenhouse gas emissions through policies and measures such as carbon pricing, renewable energy incentives, and energy efficiency standards. International agreements such as the Paris Agreement aim to limit global warming to well below 2°C above pre-industrial levels, with efforts to pursue 1.5°C. Transitioning to a low-carbon economy can create new job opportunities, promote sustainable development, and reduce dependency on fossil fuels.
2. Adaptation Strategies: Adapting to climate change involves enhancing resilience to its impacts through climate-smart agriculture, disaster preparedness, and infrastructure improvements. Investing in early warning systems, flood defenses, and drought-resistant crops can reduce risks and protect communities from climate-related hazards. Integrating climate considerations into urban planning, water management, and public health strategies is essential for building adaptive capacity and fostering sustainable development.
3. International Cooperation: Addressing the social and economic impacts of climate change requires global cooperation, knowledge sharing, and financial support for vulnerable countries and communities. International organizations, donor agencies, and development partners play a crucial role in supporting adaptation efforts, capacity-building initiatives, and technology transfer to enhance resilience and promote sustainable development outcomes.
4. Community Engagement and Education: Empowering communities through education, awareness-raising, and participatory decision-making processes is vital for building resilience and fostering adaptive capacity. Local knowledge, traditional practices, and indigenous wisdom can contribute to effective climate adaptation strategies tailored to the needs and priorities of vulnerable populations. Engaging stakeholders, including civil society organizations, academia, and private sector partners, promotes inclusive and sustainable development approaches.

9.3.3. Environmental Impact of Climate Change

Climate change refers to long-term shifts in temperature, precipitation patterns, and other climatic parameters due to human activities, primarily the emission of greenhouse gases (GHGs) such as carbon dioxide (CO_2), methane (CH_4), and nitrous oxide (N_2O).

These emissions trap heat in the Earth's atmosphere, leading to global warming and alterations in weather patterns. The consequences of climate change are far-reaching and affect virtually every aspect of the environment, including biodiversity, ecosystems, water resources, and human societies.

9.3.3.1. Impacts on Biodiversity and Ecosystems

One of the most significant impacts of climate change is its effect on biodiversity and ecosystems. Many species are highly sensitive to changes in temperature and precipitation, leading to shifts in their distribution and abundance. For instance, certain species may migrate to higher latitudes or altitudes in search of suitable habitats as temperatures warm. Others may face increased competition or predation as ecosystems transform.

Ecosystems such as coral reefs, mangroves, and Polar regions are particularly vulnerable to climate change. Coral bleaching, caused by warmer ocean temperatures, threatens the existence of these diverse and productive marine ecosystems. Similarly, melting polar ice caps affect the habitat of polar bears, seals, and other ice-dependent species, jeopardizing their survival.

Changes in precipitation patterns also impact ecosystems, leading to altered fire regimes, desertification, and shifts in vegetation types. For example, extended droughts can increase the frequency and intensity of wildfires in forested regions, posing risks to both flora and fauna. Conversely, increased rainfall in some areas may promote the spread of invasive species or lead to flooding, disrupting ecosystems and ecological processes.

9.3.3.2. Ocean Acidification and Sea Level Rise

Climate change not only affects terrestrial ecosystems but also has profound implications for the world's oceans. The absorption of excess CO_2 by seawater results in ocean acidification, which poses a threat to marine organisms such as shellfish, corals, and plankton. Acidification can impair shell formation in mollusks and disrupt the marine food web, ultimately affecting fisheries and coastal economies that rely on marine resources.

Furthermore, climate change contributes to sea level rise through the thermal expansion of seawater and the melting of glaciers and ice sheets. Rising sea levels exacerbate coastal erosion, inundate low-lying areas, and increase the frequency of storm surges and flooding events. Coastal communities, including major cities, infrastructure, and valuable ecosystems such as wetlands, are increasingly vulnerable to these impacts.

9.3.3.3. Water Resources and Hydrological Cycle

Changes in the global climate also affect the hydrological cycle, altering precipitation patterns and water availability in different regions. Some areas may experience more frequent and intense droughts, leading to water scarcity and conflicts over resources. Others may face increased precipitation and flooding, which can overwhelm infrastructure and disrupt agriculture and livelihoods.

Snowpack and glacier melt contribute significantly to freshwater supplies in many regions, particularly mountainous areas. However, as temperatures rise, glaciers are retreating at unprecedented rates, threatening the stability of water sources for millions of people who depend on meltwater for drinking, irrigation, and hydropower generation.

9.3.3.4. Impacts on Human Health and Well-being

Climate change has direct and indirect impacts on human health and well-being, primarily through changes in temperature extremes, air quality, vector-borne diseases, and food security. Heatwaves and extreme weather events such as hurricanes and wildfires pose immediate risks to human health, causing heat-related illnesses, injuries, and fatalities. Vulnerable populations, including the elderly, children, and individuals with pre-existing conditions, are particularly at risk.

Poor air quality associated with increased ground-level ozone and particulate matter exacerbates respiratory diseases such as asthma and bronchitis, leading to higher healthcare costs and reduced quality of life. Changes in precipitation patterns can also influence the spread of waterborne diseases and vector-borne diseases such as malaria and dengue fever, affecting populations in both developed and developing countries.

Furthermore, disruptions to agricultural production and food supply chains due to climate change can impact food security and nutrition outcomes globally. Extreme weather events, shifts in growing seasons, and changes in pest and disease dynamics pose challenges to crop yields and livestock productivity, jeopardizing the livelihoods of farmers and food producers.

9.3.3.5. Mitigation and Adaptation Strategies

Addressing the impacts of climate change requires concerted efforts at local, national, and global levels to mitigate greenhouse gas emissions and adapt to existing and projected changes. Mitigation strategies aim to reduce GHG emissions through policies and measures such as transitioning to renewable energy sources, improving energy efficiency, and promoting sustainable land use practices. These efforts not only mitigate climate change but also offer co-benefits such as improved air quality, enhanced energy security, and economic opportunities in green technologies.

Adaptation strategies focus on building resilience to climate impacts, including enhancing infrastructure, implementing early warning systems, and diversifying livelihoods. Nature-based solutions such as ecosystem restoration, sustainable agriculture, and coastal protection can help communities adapt to changing climatic conditions while conserving biodiversity and ecosystem services. International cooperation and partnerships are essential for mobilizing financial resources, technology transfer, and capacity-building efforts to support vulnerable regions and communities in adapting to climate change impacts. The Paris Agreement, adopted in 2015 under the United Nations Framework Convention on Climate Change (UNFCCC), represents a global commitment to limiting global warming to well below 2℃ above pre-industrial levels and pursuing efforts to limit it to 1.5 ℃.

9.3.3.6. Environmental Consequences of Climate Change

The climate change risk assessment (CCRA) report outlines the current risk, the potential impact of global climate change on every risk issue, the implications for the ecosystem services sector, and therefore the implications for adaptation. Here we tend to define the salient components of risk. It ought to be noted that the CCRA considers the risks of a dynamics climate solely, not the risks to that the countries could also be exposed from world climate-related events. There's doubtless to be an impact on the country's economy from climate-related disruption elsewhere in the world, for example, through international trade, supply chains, and migration.

1. Changes in Soil Moisture Level: Most habitats and species are adapted to a rather wet environment. Climate change forecasts expect moisture levels to decrease, thus habitats that are sensitive to moisture levels can suffer if the climate becomes a lot drier. Reduction in operation of loss of habitats might impact food production, water system, and quality, and use of land for business enterprise and leisure. Reduced soil moisture is additionally connected to raise risk of wildfires, and therefore the vulnerability of species to pests and disease. Potential diversifications to the current risk are extremely species and site specific, like developing new earth areas and managing bog land, with no universal solution.
2. Extreme Flooding and Coastline Changes: A coastal setting is dynamic and with high biodiversity, supporting a vast range of species. Changes in water level will amend coastal habitats through erosion or sediment deposits, providing risks and new opportunities to plant and animal species. Some coastal options additionally play a crucial role in protecting human settlements from flooding. Usually, coastal changes occur over a long time however events like as storm surges and major flooding will result in large-scale changes during a short time.
3. Increased in Water Temperature: Nowadays water stress is increasing in several countries, in many regions, floods, and water quality problems are likely to be worse because of climate change. Approx. 75–220 million individuals face a lot of severe water shortages by 2020. A reduction in the water system because of climate change combined with a rise in demand for agricultural irrigation can have an impact on the ability of aquatic ecosystems to maintain and replenish. It's known that low water flows can lead to increases in pollution and nutrient concentrations. This will affect each wildlife and drinking water. The potential drying-up of watercourses additionally has implications for wider landscape pleasantness. Climate impacts with totally different temperature stabilization eventualities are indicated as:
    - Within 2°C: For several regions underneath water stress, the world means the temperature will increase above 1.5°C result in decreases in water supply.
    - >2°C to 2.5°C: Above 2 to 2.5°C world average temperature increase it is projected that a further 2.4 to 3.1 billion people are in danger of water stress

- >3°C: Above 2.5°C warming the extent of risk begins to saturate within the range of 3.1–3.5 billion additional persons in danger.

4. Changes in Species Migration Patterns: Several animals, particularly birds, migrate to warmer climates in winter. Dynamic climates will so impact migration patterns. This could present risk and opportunity to migratory species. Observations recommend that patterns are already changing, in terms of geographical and temporal order, behavior that itself may be thought to have associate degree of adaptation to climate change. However, concern is that it may impact breeding patterns and successes.
5. Increased Risk from Pests, Diseases, and Invasive Non-Native Species: These problems are interlinked and therefore the risk is anticipated to extend because of temperature as a result of several species are climate sensitive, and additionally as a result of typically warmer winters offer a lot of conducive environment for nonnative species and pathogens. Most of the analysis thus far has focused on human health effects; however, the risk also applies to biodiversity. There's already legislation to stop the introduction of non-native species (either deliberate or accidental), which can displace existing species and result in the introduction of latest pests and diseases additionally as disrupt agriculture.

9.3.4. Climate Change and Food Security

Global climate change has the potential to extend food insecurity. Existing threats to food security and livelihoods are going to be exacerbated by global climate change because of a mix of things that include; the increasing frequency and intensity of climate hazards, decreasing agricultural yields and reduced production, rising sanitation and health risks, increasing water scarceness, and exasperating conflicts over scarce resources. Owing to an increase in extreme weather events and variability of weather patterns, climate change is expected to negatively affect food security outcomes, potentially affecting all four dimensions of food security as shown in Figure 9.1.

Figure 9.1. Climate change impact on food security.

Source: https://www.researchgate.net/publication/363739433_Impact_of_Climate_Change_on_Environment_and_Society.

1. Availability: Global climate change will scale back agricultural production, through environmental degradation, dynamic agro-ecological conditions, and shifting growing seasons. Changes in precipitation and higher temperatures will also affect crop productivity, reducing food availability.
2. Access: Food prices can increase as a result of reduced production, reducing access to food. For the foremost vulnerable, reduced agricultural production also will mean lower income.
3. Utilization: climate change is anticipated to extend the incidence of diseases, therefore affecting utilization of food, significantly among children. Moreover, reducing agricultural production implies that the foremost vulnerable households can have less food available, increasing the danger of malnutrition.
4. Stability: Extreme weather events disrupt the steadiness of food production and access likewise people's livelihoods (Lambeck et al., 2003; Markson et al., 1978).

CASE STUDY

IMPACT OF CLIMATE CHANGE ON COASTAL COMMUNITIES

Background:

Coastal communities around the world are experiencing significant changes due to climate change. Rising sea levels, increased frequency of extreme weather events, and changes in precipitation patterns are among the key challenges these communities face. This case study focuses on a small coastal town, Rivertown, which is grappling with the impacts of climate change on its environment and economy.

Methodology

1. Climate Data Analysis: Historical climate data for Rivertown, including temperature trends, precipitation patterns, and sea level rise, are analyzed. This data helps establish baseline conditions and identify trends over the past few decades.
2. Impact Assessment: Stakeholder interviews and field surveys are conducted to understand how climate change is affecting Rivertown. Key impacts identified include:
    - Increased frequency of storm surges leading to coastal erosion and property damage.
    - Shifts in fish populations affecting local fisheries.
    - Changes in tourism patterns due to altered beach conditions.
    - Infrastructure vulnerabilities to flooding and sea level rise.
3. Modeling Future Scenarios: Using climate projection models, future scenarios are developed to predict potential changes in temperature, precipitation, and sea levels in Rivertown over the next 50 years. These scenarios consider various emission scenarios and their implications for local climate.
4. Adaptation Strategies: Based on the impact assessment and future scenarios; adaptation strategies are proposed:
    - Ecosystem-based approaches: Restoring coastal wetlands and mangroves to provide natural flood protection.
    - Infrastructure upgrades: Elevating vulnerable buildings and roads, and improving drainage systems.
    - Community awareness and education: Campaigns to inform residents about climate risks and encourage sustainable practices.

- Policy recommendations: Advocating for coastal zoning regulations and building codes that account for future climate risks.

5. Implementation and Monitoring: The proposed adaptation strategies are implemented in collaboration with local government, NGOs, and community groups. Monitoring programs are established to track the effectiveness of these strategies over time and make adjustments as needed.

Results:

The implementation of adaptation strategies leads to several positive outcomes:

- Reduction in flood damage and infrastructure losses during extreme weather events.
- Enhanced resilience of local ecosystems and fisheries.
- Improved community awareness and preparedness for climate impacts.

Conclusion:

This case study highlights the importance of proactive climate adaptation planning for coastal communities like Rivertown. By understanding current impacts, modeling future scenarios, and implementing targeted strategies, communities can effectively mitigate the adverse effects of climate change and build resilience for the future. Continued monitoring and adaptation will be essential as climate change continues to evolve.

CLASS ACTIVITY

CLIMATE CHANGE IN A JAR

Objective: To demonstrate the greenhouse effect using a simple model.

Instructions:

1. Fill a clear jar with water and add a few drops of food coloring.
2. Place the jar in direct sunlight.
3. Cover the jar with a clear plastic wrap and secure it with a rubber band.
4. Observe how the water heats up and discuss how this relates to Earth's atmosphere and climate change.

Discussion Points:

- Explain how the plastic wrap acts like the atmosphere, trapping heat.
- Discuss the impact of increased greenhouse gases on global temperatures.
- Brainstorm actions individuals can take to mitigate climate change.

SUMMARY

- Climate change refers to long-term shifts in temperatures and weather patterns. Such shifts can be natural, due to changes in the sun's activity or large volcanic eruptions. But since the 1800s, human activities have been the main driver of climate change, primarily due to the burning of fossil fuels like coal, oil and gas.
- The main greenhouse gases that are causing climate change include carbon dioxide and methane. These come from using gasoline for driving a car or coal for heating a building, for example. Clearing land and cutting down forests can also release carbon dioxide. Agriculture, oil, and gas operations are major sources of methane emissions. Energy, industry, transport, buildings, agriculture, and land use are among the main sectors causing greenhouse gases.
- Climate change has manifested in numerous impactful ways globally. Rising temperatures have led to hotter days and more frequent heatwaves, causing health risks and altering natural systems. Melting ice caps and glaciers contribute to rising sea levels, threatening coastal cities and small island nations with flooding and erosion. Furthermore, the increased frequency and intensity of extreme weather events, such as hurricanes, droughts, and wildfires, have devastating effects on communities, economies, and ecosystems.
- Climate change profoundly impacts societies worldwide. It exacerbates social inequalities as vulnerable communities, often with fewer resources, are disproportionately affected by extreme weather events, rising sea levels, and food insecurity. Health risks increase due to heatwaves, poor air quality, and the spread of vector-borne diseases, placing additional strain on healthcare systems. Moreover, climate-induced displacement forces millions to migrate, leading to social and political tensions. These social impacts highlight the need for inclusive and equitable climate policies to protect and support all communities.
- Climate change significantly impacts the environment. It accelerates the melting of polar ice caps and glaciers, leading to rising sea levels and habitat loss for species dependent on ice. Increased temperatures and changing precipitation patterns disrupt ecosystems, resulting in shifts in species distributions and threatening biodiversity. Additionally, ocean acidification, driven by higher CO_2 levels, harms marine life, particularly coral reefs and shellfish, disrupting marine ecosystems and food webs. These environmental changes underscore the urgency of addressing climate change to preserve ecological balance.

REVIEW QUESTIONS

1. What are the primary natural factors that influence climate change?
2. How do volcanic eruptions contribute to climate variability?
3. Explain the role of solar radiation variations in driving natural climate changes.
4. What is the impact of deforestation on atmospheric CO_2 levels and climate change?

5. Discuss how industrial processes contribute to the increase in greenhouse gases.
6. What are the key environmental impacts of climate change on biodiversity and ecosystems?
7. How does climate change exacerbate social inequalities and affect vulnerable communities?

REFERENCES

1. Dansgaard, W., & Oeschger, H. (1989). In Oeschger, H., & Langway, C. C. (Eds.), The Environmental Record in Glaciers and Ice Sheets (pp. 287–318). Chichester, UK: Wiley.
2. Dansgaard, W., et al. (1993). Evidence of general instability of past climate from a 250-kyr ice-core record. Nature, 364, 218–220.
3. Haug, G. H., et al. (2003). Climate and the collapse of Maya civilization. Science, 299, 1721–1725.
4. Lambeck, K., Esat, T. M., & Potter, E. K. (2003). Links between climate and sea levels for the past three million years. Nature, 419, 199–206.
5. Markson, R. (1978). Solar modification of atmospheric electrification and possible implications for the Sun-weather relationship. Nature, 244, 197–200.
6. Paillard, D. (1998). The timing of Pleistocene glaciations from a simple multiple-state model. Nature, 391, 378–381.
7. Richerson, P. J., Boyd, R., & Bettinger, R. L. (2001). Was agriculture impossible during the Pleistocene but mandatory during the Holocene? A climate change hypothesis. American Antiquity, 66, 1–50.
8. Shoemaker, E. M. (1983). Asteroid and comet bombardment of the Earth. Annual Review of Earth and Planetary Science, 11, 461–494.
9. Tengen, I., Lacis, A. A., & Fung, I. (1996). The influence on climate forcing of mineral aerosols from disturbed soils. Nature, 380, 419–422.
10. Tucker, C. J., Dregne, H. E., & Newcomb, W. W. (1991). Expansion and contraction of the Sahara Desert from 1980 to 1990. Science, 253, 299–301.
11. Weaver, A. J., & Hughes, T. M. C. (1994). Rapid interglacial climate fluctuations are driven by North Atlantic Ocean circulation. Nature, 367, 447–450.
12. Zachos, J., et al. (2001). Trends, rhythms, and aberrations in global climate 65 Ma to present. Science, 292, 686–693.

CHAPTER

10 Applied Climatology and Future Projections

LEARNING OBJECTIVES

After studying this chapter, you will be able to:

- Define applied climatology;
- List the emerging objects and practices in applied climatology;
- Explain predictions of future global climate;
- Describe the climate to evolve in the future;
- Elaborate the extreme events changing.

KEY TERMS FROM THIS CHAPTER

Climate change	Climate data	Cyclones
Greenhouse gas emission	Heatwaves	Humidity
Irrigation	Meteorology	Permafrost
Precipitation	Storms	Thermal expansion
Thermohaline circulation	Urban climatology	Water vapor

INTRODUCTION

Applied climatology is the practical use of climatic data and principles to address specific real-world problems. It involves analyzing past and current climate patterns to understand their impacts on various sectors such as agriculture, water resources, health, and urban planning. This field plays a critical role in helping societies adapt to climate variability and change, and in mitigating the adverse effects of climate extremes.

One key aspect of applied climatology is its focus on practical applications. For instance, in agriculture, climatologists analyze temperature and precipitation trends to predict crop yields, inform irrigation practices, and develop strategies to cope with droughts and floods. In the water resources sector, climate data are used to manage water supplies, forecast river flows, and plan for future water needs under changing climatic conditions. Similarly, in public health, climatologists study the relationship between climate and disease outbreaks to improve health preparedness and response strategies.

Urban planning is another area where applied climatology is essential. Cities face unique challenges from climate change, such as increased heatwaves, changing precipitation patterns, and sea-level rise. Applied climatology provides critical data for designing resilient infrastructure, developing sustainable urban policies, and mitigating the urban heat island effect.

Looking to the future, climate projections are a fundamental tool for understanding potential changes in the Earth's climate system. These projections, based on complex climate models, provide insights into how temperature, precipitation, and other climate variables may evolve under different greenhouse gas emission scenarios. They are essential for planning long-term adaptation and mitigation strategies. For instance, future projections help policymakers and planners design infrastructures that can withstand future climate extremes and develop policies to reduce carbon emissions.

The application of climate projections is not without challenges. Uncertainties in climate models and the inherent variability of the climate system mean that projections must be interpreted with caution. However, advances in climate science and modeling techniques continue to improve the accuracy and reliability of these projections.

In summary, applied climatology bridges the gap between climate science and practical applications, providing valuable insights and tools for addressing the impacts of climate variability and change. Future climate projections further enhance this field by enabling forward-looking planning and decision-making, helping societies to prepare for and adapt to the changing climate. This integration of past, present, and future climate knowledge is crucial for building a resilient and sustainable future.

10.1. APPLIED CLIMATOLOGY

Applied climatology is the scientific discipline that involves the practical application of climate data and principles to solve real-world problems across various sectors. It translates theoretical climatological knowledge into actionable insights and strategies, helping to address the impacts of climate variability and change on human activities and natural systems.

10.1.1. Key Areas of Application

1. Agriculture:

- Crop Management: Applied climatology assists farmers in selecting suitable crops based on climatic conditions and forecasting weather patterns that affect crop growth. This includes predicting droughts, frosts, and heatwaves that could impact yields.
- Irrigation Planning: Climate data help optimize irrigation schedules and water usage, crucial for regions facing water scarcity or erratic rainfall patterns.

2. Water Resources Management:

- Flood Forecasting: Climatologists use rainfall data and hydrological models to predict river flows and flood risks, enabling early warning systems and disaster preparedness.
- Water Supply Planning: Long-term climate data inform the management of reservoirs and water supply systems to ensure sustainability under changing climatic conditions.

3. Urban and Regional Planning:

- Infrastructure Design: Applied climatology guides the design of resilient infrastructure, such as stormwater drainage systems and buildings that can withstand extreme weather events.
- Urban Heat Island Mitigation: Strategies are developed to reduce the urban heat island effect through green spaces, reflective building materials, and efficient urban layouts.

4. Public Health:

- Disease Prevention: Climate data are used to predict and mitigate the spread of climate-sensitive diseases like malaria and dengue fever, which are influenced by temperature and precipitation patterns.
- Heatwave Response: Applied climatology helps in planning and implementing measures to protect vulnerable populations from heat-related health risks.

5. Energy:
- Renewable Energy Planning: Climatologists assess wind, solar, and hydropower potential based on long-term climate data to optimize the location and efficiency of renewable energy installations.
- Energy Demand Forecasting: Climate information is crucial for predicting energy demand, especially for heating and cooling needs, which are influenced by temperature variations.

6. Disaster Management:
- Risk Assessment: Applied climatology provides critical data for assessing risks associated with natural disasters such as hurricanes, tornadoes, and wildfires.
- Emergency Preparedness: Climate data help in planning for emergency responses and recovery efforts, minimizing the impact of disasters on communities and infrastructure.

10.1.2. Tools and Techniques

1. Climate Models: These simulate the Earth's climate system and project future climate scenarios, providing valuable information for planning and decision-making.
2. Geographic Information Systems (GIS): This technology is used to map and analyze climatic data spatially, aiding in the visualization of climate impacts and the development of localized adaptation strategies.
3. Remote Sensing: Satellite data provide real-time monitoring of climatic variables such as temperature, precipitation, and vegetation cover, enhancing the accuracy of climate assessments.
4. Statistical Analysis:
   - Statistical methods are employed to analyze historical climate data and detect trends, variability, and correlations with various climatic impacts.

10.1.3. Applied Climatology, Where Did It Come From, and Where Is It Going?

This takes a historical perspective on the evolution of "application" discourses in meteorology, climatology, and geography to address the issue of applied climatology. By doing this, we record the ways in which applied climatologists have conceptualized their positions and duties in society and the ways in which the term "applied climatology" has come to be associated with specific social actors (and industries). We draw attention to the ways that applied climatology diverged from mainstream climatology and geography due to disagreements over "relevance" and environmental determinism, respectively, and how these developments shaped the field's "downstream" political orientation.

Although a complete history of the field's development is unfeasible, we highlight significant historical turning points that help us understand the direction and history of applied climatology as well as the reasons behind the failure of some ideas (like a notional Marxist climatology). It is divided into three major periods for heuristic purposes: pre-1945, post-war emergence and consolidation (1945–1980), and the "golden age" (1980, onward).

10.1.4. Early Foundations of Applied Climatology

Since ancient times, climate has been considered a valid subject of study (Fleming and Jankovic, 2011; Heymann, 2010). By the late 19th and early 20th centuries, the field of climatology—the study of the average properties of the atmosphere—became established in the newly formed field of physical geography. The task of spatial description dominated geographical study during this time (Castree, 2005). This meant, in terms of climatology, compiling average weather data, recording climate change, characterizing climates regionally, and categorizing the relationships between vegetation and climate (Aspinall, 2010; Skaggs, 2004). The groundwork for the development of applied climatology in the field of geography was established by this work. By 1940, a wealth of climate knowledge had been amassed that would prove indispensable to applied endeavors, including baselines for water resources, crop-climate suitability, and the fundamentals of structural design (Changnon, 2005a).

Two other significant events that shaped the direction and nature of applied climatology occurred during this time. First, meteorology—a mostly mathematical field of study—was split off from climatology, a field of study focused on geography. Meteorology emerged as a reductionist science based on mathematical formulation and prediction after dynamic meteorology was formed in the 1920s and atmospheric inquiry was realigned towards the form of air masses, fronts, and physical movements of the atmosphere. This led to the emergence of the weather forecasting industry. Because of this, meteorologists were primarily the purview of physicists, chemists, and mathematicians. This led to a long-lasting conflict that still exists today regarding the institutional ownership and relative "scientific status" of meteorology and climatology (Carleton, 1999).

Secondly, during this time, an increasingly maligned interest in environmental determinism emerged from the foundational ideas of geography as an integrative social-environmental discipline (Skaggs, 2004). Such methods were common in all fields of science at the time and were not limited to geography. Climate science trailed suit, with influential writings by authors like Ellsworth Huntington and Ellen Churchill Semple attempting to attribute social characteristics to climate, thus legitimizing the superiority of the white race (Castree, 2005). For example, Huntington (1924) compiled data on economic output and climate to support his case for the ideal ambient temperature for human productivity. He also attempted to explain the "advance" of Anglo-American civilizations by drawing historical comparisons. Reactions to this application of climatological determinism varied from tacit acceptance to outright outrage (Kobayashi, 2010). Major geography scholars such as Carl Sauer and Harlan Barrows offered a variety of solutions, ranging from giving up on studying physical environmental processes in favor of "human ecology" to giving up on just the explanatory parts of physical geography (Skaggs, 2004). Physical geographers had a noticeable sense of alienation from issues about human-environment relationships, which some argue has needlessly impeded the advancement of applied climatology (Hare, 1966; Sewell et al. (Skaggs, 2004; 1968).

Given these two divergences in climatology, originating from meteorology and human-environment interactions, it is pertinent to consider the objective that D. Brunt, the President of the Royal Meteorological Society in 1944, had envisioned for what he perceived to be a developing field of human biometeorology. Following a discussion of the paper "The Poor White Problem in South Africa," Brunt proposed that:

The Colonial Office ought to survey the entire British Empire right away to decide which regions are acceptable for white people to settle, multiply, and live in. The meteorologist, the medical man, and the economist should all be involved in the work if such a survey is ever conducted—which it should, of course, be done as soon as possible. (Brunt, 1944).

The idea behind this project, which was intended for an applied climatologist, was to use physics and mathematics to optimize the climate for human habitation while avoiding or constructively addressing issues related to race, colonialism, and politics. Was applied climatology to be the field responsible for this?

10.1.5. Post-War Emergence of the Subdiscipline

The Second World War had a significant impact on applied climatology, as it did on many other fields in geography. However, compared to other contexts, the influence in applied climatology was far more direct and formative. Meteorologists and climatologists were called upon during the war to forecast weather that would affect soldiers, ships, and aircraft as well as to worry about operational activity patterns and logistical flows (Jacobs, 1947). In reality, a number of the most well-known applied climatologists of the ensuing decades participated in military operations (e.g., Skaggs (2004). Landsberg, 1958; Landsberg and Jacobs, 1951; Lee, 1968). Famously defining the field as the "scientific analysis of [a statistical collective of individual weather conditions] in light of a useful application for an operational purpose," Landsberg and Jacobs (1951) made a seminal contribution to applied climatology. This definition held up well for the remainder of the century (Hobbs, 1997; Mather, 1974), and Changnon (2005b) is the only revision that has occurred recently.

The idea of applied climatology was just getting started in geography at a time when physical and human geographers were strongly opposing regional descriptive approaches and organizing the field to look for general environmental, spatial, and social laws (Leighly, 1955; Schaefer, 1953). Additionally, as climatologists became more interested in scientifically explanatory endeavors, several new specializations and subfields emerged around scales and types of emphasis (Gregory, 2000; Skaggs, 2004). For instance, statistical climatology and data collection gained traction as an institutional endeavor and new companies were established to gather, standardize, ensure quality, and analyze meteorological data for various uses (Changnon, 2005a). Within the atmospheric boundary layer, energetic flows and distributions among environments were investigated by energy balance climatology. These studies developed conceptual tools that primarily advanced the applied fields of land-surface climatology (including agriculture), physiological climatology or biometeorology, and urban climatology by integrating physical principles into understandings of meso- and micro-scale environments (Chang, 1968; Oke, 1978; Sellers, 1965). As the possibility of human-induced global warming started to gain policy currency, climate change also started to emerge as a major organizing research theme, both in the context of historical change and contemporary vulnerabilities (Hare, 1966).

Alongside these geographic advancements, meteorology's synoptic and dynamic climatology continued to advance. With the help of new computing technologies, these fields made enormous progress in understanding and modeling atmospheric dynamics across a variety of time and space scales (Barry and Perry, 1973; Hare, 1955). These advancements opened up new avenues for weather forecasting and applications related to the global climate, ranging from simulating nuclear fallout and global climate change to modeling the effects of acid rain and air pollution. An increasing number of elements, from human pollution and land-cover change to atmosphere-ocean and atmosphere-vegetation coupling, could be modeled thanks to the representation of complexity made possible by these models (Skaggs, 2004). Other applied fields include the creation of climate-effect models for various public and private purposes (e.g., infrastructure, water supply, and agriculture) made it possible to update decision-making processes in real-time and create new markets for meteorological "products" (Changnon, 2005a; Mason, 1970).

As climatologists proceeded with their work in these and other subfields, persistent disagreements regarding applied science and its disciplinary role were made evident by ongoing discussions surrounding a geographical-climatological imperative. Bryan (1944), Leighly (1955), and Thornthwaite (1961) all opposed distinguishing (geographical) climatology and meteorology and argued that all physical geographers should have a solid background in mathematics and physics to further solidify the process/quantitative revolution in physical geography. However, Thornthwaite (1961) contended that "topo climatology," which aimed to comprehend the connections between land and atmosphere, offered a compelling new story for geographer-climatologists that resonated with Hare's (1966) process-based "environmental climatology." In contrast, Stringer (1958) pushed for the acceptance of geographical meteorology, which attempted to explain the observed world by bridging the fields of regional geographical "climatography" and upper air meteorology's universal physics. The local intricacies of elements and factors, which annoy mathematicians so much, are, he claimed, an essential aspect of nature and humankind. Geographers were supposed to "keep meteorology's feet on the ground." As students studied the social sciences or the physical sciences (meteorology), these narratives became self-reinforcing, with geography remaining one of the few disciplines to uncomfortably straddle the nature/culture divide. To integrate physical and human geography, Terjung (1976) proposed a geographical climatology that was based on human-environment concerns and operationalized through systems theory using quantitative mathematical models. Under all of these threads, though not overtly so, a subtle geographical climatology was being carved out, emphasizing the study of environmental (rather than just atmospheric) processes related to understanding climatic environments (Hare, 1966).

Most of the applied climatology that emerged during this period easily fit into the category of "environmental climatology," resulting in groundbreaking books that combined advances in climatological theory with descriptive research that linked climate variables to human activity (Hobbs, 1980; Mather, 1974; Maunder, 1970; Oliver, 1973; Smith, 1975). Significant work streams on physiological climatology (Lee, 1953, 1968) and economy-climate linkages (Curry, 1952; Maunder, 1970; Perry, 1971) emerged in the immediate post-war period. According to both workstreams, people could: (1) use technology to protect themselves and their activities from climatic variability; and (2) change the type of activities they engage in to reduce loss or disturbance. Therefore, the goal of this applied work was to free human action from weather-related limitations. In the opinion of Hare (1966):

We are entering a time when environmental adaptations won't just be numerous people overcoming or submitting to their local circumstances; instead, they will be deliberate, planned efforts possibly entailing extensive, purposeful manipulation or control. Of course, advanced societies already have a great deal of environmental control. We can anticipate that the process will continue to progress. Our attitude as we reconcile with nature in the future will be one of self-conscious awareness. (Hare, 1966).

According to different accounts, the goal of the applied climatology texts from this era (which are strikingly similar) is to assist communities in "living rationally within their climatic environment" (Mather, 1974). According to these authors (Hobbs, 1980; Maunder, 1970; Smith, 1975), the atmosphere is a resource that should be utilized and optimized through climatological methods and determination. "[I]t is almost axiomatic that the user himself does not understand the real influence of climate and weather on his operation," according to Mather (1972), and it is the applied climatologist's responsibility to reframe the issue accordingly. As a result, the subjects of "application" were mostly determined by the "users'" self-selection.

There is one more thing to consider, which could be viewed as a missed chance. Notwithstanding the substantial body of research in human geography regarding the interactions between people and their surroundings (e.g., The ontological underpinnings upon which applied climate studies were compiled (Glecken, 1967; Tuan, 1977) are still mainly unknown and limited to the level of empirical correlation. In other words, research from both "sides" of geography provided shallow descriptions but no deep biophysical or social explanation. Physical approaches reduced atmospheric processes to "events," while human processes producing social formations were locked off at the level of industry inputs (Johnston, 1983; Thornes, 1983). However, this was not applied climatology's goal; rather, it was to maximize economic returns and biophysical indicators rather than to explain social life. Applied climatology thus tracked meteorological advancements more closely than geography did. Thus, by asserting a politics of the biophysical (to the exclusion of the social, cultural, and/or political), applied climatology effectively sidestepped arguments about ontology in geography. After all, who could argue against the influence of climate on agricultural productivity or the fact that heat stresses human circulatory systems? Similarly, approaches in cultural ecology succeeded in reducing social processes to mechanical "responses" to environmental stimuli (Braun, 2004), and the applied climatology case makes no claims about its subjects other than the fact that they suffer biophysical economic losses or benefits from specific events.

Nonetheless, some in applied climatology demonstrated a desire to consider the larger stories surrounding their research. Lee (1969) offered a straightforward yet impactful illustration of a spectrum of meanings in biometeorological research.

At [one] end is the study of an animal's reaction to environmental stressors only for academic purposes, with no consideration for any potential broader significance (and the milk cow was chosen for at least as many reasons as, say, the salamander!). A study that looks at how it can help establish general principles of environmental adaptation would be a little more liberal. If the study's goal is to increase animal production's economics, then human welfare—at least limited human welfare—comes into play. If the researcher or his sponsor is interested in the production of milk in developing or agriculturally disadvantaged nations, then there are undoubtedly social ramifications involved. And from here, it's not far to become preoccupied with desire as a peace-threatener. (Lee, 1969).

Lee dared to wonder what kinds of effects such an approach might have on the world, and whether or not it would ultimately advance his vision of the field as one that facilitates the "reduction of social ills" (Lee, 1969), rather than viewing the applied climatology project as one of simply "optimizing the resource." Ten years later, there was "a swing of our profession towards taking responsibility towards society very much more seriously than we have in the past," according to Hare (1979: 1174). Noting that the first World Climate Conference would bring together a large number of scientists who are not in the atmospheric sciences, he proposed that:

Beyond its predictability, which is our concern, climate also affects the peoples of the world. This is why it matters. We aren't capable of handling that. Our education does not qualify us as specialists in the field of climatic impact interpretation (Hare, 1979).

In keeping with these two ideas, Thornes (1981) considered the "ideology" of applied climatology by drawing on a recent provocative statement made by human geographer Harvey (1974). Noticing the dearth of philosophical discourse in the atmospheric research community, he bemoaned the "much too restrictive positivist paradigm." It does not specify for us how we are to use the knowledge that its procedures generate (Thornes, 1981). Although he acknowledged that climate researchers and their audiences could benefit from the systems frameworks for human-environment interactions proposed by Kennedy and Chorley (1971), Terjung (1976), and others, he was concerned about the discussions that were going on that weren't being had:

To fully utilize the value of the atmosphere to man, it will be necessary to ask and answer some of the following questions in the coming years: What other paradigms might we consider to stress the interactions between man and the atmosphere? Do such approaches as Marxist meteorology, humanistic climatology, and radical meteorology, exist? What is a phenomenological approach to the atmosphere? Thornes, 1981.

Despite the emergence of political ecology in geography, which raised new concerns about the role of environmental science in the exercise of political power, these questions were sadly unanswered in the ensuing years (Braun, 2004). Applied climatology, on the other hand, mostly followed a politics of the biophysical, creating quantitative and numerical methods for systems-based formulations of climate adaptation and only partially investigating "society" to derive new mechanisms for comprehending the "effect" of climate change (Warrick and Riebsame, 1981).

10.1.6. The 'Golden Age'

There has only been one significant collection of work published since the spate of influential texts on applied climatology that were released in the 1970s and early 1980s (Thompson and Perry, 1997; however, see Ebi et al. 2009). According to Legates (2012) and Perry (1995), the increased focus on global climate change has positive and negative effects on applied climatology. While interest in and investment in fields related to climate-sensitive activities have never been higher, there is also a tighter mandate and scope for the practice of these studies. Current advancements in regional and global climate modeling, meteorological forecasting, and the recently developed field of climate impact assessment (Kates et al., 1985) have developed a comprehensive institutional paradigm centered on the new field of climate change impacts, incorporating Terjung's (1976) theoretical physical-human process-response systems approach.

Notwithstanding the change in focus, climate applications are now more ingrained in human organizations than they have ever been. To address these issues, the sectors with a long history of applying climate change, such as agriculture, urban planning, and tourism, have created their communities and institutions both inside and outside of applied climatology. For instance, weather derivatives and crop insurance offer farmers a means of reducing the risk of a subpar crop yield, while energy and water companies can address the issue of drought; civil engineers work in the infrastructure and construction sectors, and architects work in the design sector. Applied climatologist positions within organizations have essentially evolved into portfolios. The depressing nature of the field may be attested to as much by the paucity of textbooks in applied climatology as by any loss of "coherence" in definition or application. A broader definition of the field, "applied climatology describes, defines, interprets, and explains the relationships between climate conditions and countless weather-sensitive activities," was outlined by Changnon (2005b) during this expansion.

Changnon (2005a) refers to this era as the "golden age" of applied climatology because of the following: (1) growing awareness of human vulnerability to climate extremes, led by major national and international investments in climate data and research programs; (2) a growing reliance on agriculture and other climate-sensitive industries, especially in light of population growth and globalizing economic flows; and (3) the low cost of information production, transmission, and analysis made possible by the internet and modeling applications that could now be fed with real-time data. Perhaps the most distinctive aspect of the "golden age" is its increasingly global nature, given the growing developmental and international responsibilities attributed to climatology (among others) (Glantz and Adeel, 2000).

10.2. EMERGING OBJECTS AND PRACTICES IN APPLIED CLIMATOLOGY

Our understanding of the world is intricately linked to the concepts, tools, and behaviors we employ to conceptualize action (Comrie, 2010; Jasanoff, 2004). This section examines five recent examples of applied climatology—"applied" because they are directly related to recommendations about human action, and "climatology" because scientific understandings of atmospheric processes are central—as a step toward thinking about the larger social structures that shape our questions and the climates that we produce. These examples were selected to contrast with the well-documented concerns surrounding global climate modeling and geoengineering (e.g., to highlight the diversity of scientific and institutional contexts within which climatologists operate) and to illustrate a shared empirical focus. g. Hulme, 2012a; Edwards, 2010).

We contend that critical readings of these examples—both positive and negative—provide insight into how the application of climate science can be understood as a cultural intervention as much as any "pure" representation of the "truth," even though our summary is inevitably succinct and incomplete. We can start to rethink how the study of atmospheric processes is connected to the creation of social-environmental outcomes by developing some of the cultural "work" of applied climatology.

10.2.1. Climate-Change Adaptation

A common theme in a lot of research on climate change adaptation is the understanding and prediction of climate variability and change. Climates and costs are projected into the future in order to logically allocate funding for adaptation in both international and local development contexts, and to determine what infrastructure adaptations may be required for future equilibrium climates (Hulme et al. 2011). Referred to as "scientific management" (Brunner and Lynch, 2010), this top-down planning approach has drawn criticism from the scientific and distributive communities.

According to scientific analysis, the climate is a complex system, and because there are still unknowns about how it will function, projections could lead to large-scale investments being made in the wrong place and wasting precious financial resources (Dessai et al. 2009). Furthermore, even if projections come to pass, a significant sunk investment in specific infrastructures is created, guaranteeing specific future path-dependencies (e.g., on specific sectors). Climate projections are the primary determinant of social investment, and as such, climate-change studies continue to frame their relevance using a determinative "predict and adapt" rationale (Mahony and Hulme, 2012).

In a more political or "distributive" analysis, national or international actors may use the scientific management approach to adaptation as a justification for the top-down distribution of resources and as a means of forcing their ideas of "adaptation" on local communities (Barnett and Campbell, 2010; Brunner and Lynch, 2010). Reductionist interpretations of adaptation, therefore, have the potential to direct resources through centralized decision-making on authorized "adaptation" projects, as opposed to giving communities the power to allocate funds among various projects and investments (and reducing path dependency on specific industries, infrastructure, etc.).) (Thomsen and colleagues. 2012). How can we start to navigate the complex local politics of multiple threats and stressors of environmental and socio-economic change, moving beyond the understanding that the "problem" of climate adaptation requires a reductionist scientific solution, which is what climate projections and economic analysis claim to provide (Wilbanks and Kates, 2010)? How can climatological work be practiced within broader conversations about environmental (and nonenvironmental) community values and futures (Adger et al., 2009) in ways that do not "reduce the future to climate" (Hulme, 2011)? What kinds of methodologies, categories, institutions, and practices might we create to achieve these goals, as well, as if they are also pertinent to our discussion here?

10.2.2. Commercialization of Climate Science

In many respects, climatology as a field of study and climate as a service have been bundled together as trade goods. Climate analyzes have been incorporated into private companies' overall operations as a resource and hazard. Climate risk is now more calculable than ever thanks to developments in synoptic and dynamic meteorology, new observational technologies, and weather forecasting applications. In the early 1990s, weather derivatives were born out of the weather insurance markets and quickly expanded to become a global industry transferring billions of dollars annually (Changnon, 2005a). These new commodities enable the development of more industry-specific technologies and climate packages, and they give the private sector a means of controlling climate risk to industry (Randalls, 2010).

While some (Changnon, 2005b; Hunt, 2013; Legates, 2012) have welcomed the emergence of weather commodity markets as a successful illustration of the applicability and success of applied climatology, others have voiced concern about the types of social organization that these commodity systems foster (Pollard et al., 2008; Randalls and Thornes, 2007). These worries are based on straightforward queries. What other forms of organization (e.g., should the benefits of climate expertise be restricted to those who can afford to pay for it through the private sector, and if not, how should climate expertise be distributed)? Is this the most effective and empowering way to practice climate knowledge for the creation of a just and happy society? How do these formations distribute the costs and benefits of participation, and how is participation determined? Other commodity markets), are these formations valid, and are the projects that these formations wish to support worthy of our support (Stanley, 2013)?

In addition, these advancements have given rise to several conflicts that go beyond business use. Thornes and Randalls (2007) examine the growing representation of the atmosphere as an economic resource that can be subjected to analyzes and criteria for economic decision-making. The creation of emissions trading programs, the idea of "airspace" or atmospheric property, and cost-benefit analyzes carried out at various atmospheric scales are examples of commodity discourses that need to be scientifically articulated to be concrete and understandable (Hulme, 2010; Randalls, 2011; Thornes and Randalls, 2007).

Some academics have been working to create an alternative set of conceptual frameworks around the idea of atmospheric services to move beyond commodity discourses related to climate change (Cooter et al. Thornes, 2010; 2013). These approaches embrace broader thinking in the "ecosystem services" literature and aim to highlight how atmospheric processes both quantitatively and qualitatively affect things humans care about. This helps to articulate (and value) the nonfinancial and public-good benefits of climatic processes (Anderson-Teixeira et al. Thornes (2010);, 2012). The strategy aims to represent a more comprehensive (e.g., greater than economic/ quantitative) approach to managing land use and the atmosphere by giving voice to the relationships and meanings that are typically ignored in "resource" assessments. But there are questions about whether an ecosystem services approach is just commodification under a different name. To address these questions, the approach must address its own implicit method and quantification hierarchies (Robertson, 2012). In summary, the underlying frameworks and lenses through which "applied climatological" research is conducted have significant implications for the physical scientists conducting the research, regardless of the outcome of debates about "atmospheric services." However, few physical scientists have accepted responsibility for the effects that these categories and practices are having on the world.

10.2.3. Climate Services for Development

In national and international climate knowledge circuits, the idea of institutionalizing and growing public climate services is gaining traction (Dutton, 2002; Semazzi, 2011). A statement from participating heads of state and government at the World Meteorological Organization's 3rd World Climate Conference in 2009 stated that "a global framework for climate services (henceforth referred to as "the Framework") should be established to strengthen production, availability, delivery, and application of science-based climate prediction and services."

The structure was built upon the understanding that developing countries struggle with producing and applying climate information for development.

Researchers and practitioners from around the world are being called upon to participate in what could be the largest climate development project in history. This project involves building a user interface program, reorganizing existing international climate networks to make it easier to implement the Framework, and developing international infrastructure for data sharing and collaborative research and modeling. Like other national and international projects aimed at climate development, practitioners will face two main obstacles to overcome.

First, efforts to develop a climate more directly must address user-subjectivities, which are not always neatly aligned with outputs or "products" of climate science that climatologists are familiar with or interested in producing (Dow and Carbone,

2007; Kandlikar et al. 2011). Numerous studies conducted in Pacific island nations and other locations have exposed the shortcomings of "loading dock" approaches to climate forecasts, in which producers of climate knowledge merely export their outputs to end users without considering their applicability or relevance (Brunner and Lynch, 2010; Cash et al. 2006). Climate scientists can consider new pathways, outputs, and techniques that become relevant and feasible when they engage with the needs and concerns of the communities they serve (Dilling and Lemos, 2011). By having conversations with end users, climatologists can avoid the formation of conditionalities that lead to path dependencies, such as the requirement for large-scale adaptation infrastructure plans to demonstrate "using" climate information packages. This can help climatologists reconsider and reframe what climate is actually about in these contexts.

Second, the possibility that the creation of climate knowledge amounts to a political intrusion into intricate development processes is being faced by practitioners. It is no longer possible to view merely offering someone useful information or forecasts as ethically neutral because community elites frequently use these forecasts, which can worsen inequality, support specific classes or groups, foster pathologies, and marginalize those who are most at risk from climatic variability and change (Lemos and Dilling, 2007). Climate practitioners must take into account and interact with the social formations in which they operate if the goal of offering climate services for development is to liberate the most vulnerable and impoverished members of society, as opposed to some privileged groups (Barnett and Campbell, 2010; Brunner and Lynch, 2010; Lemos and Dilling, 2007). It is possible to expand on ideas of progressive politics for applied climatology and potential avenues of action by considering climate science as an intervention into institutional formations of development and sociopolitical meanings.

10.2.4. Sustainable Architecture

Although much theory has been developed, there is still a noticeable lack of practical applications in the fields of building design and urban planning, which have long been considered promising areas to demonstrate the relevance of urban climatology (Mills, 2006; Oke, 1984). According to Mills (2006), prescriptive applications for matching building energy regimes with climatic processes to minimize energy consumption and waste can be developed by urban climatologists, who can then contribute to the advancement of the sustainable design project. Climatologists can aid in rephrasing questions about climate design and support sustainability initiatives by accepting specific pre-defined parameters, such as building temperature and humidity.

This formulation, however, fundamentally depends on climatologists optimizing around a set of externally defined parameters (temperature, energy use, etc.).). Baek (2010) makes a more thorough case for a "cultural climatology" that challenges the notions of "sustainability." Western concepts for building design (e.g., The homogenization of human-home relations from climate (due to thick interior walls) and the framing of indoor climate as an optimization problem represent a significant historical shift in Japanese practice. The idea of privacy did not exist in traditional Japanese homes, which had thick outer walls and paper partitions between rooms, until the American occupation of Japan after World War II (Baek, 2010). The hibachi (woodburner) was the center of family gatherings for warmth, and the courtyard was intentionally created as a necessary barrier that encouraged relationships with the sky. Climate regulation within the home was a social process. Therefore, designers' concerns should have been about reproducing and articulating social values and enabling specific social practices rather than about optimizing climate parameters in and of themselves.

To progress towards an architecture that is both environmentally and socially sustainable, it might be worthwhile to reconsider how applied climate pursuits are framed. e. wall thickness, the way a house works, and how it relates to activities) as opposed to just viewing the environment as a resource that should be maximized. These frames are loaded with normative presumptions about the kinds of worlds we should create, as well as (cultural) values. We interpret Baek's proposal as an invitation to consider climatology's work in a new way, as it frames biophysical processes in ways that can reproduce and embed specific human-atmospheric relations into social formations and decision contexts, rather than as a call to abolish insulating inner walls. This can prompt us to inquire about the relationships that we hope to maintain and perpetuate through our interactions with the atmosphere; similar inquiries have already been made regarding the vulnerability and adaptation to climate change (Adger et al., 2009), and ought to be expanded to encompass various atmospheric interactions.

(Adger et al., 2009) (Anderson et al., 2012)

10.3. PREDICTIONS OF FUTURE GLOBAL CLIMATE

The Intergovernmental Panel on Climate Change is made up of scientists from all over the world (IPCC). These scientists have discovered that burning fossil fuels releases greenhouse gases into the atmosphere, such as carbon dioxide, which is why the average global surface air temperature increased by 1°C (nearly 2°F) between 1900 and 2020. Even though there hasn't been much of a change, there has been no warming like this in the last 2000 years of records. Numerous effects on the planet can result from even one degree. According to climate models, if greenhouse gas emissions continue to rise at their current rate, the average global temperature will rise by an additional 4°C (7.2°F) during the 21st century. Models predict that it may become impossible to keep the rise in global average temperatures to within a range of 1.5–2°C (2–3°F) if immediate action is not taken to reduce greenhouse gas emissions (Figure 10.1).

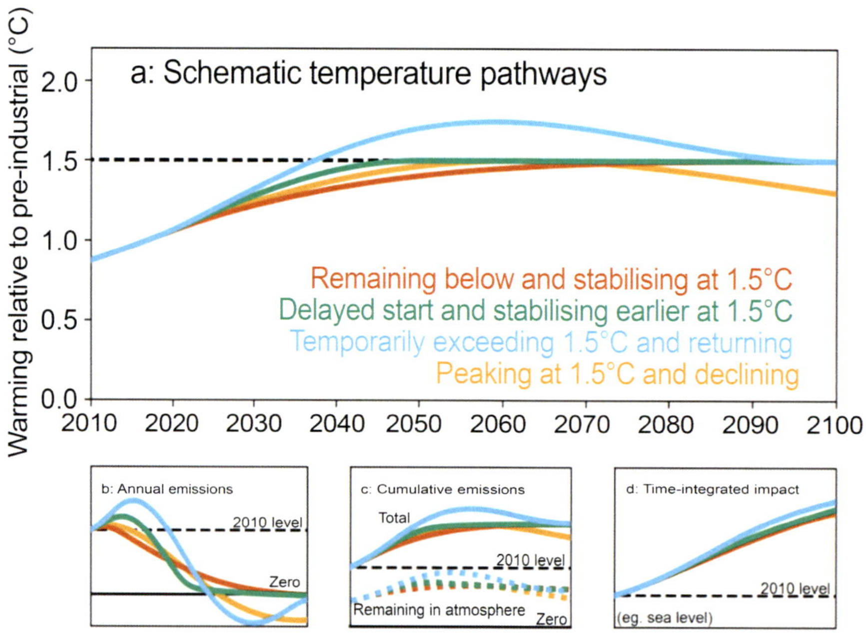

Figure 10.1. The amount of climate change by the end of the century depends on decisions we make today. If we reduce CO_2 amounts to stop increasing after 2050, the global average temperature will increase from 1–1.5°C, and this is considered a best-case scenario (blue line in the graph). If we don't reduce CO_2 and the amounts continue to increase, the worst-case scenario warming will be 4.5–5°C (red line in the graph).

Source: https://www.ipcc.ch/site/assets/uploads/sites/2/2019/01/figure-1.4.png

10.3.1. Predicted Impacts of Climate Change

Climate change is expected to have varying effects on different regions. As an illustration, it is anticipated that temperature increases will be higher on land than over oceans and higher at high latitudes than in the tropics and mid-latitudes (Figure 10.2).

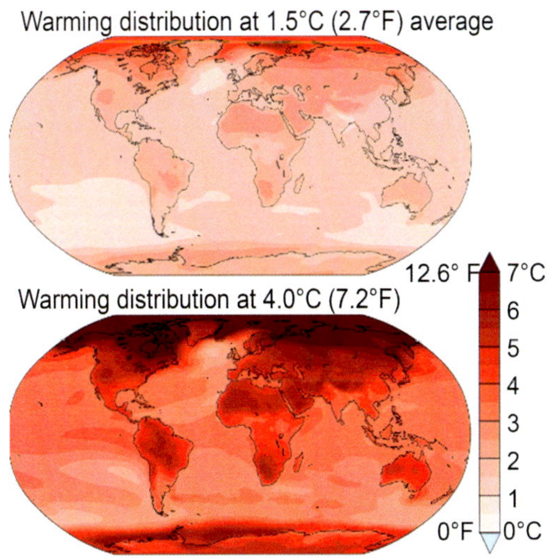

Figure 10.2. Warming is already occurring in all areas of the globe, but models of future temperatures show that the changes will not be distributed equally. Polar Regions and land areas are expected to see the largest temperature changes.

Source: https://upload.wikimedia.org/wikipedia/commons/2/29/Projected_Change_in_Temperatures.svg

Other aspects of the climate, like rain, snow, and clouds, will change as a result of and are already changing due to warmer temperatures. The ocean, life, ice, and every other component of the Earth system are all changing as a result of them.

10.3.2. Changing Precipitation

The water cycle will "speed up" as a result of a higher rate of evaporation when the average global temperature rises. Precipitation will increase as atmospheric water vapor increases. For every degree of global warming, the average amount of precipitation increases by 7% globally, meaning that future rain and snowfall will likely be significantly higher, and certain regions may be more vulnerable to flooding. It is predicted that heavy rain events will increase in likelihood by 11.7 times and intensity by 14% with a 2°C rise in temperature. On the other hand, variations in precipitation won't be dispersed equally. Certain places will receive more, while others will receive less (Figure 10.3).

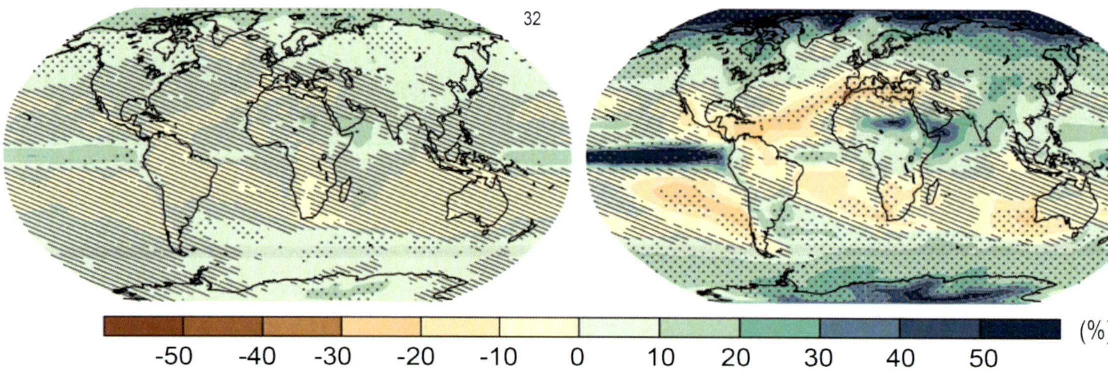

Figure 10.3. Future changes in precipitation will vary regionally, with some parts of the globe likely to become wetter and other areas projected to become drier.

Source: https://climatechange.chicago.gov/sites/production/files/2016-07/globalprecipmap-large.png

10.3.3. Melting Snow and Ice

Snow and ice melt as a result of climate change. There will likely be less snow and ice on the planet overall as a result of the melting of glaciers, ice sheets, and other snow and ice on land in the summer months continuing to exceed the amount of precipitation that falls in the winter months. The amount of permafrost in the Arctic has decreased over the past century, along with the size of mountain glaciers worldwide. The ice sheet in Greenland is also melting more quickly. It is anticipated that there will be less sea ice, or frozen seawater, floating in the Arctic Ocean and surrounding Antarctica. The Arctic's summertime sea ice thickness has already decreased by nearly half from 1950 levels. Sea ice in the Arctic is melting faster than that in the Antarctic. The circulation of the ocean may also alter as a result of ice melting. By the end of the century, summers in the Arctic Ocean will probably be free of ice, despite some uncertainty regarding the rate of melt.

10.3.4. Rising Sea Level

One way that a warmer climate raises sea levels is through the melting of glaciers and ice sheets, which add water to the oceans. Another way is through the expansion of ocean water due to warming, which also raises sea levels. Depending on the region and location, sea levels have risen by approximately 0–10–0–20 meters (0–3–0–75 feet) since 1880. About half of the rise was caused by thermal expansion and ice melting, though the precise amount of each source's contribution is not entirely known. Models indicate that sea levels will rise by an additional 0–25–0–30 meters by 2050, and if greenhouse gas emissions are not immediately reduced, global sea levels are predicted to rise by 1–1 meter (3–5 feet) by 2100. Even higher levels could be experienced in some low-lying areas, endangering wetlands, coastal communities, and international trade. The current levels of greenhouse gases will probably cause a sea level rise of approximately 0.6 meters (2 feet) by the end of the century, even in the case of rapid emission reductions (Figure 10.4).

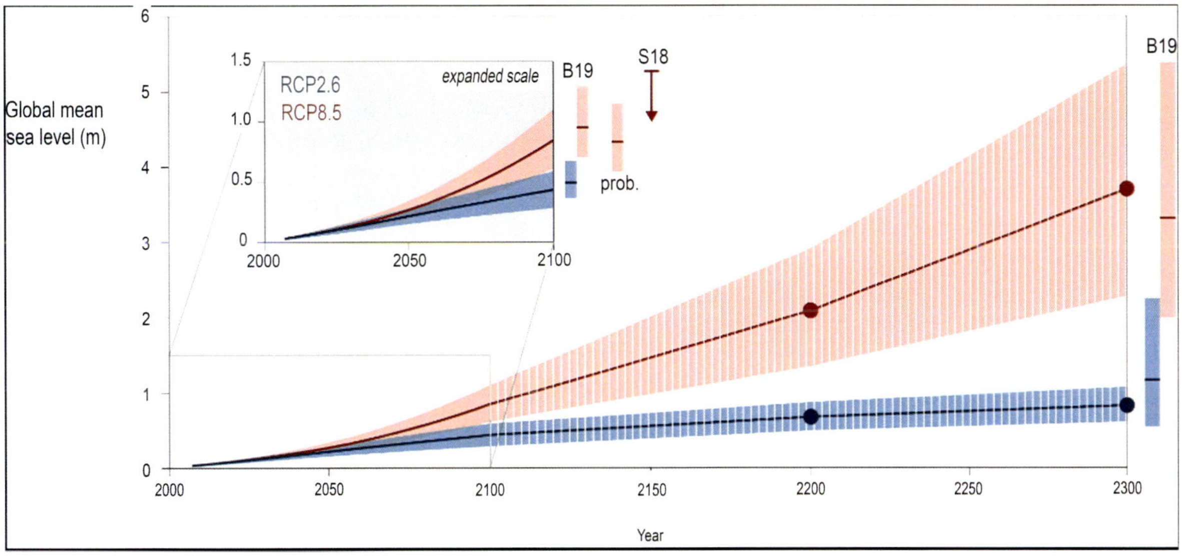

Figure 10.4. Observed (black line) and projected sea level rise through 2100, where the colored curves indicate sea level rise for different emissions scenarios.

Source: https://www.ipcc.ch/site/assets/uploads/sites/3/2019/10/IPCC-SROCC-CH_4_2.jpg

10.3.5. Acidic Ocean Water

It is anticipated that Earth's oceans will absorb some of the excess heat and carbon dioxide from the atmosphere, acting as a buffer against climate change. In the near term, this is good news, but in the long term, it presents more challenges. Weak carbonic acid is created when carbon dioxide and saltwater combine. According to scientists, this process has caused the oceans' pH to drop by roughly 0.1 pH since pre-industrial times. By the year 2100, further acidification of pH 0 to 0 is anticipated. The ocean's increased acidity harms marine life, including coral reefs.

10.3.6. Changes to Ocean Currents

As the climate warms, large-scale ocean currents known as thermohaline circulation—which are influenced by variations in salinity and temperature—may also be disturbed. Salinity can be changed by variations in precipitation patterns and the inflow of fresh water into the oceans as a result of ice melting. Rising water temperatures and shifting salinities have the potential to sabotage the currents. In a worst-case scenario, thermohaline circulation might be severely hampered or possibly stopped in some oceanic regions, which could have a significant impact on climate.

10.3.7. Changing Severe Weather

According to some climate scientists, global warming will cause changes in hurricanes, typhoons, and other tropical cyclones. The energy that propels these enormous storms comes from warm ocean surface waters. It is anticipated that in the future, warmer oceans will exacerbate these kinds of storms. There might not be an increase in tropical cyclones globally in the future, but some scientists predict a greater frequency of the strongest and most catastrophic storms. According to some scientists, there is already evidence of an increase in the frequency of the strongest storms. Some people are not as sure.

10.3.8. Changing Clouds

A bit of a wild card in global climate models is clouds. There is more water vapor in the atmosphere as a result of faster overall evaporation rates brought on by warmer global temperatures, and additional clouds. The effects of various cloud types on climate vary depending on the location. Some cool the climate by shading the Earth. Others increase the greenhouse effect by trapping heat in their water droplets and vapor. Although more clouds are predicted to accompany warming temperatures, scientists are unsure of how this increased cloud cover will affect the climate system overall. Scientific research on cloud influence modeling in the climate system is ongoing.

10.3.9. Risks to Marine Life

As sea surface temperatures rise, ocean ecosystems will shift. Fish and other animals can relocate to different ecosystems at higher latitudes where the water is cooler. However, a great deal of marine life, including corals and kelp, is in danger because they can't swim in other places. Approximately 25% of the world's coral reefs have died in the last few decades due to warmer shallow ocean waters. After being weakened by bleaching, a process linked to warmed waters, many of the coral animals perished.

10.3.10. Risks to Life on Land

Numerous plant and animal species will have their geographic ranges altered by variations in temperature, precipitation, and seasonal timing. Many species will go extinct if their range of possible habitats shrinks because they can only live in environments that meet their needs. 18% of insects, 16% of plants, and 8% of vertebrate animals are expected to lose more than half of their geographic range if global warming is limited to 2°C. However, the loss of habitat for insects, plants, and vertebrates is reduced by nearly half if we can limit global warming to 1.5°C. However, some species' ranges may expand as a result of climate change, such as mosquitos that spread various diseases. It is also anticipated that rising surface temperatures will lead to more heat waves and droughts, which could hurt crop yield, raise the danger of wildfires, and even harm people's health (Figure 10.5).

Figure 10.5. Climate change is causing many other aspects of Earth to change, including the examples noted in this graphic. Using models, scientists can project how these aspects of Earth are likely to change in the future as the climate continues to warm.

Source: https://static01.nyt.com/images/2021/04/16/climate/00cli-faq-promo/00cli-faq-promo-superJumbo.jpg

10.3.11. Abrupt Changes are Also Possible as the Climate Warms

While some climate changes are more abrupt and unpredictable, others are more gradual and predictable. The latter are frequently known as "tipping points." A tipping point is a significant, sudden shift that is difficult to reverse at the last minute, not even with extreme means. Some potential turning points are as follows.

- The collapse of major ice sheets in Greenland and Antarctica

  The process of these ice sheets melting is continuous. Nonetheless, there are indications that moderate melting could quickly spiral out of control and result in the abrupt loss of significant amounts of ice. A collapse of this kind might affect ocean circulation and cause drastic changes in sea level.

- Disruption of thermohaline circulation

  The transfer of heat in the climate system would be significantly altered if the ocean's circulation were to drastically alter or even cease entirely.

- A sudden release of methane

  Should methane, a powerful greenhouse gas, be suddenly released from the Arctic permafrost and the methane hydrates beneath the ocean floor, the pace of warming would accelerate. This rapid release of methane would initiate a feedback loop, intensifying greenhouse warming due to methane and triggering further methane emissions. Some researchers propose that such abrupt increases in methane may have contributed to past mass extinctions.

- Ocean uptake of carbon

  Currently, CO_2 that would normally remain in the atmosphere is being absorbed by the ocean. Seawater will eventually reach a point where it is saturated with CO_2 and cannot absorb anymore. At that point, anthropogenic emissions of CO_2 would all land in the atmosphere, increasing the rate of greenhouse warming. Ocean acidification may also harm marine life by killing photosynthesizing plankton and stopping them from taking up carbon dioxide from the atmosphere. In the presence of acidic waters, the shells of a wide variety of marine organisms may start to dissolve and release the carbon they have stored back into the environment.

  It is thought that none of these tipping points will materialize in the ensuing decades. However, we must consider them when assessing the total risks associated with climate change because the effects of any one of them are so dire and because it is so difficult to stop them once they start.

10.4. CLIMATE TO EVOLVE IN THE FUTURE

Much more warming is anticipated if CO_2 emissions continue to rise rapidly. If society keeps depending on fossil fuels to the same degree as it does now, atmospheric concentrations of carbon dioxide (CO_2) are predicted to double from pre-industrial levels by 2050 and triple by 2100. When combined with increases in other greenhouse gases, the "high emissions" pathway for CO_2 is predicted to cause an average global warming of approximately 4.5°C by 2100, though it could be as low as 3°C or as high as 6°C. Warming would be drastically decreased later this century and beyond under a "low emissions" pathway that is predicated on a swift transition away from the use of fossil fuels over the ensuing few decades (Figure 10.6).

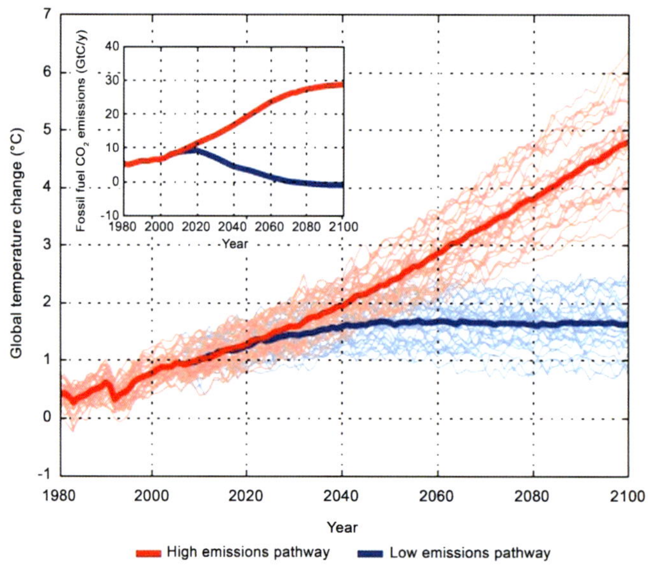

Figure 10.6. Future projected climate change depends on net emissions of greenhouse gases. Retrospective and future projected global surface air temperature changes (°C; relative to 1861–1880) under both high and low emissions pathways. Individual model simulations are shown as faint lines, with bold lines indicating the multi-model average. The corresponding two emissions pathways, including all industrial sources, are included in the inset. Emission units are gigatons (billion tones) of carbon per year (GtC/y).

Source: https://www.science.org.au/files/userfiles/learning/images/climate-change/climate-figure-4–1.jpg

Beyond the next few decades, more increases in atmospheric moisture content, more intense heatwaves, fewer frosts, more melting of mountain glaciers and ice sheets, changes in rainfall patterns (with most tropical and high-latitude regions experiencing increases and many subtropical and mid-latitude regions experiencing decreases), increased ocean warming, and increased sea level rise are all predicted effects of global warming. The amount of anticipated change is contingent upon both climate feedback and future greenhouse gas emissions.

Future forecasts, which are based on climate models run by numerous research facilities across the globe, generally concur on the patterns of global warming: higher temperatures are recorded over land than over seas, and higher temperatures are recorded at high northern latitudes than in the tropics and Southern Ocean. Future changes will be smaller in the event of reduced emissions than in the case of increased emissions, and they will depend on the emissions pathway. On finer, more localized regional scales, the models can yield disparate outcomes. For instance, certain models predict significant modifications to natural phenomena like El Niño or abrupt alterations to vegetation, while regional precipitation forecasts differ amongst models (Figure 10.7).

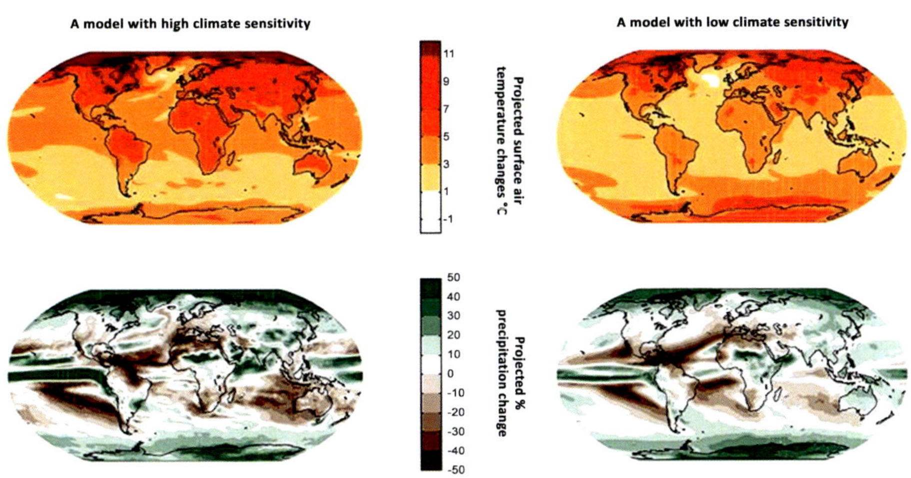

Figure 10.7. Projections of temperature and rainfall show consistent features at large scales but differ regionally, especially for rainfall. Projected global distributions of surface air temperature changes (top) and percentage precipitation change (bottom) averaged for the years 2081–2100 (relative to 1981–2000), under a high emissions pathway for two particular climate models, one with relatively high sensitivity to an initial disturbance to the climate system (left-hand panels) and one with relatively low climate sensitivity (right-hand panels). The projections have many similar patterns but differ in regional details, as is typical of climate projections from different models.

Source: https://www.science.org.au/files/userfiles/learning/images/climate-change/climate-figure-4–2.jpg

10.4.1. Long-term Climate Change is Effectively Irreversible

The choices we make in the upcoming decades regarding carbon emissions will have a long-term impact on our climate because emissions will significantly influence the rate of future climate change, especially after 2030 (. We will have to endure a warmer climate for centuries to come, even if greenhouse gas emissions are cut to almost zero this century. The change will persist for a very long time in the climate system's slower-responding components, such as the deep Ocean, ice sheets, and permafrost. Long after emissions are stopped, there will still be a lot of related effects, like sea level rise, and processes that make climate change worse, like methane and CO_2 emissions from thawing permafrost soils.

Because of these features of the climate system, reducing net greenhouse gas emissions to almost zero levels is the only way to halt climate change caused by humans (absent the use of "geoengineering," the purposeful, extensive alteration of the climate. Future climate change will be more significant the longer it takes to accomplish this and the more greenhouse gases released in the interim. Corresponding to any given threshold, the total amount of carbon dioxide emissions cannot be allowed to rise above.

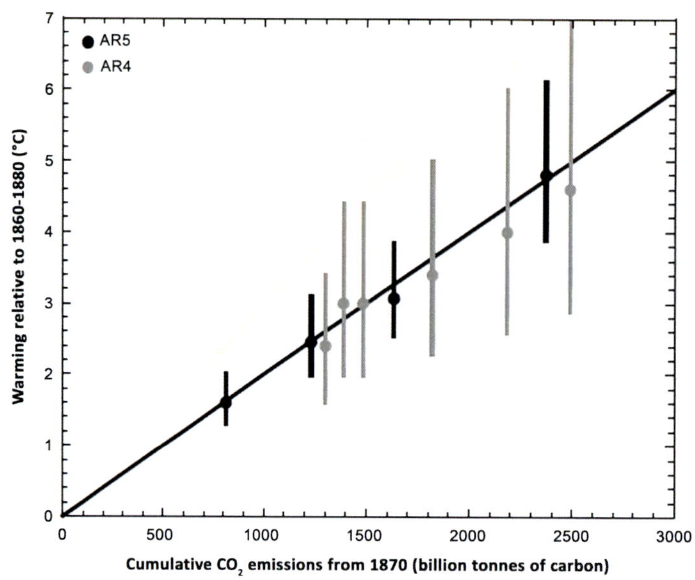

Figure 10.8. Global warming is closely related to cumulative CO_2 emissions. Points represent Intergovernmental Panel on Climate Change projections from the Fourth and Fifth Assessments (IPCC AR4, AR5); colored bands represent uncertainty, by showing the relationship if the climate were more (red) or less (blue) sensitive to disturbance than the current best estimates.

Source: https://www.science.org.au/files/userfiles/learning/images/climate-change/climate-figure-4–3.jpg

The total amount of CO_2 emissions has a direct correlation with the amount of future global warming. For example, the total amount of CO_2 emitted from human activities (taking into account the effects of other gases) would need to remain below a "carbon quota" between 820 and 950 billion tons of carbon to have a 50:50 chance of keeping global average temperatures to no more than 2°C above preindustrial levels. Humanity has already surpassed half of this cap in terms of emissions; total emissions between 1870 and 2013 were 530 billion tons. The amount of the remaining quota is equal to the current emissions for about thirty years. Long-term global emissions reductions would need to average between 5.5 and 8% annually to stay within such a carbon quota, taking into account the time needed to reverse the current emissions growth (Figure 10.8).

(Aspinall et al., 2010) (Baek et al., 2010)

10.5. EXTREME EVENTS CHANGING

10.5.1. Human-Induced Climate Change is Super Imposed on Natural Variability

Extremely cold and hot days happen more frequently in a warming climate. These alterations have already been noted. For instance, most parts of the world have seen a decline in cold days and nights over the past few decades, while hot days and nights have increased in frequency, intensity, and duration. Over much of Australia, heatwaves have become more frequent, longer, and more intense since records began, with trends picking up speed since 1970 (Figure 10.9).

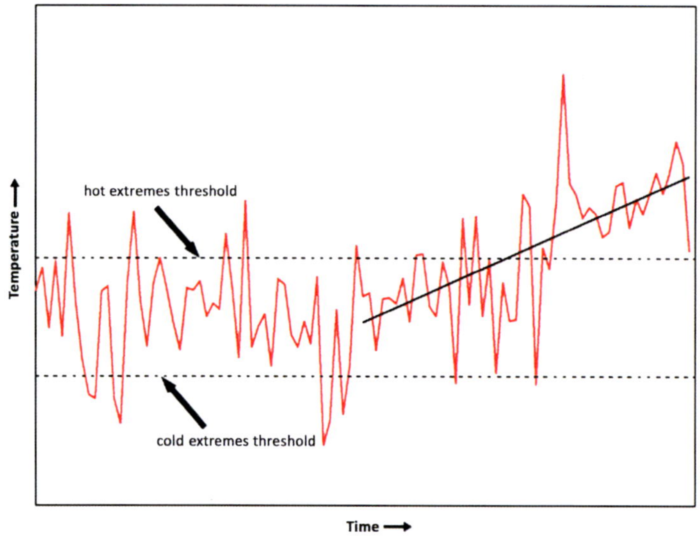

Figure 10.9. Temperature extremes change as average temperature increases. In this schematic illustration, the increase in average temperature is shown by the sloping line on the right. The idealized temperature time series has similar variability throughout the whole record. In the latter part of the record, the hot extremes threshold is exceeded progressively more frequently.

Source: https://www.science.org.au/files/userfiles/learning/images/climate-change/climate-figure-5–1.jpg

As global average temperatures rise, extreme rainfall events are predicted to grow in frequency and intensity due to the increased moisture content in a warmer atmosphere. Globally, there has already been a noticeable increase in the frequency and intensity of heavy rainfall events over most land areas in recent decades, albeit with notable regional and seasonal variations. For instance, there are fewer seasons with significant rainfall in southern Australia. Although there isn't a discernible pattern in the frequency of droughts worldwide, there are signs that since the mid-1900s, the frequency of droughts has increased in some areas (like southwest Australia) and decreased in others (like northwest Australia).

There are currently insufficient high-quality observational data on other extreme weather events, like tropical cyclones, to draw firm conclusions about historical long-term trends. On the other hand, increased rainfall from tropical cyclones is anticipated as the climate warms more.

We can now start linking changes in the climate system to a number of underlying natural and human causes thanks to recent scientific advancements. For instance, it is now feasible to calculate how much the likelihood of some types of extreme events is influenced by human-caused global warming. The recorded rise in exceptionally hot days and heatwaves is a result of human activity. Although the record high temperatures of the Australian summer of 2012–2013 might have happened by chance, human influences on the climate made it much more likely. However, there is still far less certainty about how human influences are affecting other extremes, like rainfall or tropical cyclones, due to their high degree of natural variability.

10.5.2. Extremes are Expected to Change in the Future

High temperature extremes will get hotter and cold temperature extremes will get less cold as a result of continued greenhouse gas emissions. Future emission levels will determine how quickly Australia's temperature extremes change; higher emissions will result in increasingly more frequent high extreme temperatures. According to climate model projections, there is a significant degree of uncertainty, but it is suggested that under a high emissions pathway, Australia's heavy rainfall events may tend to increase over the next few decades. Though there is significant regional variance in these trends, projections generally indicate an intensification of the wettest days and a reduction in the return time of the most extreme events worldwide. In Australia, increased temperatures are expected to result in increased frequency and intensity of extreme precipitation, punctuated by extended dry spells and significant regional variations (Figures 10.10–10.12).

A high-temperature event that is predicted to occur once every 20 years at the end of the 21st century is likely to be over 4°C hotter than it is now on many continents, including Australia. Furthermore, by the end of the twenty-first century, in many regions, a temperature that occurs once every twenty years would instead become an annual or once every two years occurrence.

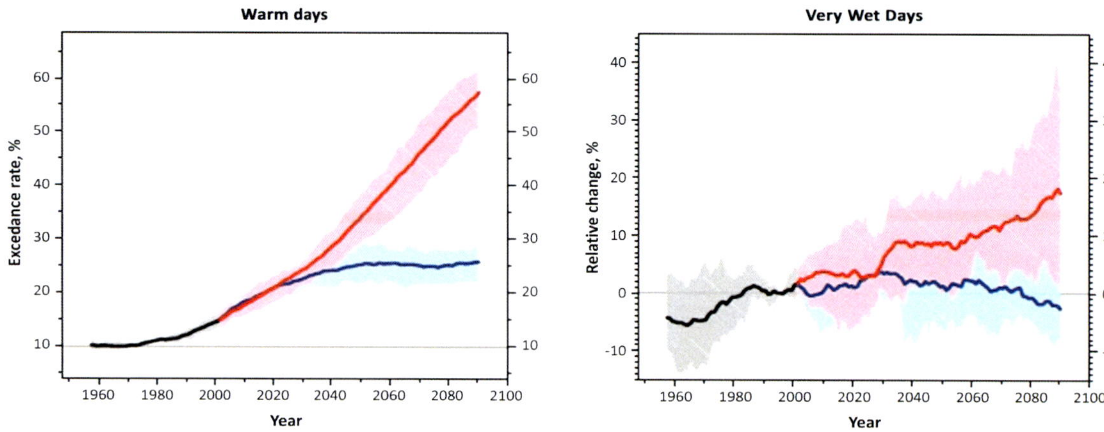

Figure 10.10. Future increases in extreme temperatures in Australia are strongly linked to global greenhouse gas emissions. But future changes in heavy rainfall are much less certain. Plots show Australia-wide changes in the percentage of days annually with daily maximum surface air temperature warmer than the temperature exceeded by the hottest 10% of days during 1961–1990; and (right) the percentage change in annual precipitation from the wettest 5% of rainfall days (relative to 1986–2005). Red and blue lines represent outcomes under high-emissions and low-emissions pathways.

Source: https://www.science.org.au/files/userfiles/learning/images/climate-change/climate-figure-5–2.jpg

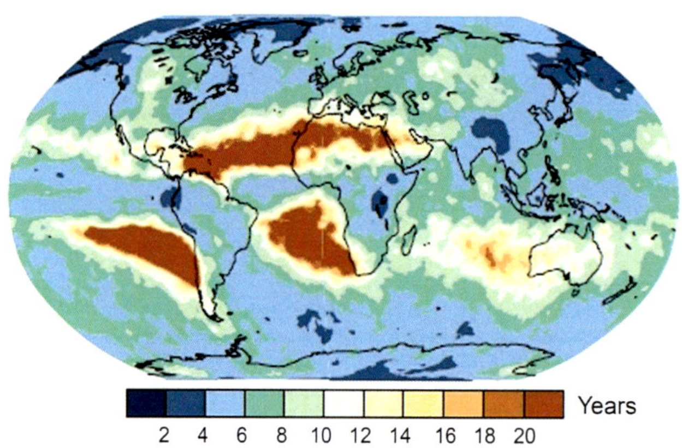

Figure 10.11. Over most continents, a heavy rainfall event that occurs only once in 20 years today is expected to occur at least twice as often by the end of the 21st century. The map shows projections, under a high emission pathway, of the return period during 2081–2100 for daily precipitation values that have a 20-year return period during 1986–2005.

Source: https://www.science.org.au/files/userfiles/learning/images/climate-change/climate-figure-5–3.jpg

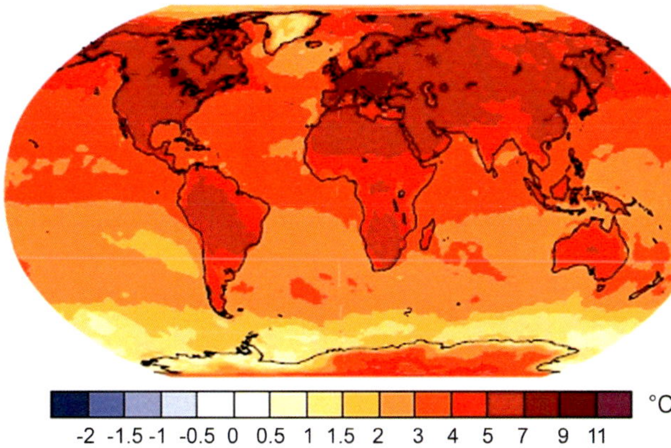

Figure 10.12. The maximum temperature in any 20 years is expected to increase with time, being substantially higher at the end of the 21st century than today. The map shows projections under a high emissions pathway of the change from 1986–2005 to 2081–2100 in 20-year return values of daily maximum temperatures.

Source: https://www.science.org.au/files/userfiles/learning/images/climate-change/climate-figure-5–4.jpg

Future changes in other extreme weather events are less certain. Evidence suggests there will be fewer tropical cyclones, but that the strongest cyclones will produce heavier rainfall than they do currently.

(Barnett et al., 2010) (Barry et al., 1973)

CASE STUDY

ADAPTING TO HEAT STRESS IN ANTWERP (BELGIUM) BASED ON DETAILED THERMAL MAPPING

The city of Antwerp, facing increasing heat stress, has adopted adaptation measures at the city-wide, local, and citizen scale. This includes the development of a heat forecast and warning system, which has raised awareness at the political level. However, challenges remain for communication and technical integration.

The city of Antwerp, in order to better understand the problem of heat stress, commissioned the research organization VITO to map the current and future temperatures and thermal comfort in the city. The research results indicate that the urban heat island of Antwerp exacerbates the impact of climate change on the urban population as the amount of heatwave days in the city rises twice as fast as in the rural surroundings. To tackle the problem of heat stress in the city, adaptation measures at three different scales (city-wide, local, and the individual citizen) are put forth. At the city-wide scale, the installation of green roofs is made mandatory for new or renovated buildings with a suitable roof, as are permeable and green parking lots. The regulations also aim to increase the albedo of public buildings. At the local scale, the thermal comfort is improved by installing fountains and ponds, planting trees and creating parks in public spaces that are renovated, while involving inhabitants through citizen science measurement campaigns. Finally, a dedicated heat forecast and warning system is put in place to minimize the health impacts on individual citizens.

Challenges

In the framework of the Copernicus European Health service, VITO applied its urban climate model UrbClim to map the air temperatures and urban heat island (UHI) extent of 100 European cities (including Antwerp) with a horizontal resolution of 100m. In the case of Antwerp, the results reveal the presence of an urban heat island, with a yearly average of 2°C in the center of the city, which can reach up to 9°C during summer evenings and nights. Due to UHI, Antwerp experienced in 2008–2017 twice as many heatwave days (defined as days with a maximum temperature over 30°C and a minimum temperature over 18°C) than the rural surroundings, exposing urban residents to much higher levels of heat stress compared to people living in the nearby rural areas.

The analysis of future climate projections (done under the FP7 RAMSES and the H2020 Climate-fit. city projects), suggests that the number of heatwave days in Antwerp is expected to increase by a factor of nearly ten towards the end of the century under the RCP8.5 scenario. Without eventual land use changes, the urban heat island intensity is expected to remain more or less at the same level, increasing the heat stress in the urban areas on top of the climate change effect.

Objectives

Motivated by the research results, the city of Antwerp decided to implement adaptation measures to tackle the problem of heat stress in the city. The goals of the identified set of measures are to: (i) reduce the local heat stress as much as possible through changes in the built environment, (ii) inform citizens about the problem, (iii) engage them through citizen science campaigns, and (iv) minimize the health impacts with a heat forecast and warning system, targeting vulnerable groups.

Ø Adaptation Options Implemented in This Case

- Climate proofing of buildings against excessive heat
- Establishment of early warning systems
- Using water to cope with heat waves in cities
- Awareness-raising campaigns for stakeholders' behavioural change

Solutions

To achieve optimal results, the adaptation measures are implemented simultaneously on three scales: 1) city-wide, 2) local, and 3) the person. The implementation of the adaptation measures described here just started or has been planned. Full implementation and city-wide results will take a long time and are only expected to be completed by 2030.

City-wide scale

The construction of buildings in the city of Antwerp is regulated by a building code, which all inhabitants and developers need to adhere to when renovating or constructing a building. In this code, specific instructions were added (9/10/2014) to help reduce heat stress in the city over time:

- For all new or renovated roofs with a slope of less than 15% and a surface area of more than $20m^2$, it is compulsory to install a green roof on top. This drastically lowers the temperature of the roof and cools air temperature by retaining and evapo-transpiring rainwater. Additionally, green roofs provide extra thermal isolation for the building reducing the need for heating and cooling.
- All new installed private gardens and open parking lots need to be green and permeable. Only $20m^2$ can be paved in gardens $<60m^2$ and only 1/3 in gardens $>60m^2$. All outdoor private parking lots need to have a permeable grassed surface.
- The majority of the buildings in the city center have historical plaster facades. When renovated, these building fronts need to be painted in the original light, preferably white color. White buildings reflect more sunlight and will not warm up as easily as dark buildings, thereby reducing the heat radiation from these buildings.

Local-scale

Regularly, large squares, parks, and neighborhoods in the city are renovated. During the planning phase, the city administration included the optimization of the thermal comfort situation as a new factor to be considered. To enable targeting actions, detailed information is needed on the local microclimate. Advised by VITO, the city of Antwerp has decided to use the Wet Bulb Globe Temperature (WBGT) indicator when assessing and optimizing the heat stress impact of renovation plans. WBGT, in contrast to simple temperature measurements, takes into account the radiation load (both shortwave and longwave), humidity, and wind speed, which all influence human thermal comfort. Several detailed (1m resolution) modeling studies were performed by VITO to quantify the local WBGT values and assess the potential impact of planned adaptation measures. This led to the inclusion of green-blue infrastructure measures (e.g., trees, permeable surfaces, water ponds, fountains) in renovation plans. The modeling was complemented by a citizen science measurement campaign during the summer of 2018 in the framework of the H2020 Ground Truth 2.0 project. About 20 residents of the Sint-Andries neighborhood were engaged in measuring the WBGT in various types of locations. Besides validation of the model results, this campaign raised awareness of the heat stress problem and stimulated a discussion about the possible adaptation measures.

Individual scale

In Belgium, 'heat health action plans' are triggered based on temperature forecasts in rural settings. This leads to an underestimation of heat stress in cities like Antwerp, where a considerable urban heat island effect causes twice as many heat wave days in urban areas than in rural surroundings. To provide a more accurate heat stress forecast for Antwerp, a short-term (5-day) heat forecast system, based on a combination of the regular European forecast model by ECMWF and the UrbClim model, was set up by VITO. The system delivers a forecast for each neighborhood of Antwerp, taking into account the urban heat island effect. This allows efficient deploying of aid resources, targeting mostly the vulnerable elderly and children, to the places where they are needed most. Furthermore, a web platform was developed by the city of Antwerp to issue heatwave warnings to health aid workers and other relevant stakeholders, including advice on what to do in case of a heatwave. The system is active during the warm months of the year in Belgium (April-September) and is managed by the city administration.

CLASS ACTIVITY

Students will gather relevant climate data for their scenario. They will use online databases and tools to collect data on temperature, precipitation, extreme weather events, and climate projections.

SUMMARY

- Applied climatology is the scientific discipline that involves the practical application of climate data and principles to solve real-world problems across various sectors.
- Climate as an object of analysis and climatology as an economic service have come to be packaged as commercial commodities.
- Climate change is predicted to impact regions differently. For example, temperature increases are expected to be greater on land than over oceans and greater at high latitudes than in the tropics and mid-latitudes.
- A warmer average global temperature will cause the water cycle to "speed up" due to a higher rate of evaporation.
- As the climate warms, snow and ice melt. It is predicted that the melting of glaciers, ice sheets, and other snow and ice on land in the summer will continue to be greater than the amount of precipitation that falls in the winter, which means a decrease in the total amount of snow and ice on the planet.
- Earth's oceans are predicted to act as a buffer against climate change by taking up some of the excess heat and carbon dioxide from the atmosphere.
- Clouds are a bit of a wild card in global climate models. Warmer global temperatures produce faster overall evaporation rates, resulting in more water vapor in the atmosphere and more clouds.
- In a warming climate, extremely cold days occur less often, and very hot days occur more often. These changes have already been observed.
- As the climate continues to warm in response to further greenhouse gas emissions, high temperature extremes will become hotter, and cold extremes will become less cold.

REVIEW QUESTIONS

1. What do you understand by applied climatology?
2. Describe the key areas of application of applied climatology.
3. What are the emerging objects and practices in applied climatology?
4. Explain the predicted impacts of climate change.
5. What are the extreme events changing?

REFERENCES

1. Adger, W. N., Lorenzoni, I., & O'Brien, K. L. (2009). Adapting to Climate Change: Thresholds, Values, Governance. Cambridge: Cambridge University Press.
2. Anderson-Teixeira, K. J., Snyder, P. K., Twine, T. E., et al. (2012). Climate-regulation services of natural and agricultural ecoregions of the Americas. Nature Climate Change, 2, 177–181.

3. Aspinall, R. (2010). A century of physical geography research in the Annals. Annals of the Association of American Geographers, 100, 1049–1059.
4. Baek, J. (2010). Climate, sustainability and the space of ethics: Tetsuro Watsuji's cultural climatology and residential architecture. Architectural Theory Review, 15, 377–395.
5. Barnett, J., & Campbell, J. (2010). Climate Change and Small Island States: Power, Knowledge, and the South Pacific. Washington, DC: Earthscan.
6. Barry, R. G., & Perry, A. H. (1973). Synoptic Climatology: Methods and Applications. London: Methuen.
7. Bickerstaff, K., & Walker, G. (2003). The place(s) of matter: Matter out of place–public understandings of air pollution. Progress in Human Geography, 27, 45–67.
8. Braun, B. (2004). Nature and culture: On the career of a false problem. In Duncan, J. S., Johnson, N. C., & Schein, R. H. (Eds.), A Companion to Cultural Geography (pp. 151–179). Malden, MA: Blackwell.
9. Briggs, D. J. (1981). The principles and practice of applied geography. Applied Geography, 1, 1–8.
10. Brunner, R. D., & Lynch, A. H. (2010). Adaptive Governance and Climate Change. Boston, MA: American Meteorological Society.
11. Brunt, D. (1944). Progress in meteorology. Quarterly Journal of the Royal Meteorological Society, 70, 1–11.
12. Bryan, K. (1944). Physical geography in the training of the geographer. Annals of the Association of American Geographers, 34, 183–189.
13. Carleton, A. M. (1999). Methodology in climatology. Annals of the Association of American Geographers, 89, 713–735.
14. Cash, D. W., Borck, J. C., & Patt, A. G. (2006). Countering the loading-dock approach to linking science and decision making: Comparative analysis of El Niño/Southern Oscillation (ENSO) forecasting systems. Science, Technology and Human Values, 31, 465–494.
15. Castree, N. (2005). Nature. London: Routledge.
16. Castree, N. (2012). Progressing physical geography. Progress in Physical Geography, 36, 298–304.
17. Chang, J. H. (1968). Climate and Agriculture: An Ecological Survey. Chicago, IL: Aldine.
18. Changnon, S. A. (2005a). Applied climatology. In Oliver, J. E. (Ed.), Encyclopedia of World Climatology (pp. 54–58). Dordrecht: Springer.
19. Changnon, S. A. (2005b). Applied climatology: The golden age has begun. Bulletin of the American Meteorological Society, 86, 915–919.
20. Clark, N., Massey, D., & Sarre, P. (2006). A World in the Making. Milton Keynes: Open University.
21. Coe, N. M., Cochrane, A., Randalls, S., et al. (2012). 'State, science and the skies:

Governmentalities of the British atmosphere', by Mark Whitehead (2009): A critical review. Geoforum, 43, 1057–1064.

22. Comrie, A. C. (2010). Nietzsche's challenge to physical geography. ACME: An International E-Journal for Critical Geographies, 9, 34–46.
23. Cooter, E. J., Rea, A., Bruins, R., et al. (2013). The role of the atmosphere in the provision of ecosystem services. Science of the Total Environment, 448, 197–208.
24. Cupples, J. (2009). Culture, nature and particulate matter–hybrid reframing's in air pollution scholarship. Atmospheric Environment, 43, 207–217.
25. Curry, L. (1952). Climate and economic life: A new approach with examples from the United States. Geographical Review, 42, 367–383.
26. Demeritt, D. (1996). Social theory and the reconstruction of science and geography. Transactions of the Institute of British Geographers, 21, 484–503.
27. Demeritt, D. (2009). From externality to inputs and interference: Framing environmental research in geography. *Transactions of the Institute of British Geographers, 34*, 3–11.
28. Dessai, S., Hulme, M., Lempert, R., et al. (2009). Climate prediction: A limit to adaptation? In Adger, W. N., Lorenzoni, I., & O'Brien, K. L. (Eds.), *Adapting to Climate Change: Thresholds, Values, Governance* (pp. 64–78). Cambridge: Cambridge University Press.
29. Dilling, L., & Lemos, M. C. (2011). Creating usable science: Opportunities and constraints for climate knowledge use and their implications for science policy. *Global Environmental Change, 21*, 680–689.
30. Dirks, K. N., Sharma, P., Salmond, J. A., et al. (2012). Personal exposure to air pollution for various modes of transport in Auckland, New Zealand. *The Open Atmospheric Science Journal, 6*, 84–92.
31. Dixon, D. P., & Jones, J. P. III. (1996). For a supercalifragilisticexpialidocious scientific geography. *Annals of the Association of American Geographers, 86*, 767–779.
32. Douglas, I. (1987). The influence of human geography on physical geography. *Progress in Human Geography, 11*, 517–540.
33. Dow, K., & Carbone, G. (2007). Climate science and decision making. *Geography Compass, 1*, 302–324.
34. Dutton, J. A. (2002). Opportunities and priorities in a new era for weather and climate services. *Bulletin of the American Meteorological Society, 83*, 1303–1311.
35. Ebi, K. L., Burton, I., & McGregor, G. R. (2009). *Biometeorology for Adaptation to Climate Variability and Change*. Dordrecht: Springer.
36. Edwards, P. N. (2010). *A Vast Machine: Computer Models, Climate Data and the Politics of Global Warming*. Cambridge, MA: The MIT Press.
37. Fleming, J. R., & Jankovic, V. (2011). Revisiting Klima. *Osiris, 26*, 1–16.
38. Glacken, C. J. (1967). *Traces on the Rhodian Shore: Nature and Culture in Western Thought from Ancient Times to the End of the Eighteenth Century*. Berkeley, CA: University of California Press.

39. Glantz, M. H. (1990). On the interactions between climate and society. *Population and Development Review, 16*, 179–200.

40. Glantz, M., & Adeel, Z. (2000). Climate Affairs as a next generation environmental science. *Global Environmental Change, 10*, 81–85.

41. Gregory, D. (1978). *Ideology, Science and Human Geography*. London: Hutchinson and Co.

42. Gregory, K. J. (2000). *The Changing Nature of Physical Geography*. London: Arnold.

43. Gulliver, J., & Briggs, D. J. (2004). Personal exposure to particulate air pollution in transport microenvironments. *Atmospheric Environment, 38*, 1–8.

44. Harden, C. P. (2012). Framing and reframing questions of human–environment interactions. *Annals of the Association of American Geographers, 102*, 737–747.

45. Hare, F. K. (1955). Dynamic and synoptic climatology. *Annals of the Association of American Geographers, 45*, 152–161.

46. Hare, F. K. (1966). The concept of climate. *Geography, 51*, 99–110.

47. Hare, F. K. (1979). The vaulting of intellectual barriers: The Madison thrust in climatology. *Bulletin of the American Meteorological Society, 60*, 1171–1174.

48. Harvey, D. (1974). Population, resources and the ideology of science. *Economic Geography, 50*, 256–277.

49. Heymann, M. (2010). The evolution of climate ideas and knowledge. *WIREs Climate Change, 1*, 581–597.

50. Hobbs, J. E. (1980). *Applied Climatology: A Study of Atmospheric Resources*. London: Wm Dawson and Sons.

51. Hobbs, J. E. (1997). Introduction: The emergence of applied climatology and climate impact assessment. In Thompson, R. D., & Perry, A. (Eds.), *Applied Climatology: Principles and Practice* (pp. 1–9). London: Routledge.

52. Hulme, M. (2010). Problems with making and governing global kinds of knowledge. *Global Environmental Change, 20*, 558–564.

53. Hulme, M. (2011). Reducing the future to climate: A story of climate determinism and reductionism. *Osiris, 26*, 245–266.

54. Hulme, M. (2012a). Climate change: Climate engineering through stratospheric aerosol injection. *Progress in Physical Geography, 36*, 694–705.

55. Hulme, M. (2012b). What sorts of knowledge for what sort of politics? Science, climate change and the challenge of democracy. Norwich: Science, society and sustainability (3s) research group, university of east Anglia. Retrieved from: https://mikehulme.org/wp-content/uploads/2011/09/12_05-Copenhagen-script_web.pdf.

56. Hulme, M., O'Neill, S. J., & Dessai, S. (2011). Is weather event attribution necessary for adaptation funding? *Science, 334*, 764–765.

57. Hunt, J. C. R. (2013). Meteorology in society and practical developments. *Quarterly Journal of the Royal Meteorological Society, 139*, 561–572.

58. Huntington, E. (1924). *Civilization and Climate*. New Haven, CT: Yale University Press.

59. Jacobs, W. C. (1947). Wartime developments in applied climatology. *Meteorological Monographs, 1*, 1–52.

60. Jasanoff, S. (2004). States of Knowledge: The Coproduction of Science and Social Order. London: Routledge.

61. Johnston, R. J. (1981). Applied geography, quantitative analysis, and ideology. Applied Geography, 1, 213–219.

62. Johnston, R. J. (1983). Resource analysis, resource management and the integration of physical and human geography. Progress in Physical Geography, 7, 127–146.

63. Kandlikar, M., Zerriffi, H., & Lem, C. H. (2011). Science, decision-making and development: Managing the risks of climate variation in less-industrialized countries. WIREs Climate Change, 2, 201–219.

64. Kates, R. W., Ausubel, J. H., & Berberian, M. (1985). Climate Impact Assessment: Studies of the Interaction of Climate and Society. Chichester: Wiley.

65. Kaur, S., Nieuwenhuijsen, M. J., & Colvile, R. N. (2007). Fine particulate matter and carbon monoxide exposure concentrations in urban street transport microenvironments. Atmospheric Environment, 41, 4781–4810.

66. Kennedy, B. A., & Chorley, R. J. (1971). Physical Geography: A Systems Approach. London: Prentice-Hall.

67. Kobayashi, A. (2010). People, place and region: 100 years of human geography in the Annals. Annals of the Association of American Geographers, 100, 1095–1106.

68. Landsberg, H. E. (1958). Trends in climatology. Science, 128, 749–758.

69. Landsberg, H. E., & Jacobs, W. C. (1951). Applied climatology. In Malone, T. (Ed.), Compendium of Meteorology (pp. 976–992). Boston, MA: American Meteorological Society.

70. Lee, D. H. K. (1953). Physiological climatology as a field of study. Annals of the Association of American Geographers, 43, 127–137.

71. Lee, D. H. K. (1968). Culture and climatology. International Journal of Biometeorology, 12, 317–319.

72. Lee, D. H. K. (1969). Tempora mutantur. International Journal of Biometeorology, 13, 3–5.

73. Legates, D. R. (2012). Applications of climatology. In Stoltman, J. P. (Ed.), 21st Century Geography: A Reference Handbook (pp. 67–77). London: SAGE.

74. Leighly, J. (1955). What has happened to physical geography? Annals of the Association of American Geographers, 45, 309–318.

75. Lemos, M. C., & Dilling, L. (2007). Equity in forecasting climate: Can science save the world's poor? Science and Public Policy, 34, 109–116.

76. Leviakangas, P. (2009). Valuing meteorological information. Meteorological Applications, 16, 315–323.

77. Lewthwaite, G. R. (1966). Environmentalism and determinism: A search for clarification. Annals of the Association of American Geographers, 56, 1–23.

78. Longhurst, J. W. S., Irwin, J. G., Chatterton, T. J., et al. (2009). The development of effects-based air quality management regimes. Atmospheric Environment, 43, 64–78.

79. Mahony, M., & Hulme, M. (2012). Model migrations: Mobility and boundary crossings in regional climate prediction. Transactions of the Institute of British Geographers, 37, 197–211.

80. Mason, B. J. (1970). Future developments in meteorology: An outlook to the year 2000. Quarterly Journal of the Royal Meteorological Society, 96, 349–368.

81. Mather, J. R. (1972). Applied climatology–An old discipline faces a dynamic future. The Professional Geographer, 24, 137–141.

82. Mather, J. R. (1974). Climatology: Fundamentals and Applications. New York: McGraw-Hill.

83. Maunder, W. J. (1970). The Value of the Weather. London: Methuen.

84. McGregor, G. R. (2006). Climatology: Its scientific nature and scope. International Journal of Climatology, 26, 1–5.

85. Meadows, D. H., Meadows, D. L., Randers, J., et al. (1972). The Limits to Growth: A Report for the Club of Rome's Project on the Predicament of Mankind. New York: Universe Books.

86. Mills, G. (2006). Progress toward sustainable settlements: A role for urban climatology. Theoretical and Applied Climatology, 84, 69–76.

87. Oke, T. R. (1978). Boundary Layer Climates. London: Methuen.

88. Oke, T. R. (1984). Towards a prescription for the greater use of climatic principles in settlement planning. Energy and Buildings, 7, 1–10.

89. Oliver, J. E. (1973). Climate and Man's Environment: An Introduction to Applied Climatology. New York: Wiley.

90. Pearce, J., & Kingham, S. (2008). Environmental inequalities in New Zealand: A national study of air pollution and environmental justice. Geoforum, 39, 980–993.

91. Perry, A. H. (1971). Econoclimate: A new direction for climatology. Area, 3, 178–179.

92. Perry, A. H. (1995). New climatologists for a new climatology. Progress in Physical Geography, 19, 280–285.

93. Pollard, J. S., Oldfield, J., Randalls, S., et al. (2008). Firm finances, weather derivatives and geography. Geoforum, 39, 616–624.

94. Randalls, S. (2010). Weather profits: Weather derivatives and the commercialization of meteorology. Social Studies of Science, 40, 705–730.

95. Randalls, S. (2011). Optimal climate change: Economics and climate science policy histories (from heuristic to normative). *Osiris, 26*, 224–242.

96. Robertson, M. (2012). Measurement and alienation: Making a world of ecosystem services. *Transactions of the Institute of British Geographers, 37*, 386–401.

97. Schaefer, F. K. (1953). Exceptionalism in geography: A methodological examination. *Annals of the Association of American Geographers, 43*, 226–249.

98. Sellers, W. D. (1965). *Physical Climatology*. Chicago, IL: University of Chicago Press.
99. Semazzi, F. H. M. (2011). Framework for climate services in developing countries. *Climate Research, 47*, 145–150.
100. Sewell, W. R. D., Kates, R. W., & Phillips, L. E. (1968). Human response to weather and climate: Geographical contributions. *Geographical Review, 58*, 262–280.
101. Skaggs, R. H. (2004). Climatology in American geography. *Annals of the Association of American Geographers, 94*, 446–457.
102. Smith, K. (1975). *Principles of Applied Climatology*. Maidenhead: McGraw-Hill.
103. Stanley, A. (2013). Natures of risk: Capital, rule, and production of difference. *Geoforum, 45*, 5–16.
104. Stehr, N., & von Storch, H. (2010). *Climate and Society: Climate as Resource, Climate as Risk*. Hackensack, NJ: World Scientific.
105. Stringer, E. T. (1958). Geographical meteorology. *Weather, 13*, 377–384.
106. Tadaki, M., Salmond, J., Le Heron, R., et al. (2012). Nature, culture, and the work of physical geography. *Transactions of the Institute of British Geographers, 37*, 547–562.
107. Terjung, W. H. (1976). Climatology for geographers. *Annals of the Association of American Geographers, 66*, 199–220.
108. Thompson, R. D., & Perry, A. (1997). *Applied Climatology: Principles and Practice*. London: Routledge.
109. Thomsen, D. C., Smith, T. F., & Keys, N. (2012). Adaptation or manipulation? Unpacking climate change response strategies. *Ecology and Society, 17*, 20.
110. Thornes, J. (2010). Atmospheric services. In Hester, R. E., & Harrison, R. M. (Eds.), *Ecosystem Services* (pp. 70–104). London: Royal Society of Chemistry.
111. Thornes, J. E. (1981). A paradigmatic shift in atmospheric studies? *Progress in Physical Geography, 5*, 429–440.
112. Thornes, J. E. (1983). Atmospheric management. *Journal of Geography in Higher Education, 7*, 189–198.
113. Thornes, J. E., & Randalls, S. (2007). Commodifying the atmosphere: 'Pennies from heaven'? *Geografiska Annaler: Series A, Physical Geography, 89*, 273–285.
114. Thornes, J., & McGregor, G. (2003). Cultural climatology. In Trudgill, S., & Roy, A. (Eds.), *Contemporary Meanings in Physical Geography: From What to Why?* (pp. 173–197). London: Arnold.
115. Thornthwaite, C. W. (1961). The task ahead. *Annals of the Association of American Geographers, 51*, 345–356.
116. Tuan, Y. F. (1977). *Space and Place: The Perspective of Experience*. Minneapolis, MN: University of Minnesota Press.
117. Warrick, R. A., & Riebsame, W. E. (1981). Societal response to CO2-induced climate research opportunities. *Climatic Change, 3*, 387–428.
118. Whitehead, M. (2009). *State, Science, and the Skies: Governmentalities of the British Atmosphere*. Chichester: Wiley-Blackwell.

119. Wilbanks, T. J., & Kates, R. W. (2010). Beyond adapting to climate change: Embedding adaptation in response to multiple threats and stresses. *Annals of the Association of American Geographers, 100*, 719–728.

120. Ziegler, A. D., Gillen, J., Newell, B., et al. (2013). Comprehensive research in geography. *Area, 45*, 252–254.

Index

FSC
www.fsc.org
MIX
Papier | Fördert gute Waldnutzung
FSC® C083411

Kolibri 360 GmbH
Ferdinand-Jühlke-Straße 7
99095 Erfurt, Deutschland
produktsicherheit@kolibri360.de